"Control Systems/Operational Technology (CS/OT) are being exploited due to lack of specific and integrated security. Interdepartmental collaboration, cooperation and understanding of what is connected and who has access, is paramount. While exploitation of a video source could deleteriously impact local business operations, more concerning is the potential gateway into an organization's Information Technology system or enterprise networks. In the latest edition of *Intelligent Network Video*, Fredrik Nilsson provides a primer that physical security professionals should be up to speed on to ensure Cybersecurity is taken into account for all systems."

Daryl Haegley, *GICSP, OCP, Department of Defense, Control Systems Cyber Resiliency*

"As a leading expert, Fredrik Nilsson continues to push the industry to innovate and share best practices as technology continues to evolve. He has dedicated his career to shifting the surveillance industry from traditional analog closed-circuit television to network video, and his work has made an undeniable impression on the security industry. If you're looking for a comprehensive guide to understanding video surveillance and security systems applications, this book is a must read!"

Tony Byerly, *Global President Securitas Technology,*
Securitas AB & CEO, Securitas Technology Corporation

"Fredrik Nilsson continues to crank out industry knowledge relevant to today's rapidly changing technologies and threats. The first two editions of *Intelligent Network Video* provided security professionals with fundaments of video surveillance. This updated version builds on that with advanced information on analytics, cloud-based video solutions and importantly cybersecurity. This is a must-read book for those security professionals looking to advance their skills in an ever-changing technology and threat environment."

Dan Moceri, *Executive Chairman and Co-Founder, Convergint Technologies*

"Many in the global security industry have anticipated this valuable update to *Intelligent Network Video* by Fredrik Nilsson that provides insight into the major trends shaping the future of IP-based security technologies, including the pervasive adoption of cloud-based technology and external threats to business continuity such as cybersecurity threats. Cybersecurity has always been one of the top trends in SIA's annual Security Megatrends report, with artificial intelligence, analytics and cloud and software-based systems playing growing roles in our industry. The ever-changing technology landscape makes it even more critical for industry professionals to not only stay current, but also position themselves a step ahead with technical knowledge that soars above commonly found industry information. As a former member of the Security Industry Association Board of Directors, Fredrik Nilsson provides the industry with unique insight into the future direction of networked video and what it means for the entire security ecosystem. Through his writings, presentations and commentary, Fredrik has dedicated considerable time and resources in support of all forms of industry education, from classroom instruction to online education and certification programs. This latest update to *Intelligent Network Video* is an example of his expertise and his passion for industry education that contributes to a stronger industry for all stakeholders and the people and property they work to protect."

Don Erickson, *Chief Executive Officer, Security Industry Association*

"A comprehensive, deeply informed, and clear-eyed look at the changing landscape of networked video surveillance. This new edition explores the profound importance of cybersecurity and risk management in physical security networks. It also provides the reader with a 'hype-free' look into the ways that cloud and analytics technology are reshaping the practice of physical security both today and into the future."

Pierre Racz, *Founder and CEO, Genetec Inc.*

"Video surveillance is an ever-changing industry where advances in technology drive the need for increased knowledge. Whether someone is new to the security industry, or has 30 years of experience, this book is a must read. As the author of *Intelligent Network Video*, Fredrik Nilsson brings a vast amount of knowledge and experience to the table, and he has an incredible ability to convey the topics in an understandable way to the reader. I've used the second edition of *Intelligent Network Video* for a number of years for apprenticeship and journeyman electrical training. This textbook does an amazing job teaching the concepts of video surveillance technologies. The new third edition of *Intelligent Network Video* brings the textbook to a new level! Updates to cloud storage and analytics chapters will prepare the reader for technologies that continue to grow in video surveillance. The addition of a new cybersecurity chapter is spot on and will help the reader understand the challenges associated with systems connected to the cloud."

Jim Simpson, *Director of Installer-Technician and Residential curriculum,*
Electrical Training Alliance

Intelligent Network Video

The third edition traces the trajectory of video surveillance technology from its roots to its current state and into its potential future role in security and beyond. For the reader, it is an opportunity to explore what the latest technology has to offer, as well as to gain some insight into the direction that surveillance will take us in the years ahead.

The revised edition of *Intelligent Network Video* is more comprehensive in every area than the first and second editions, printed in over 25,000 copies. There is also a new chapter on cybersecurity, as well as thoroughly revised chapters on cloud and analytics. The book takes the reader on a tour through the building blocks of intelligent network video – from imaging to network cameras and video encoders, through the IT technologies of network and storage and into video management, analytics, and system design.

Intelligent Network Video

Understanding Modern Video Surveillance Systems

Third Edition

Fredrik Nilsson
Axis Communications

CRC Press is an imprint of the
Taylor & Francis Group, an **informa** business

Designed cover image: Shutterstock

Third edition published 2024
by CRC Press
2385 NW Executive Center Drive, Suite 320, Boca Raton FL 33431

and by CRC Press
4 Park Square, Milton Park, Abingdon, Oxon, OX14 4RN

CRC Press is an imprint of Taylor & Francis Group, LLC

First edition published by CRC Press 2008
Second edition published by CRC Press 2017

Library of Congress Cataloging-in-Publication Data
Names: Nilsson, Fredrik, author. | Axis Communications.
Title: Intelligent network video: understanding modern video surveillance systems/Fredrik Nilsson, Axis Communications.
Identifiers: LCCN 2023026573 (print) | LCCN 2023026574 (ebook) | ISBN 9781032534718 (hbk) | ISBN 9781032534664 (pbk) | ISBN 9781003412205 (ebk)
Subjects: LCSH: Video surveillance.
Classification: LCC TK7882.E2 N55 2024 (print) | LCC TK7882.E2 (ebook) | DDC 621.389/28–dc23/eng/20230614
LC record available at https://lccn.loc.gov/2023026573
LC ebook record available at https://lccn.loc.gov/2023026574

ISBN: 978-1-032-53471-8 (hbk)
ISBN: 978-1-032-53466-4 (pbk)
ISBN: 978-1-003-41220-5 (ebk)

DOI: 10.4324/9781003412205

Typeset in Minion Pro
by Apex CoVantage, LLC

Contents

Acknowledgements

Writing a book, like any large project you take on in life, cannot be completed without tremendous support from colleagues, friends, and family.

I am fortunate enough to work for Axis Communications, a company that sees the value in the education this book provides. Axis not only gave me the time to work on this project but, more importantly, all the necessary internal resources to complete it. Special thanks go out to Shi-lin Chan, who acted as my shadow writer and editor for the first edition; Gunilla Burke, who helped me with the second edition; and Steve Dale and Emmanuel Prempeh, who both helped me with this third edition. Also, thanks to Mark Listewnik of CRC Press, who, like clockwork, reminds me when it's time to start working on a new edition and then patiently waits for the manuscript. Many other people should be mentioned, but the list would become too long. Everyone who has helped me with this endeavor, from proofreading to writing whole chapters, will see their invaluable contributions in these pages, and my sincere appreciation goes out to all of you.

I am also lucky to be part of an industry filled with outstanding individuals who entered the profession not only to make a living but also to help provide a safer and smarter world. Since moving to the United States over 20 years ago, my knowledge of the industry and technology has grown tremendously, thanks to so many generous partners, customers, and friends willing to share their lifelong experiences with me.

And finally, this book would not exist without the understanding and support from my wife.

Thanks for letting me once again spend so many weekends and nights on this project. I love you.

Introduction

Welcome to the third edition of *Intelligent Network Video: Understanding Modern Video Surveillance Systems*, published 15 years after the first edition was released in 2008. Since then, technology has caused a profound shift in the way we live and work. Remember when smartphones and social media were just emerging? Today, we can't imagine life without them. Technology development is not likely to slow down, and it will continue to change our lives and transform the video surveillance industry. Technologies such as cloud services and artificial intelligence (AI), along with the challenges of cybersecurity, are all part of our everyday life – and the video surveillance market.

Back in 2008, IP video was still an emerging technology. Skeptics wondered whether it was ready for primetime and if it would gain sufficient traction to ever take over the whole video surveillance market. Video analytics was viewed as the biggest opportunity for the market, with estimates that it would become a billion-dollar market in just a few years. In the context of physical security, hardly anyone was talking about cloud services. IT security and physical security were mentioned in the context of "convergence", but few considered the cyber-threat that IP video equipment might pose to corporate networks.

Fast-forwarding to 2023, the debate over which term to use – "IP cameras" or "network cameras" – has long been settled. IP cameras are a part of our world now and are known simply as "cameras". We have also seen that not only are they better than analog cameras but of course, they are also IP-connected, just like billions of other IoT devices. While many analog cameras are still in use, very few new ones are being installed, and the existing ones are either digitized via encoders or being replaced by their more capable IP counterparts. Today's cameras are very impressive in terms of resolution, light sensitivity, forensic information, and built-in analytics. They are also easier than ever to install, deploy, and maintain. And their capabilities will continue to improve, not only outperforming the human eye but in some tasks even the human brain.

The requirements on today's surveillance systems are vastly different to those from years ago, which were often installed as an "island" system and kept alive as long as possible, with no involvement from the rest of the organization except the security department. Today's systems are fully integrated into an organization's IT infrastructure, and they have visibility and interest from the C-suite. Systems sometimes contain thousands or tens of thousands of cameras, which can be spread across the globe. Surveillance systems are often integrated with other systems and include other IoT devices, such as audio devices, intercoms, body wearable cameras, and access control devices. Part of the system is installed and maintained on-premises, and other parts can be in the cloud. With the advent of deep learning capabilities at the edge, analytics (AI) are finally coming of age and bringing great value to organizations, from both security and business operations perspectives. With all the advanced devices connected to the IT networks, the threat of cybersecurity needs to be taken very seriously.

WHAT TO EXPECT IN THE FUTURE

It doesn't take a crystal ball to figure out that innovation will continue to reshape the surveillance landscape as we know it and that many of the technology trends we see in IT and consumer electronics will impact the security and surveillance market. The next section lists some of the major trends.

Greater investment in technology

The world today invests hundreds of billions of dollars every year to make citizens, customers, students, passengers, employees, and facilities safe and secure – in cities, retail locations, airports, schools, offices, data centers, etc. Around 80% of these investments are spent on manual labor, such as guarding, policing, monitoring, installation, and maintenance – only a small portion goes towards technology such as surveillance systems. As technology improves and costs for staffing increase, we can expect a greater share of investments to be made in technology.

Evolution of AI

It is hard to read the news today without finding an article about AI (artificial intelligence), and the video surveillance industry is no different. While it is hard to define what is real AI, the video surveillance industry often uses the term "analytics", which has been something of a holy grail for 20 years. While investments have been substantial, the end results were sometimes disappointing. However, the last few years have really accelerated the capabilities and accuracy of analytics. With the advent of deep learning processing at the edge, today's cameras can do amazing things when it comes to detecting, categorizing, and counting objects. Expect this trend to continue.

Cloud migration

Just as with AI, "cloud" has become a prominent term found all over the news, as well in most companies' corporate communications and business plans. It is important to realize that the cloud is both a technology and a business model. When it comes to the technology, shifting functionality to the cloud offers many advantages, such as easier system maintenance, better accessibility, and off-site redundancy backups. Video surveillance presents challenges due to the massive amounts of data that need to be transferred and stored. Therefore, so-called hybrid systems are often used, where some data still resides on the premises or even in the cameras. The proliferation of accurate and useful analytics (AI) also supports the implementation of cloud-based systems, where metadata rather than video is transferred to the cloud. The other factor driving cloud migration is, of course, the business model that revolves around the ability to pay-per-use or to pay a monthly fee for access to the system's capabilities. Most organizations do not have security as a core competency, and the trend in outsourcing security system functionality to the cloud is likely to continue.

Cameras used beyond security

With the three aforementioned trends utilizing more accurate and scalable technology to drive automation, it is easy to see how surveillance cameras will be increasingly used for applications beyond security. Unattended retail stores that enable self-service, manufacturers automating more production processes and quality controls, garages using license plate recognition for automated entry and exit, and intercoms using facial recognition for access control – all with the ability to integrate the metadata into other business analysis systems. Expect the value of surveillance cameras for both security and business operations to increase significantly in the future.

Cybersecurity awareness

Cybercrime is widespread in society today and ever increasing in the number of attacks and sophistication. For most organizations, it is not a matter of *if* but *when* their networks and data will be compromised. While security and surveillance data might not be of the greatest interest to would-be attackers, the security system itself might contain vulnerabilities and, thus a way into an organization's network. With many organizations having more IP cameras than any other type of device on their networks, it is easy to understand why cybersecurity in the physical security systems has become so important and that cybersecurity functionality in the video surveillance systems and the tools that support the process will continue to grow in importance.

Ethical use of technology

The usage of surveillance cameras in public places can be controversial, especially when using technology such as facial recognition, which may have bias in the algorithms. At the same time, privacy for citizens in most countries is becoming increasingly more important, as underscored by regulations such as the GDPR in Europe. Additionally, initiatives such as the Copenhagen Letter are endeavoring to increase awareness about the challenges of technology while at the same time seeking accountability and building trust. Expect this trend to continue, especially with more capable technology, more data uploaded to the cloud, and smarter AI.

Sustainability a major driver

Sustainability is a major trend impacting everything from the car industry to power generation and the way we consume and travel. In the video surveillance industry, there is an increased interest in more sustainable solutions, by phasing out hazardous materials such as PVC in the cameras and by finding smarter ways to produce, package, and ship cameras. Video surveillance solutions are also becoming more sustainable by means of higher light sensitivity so that cameras do not require additional light to capture colors at night, smarter compression that vastly reduces the storage and network needs, lower power consumption, and smarter cameras that generate metadata. The sustainability trend is clear in society, increasing the expectations on more sustainable video surveillance systems.

Cameras in general have become ubiquitous on street corners, in retail stores, and even built into our phones, cars, and some appliances. What is more, after the COVID-19 pandemic, cameras are now even more commonly used for remote monitoring, video conferencing, and other business operations. Over the past decade, network cameras have evolved significantly through better sensors, high-power processors, and deep learning capabilities. As a result, users are discovering greater value and versatility from IP cameras and are deploying more of them.

As for security, market research indicates the industry has reached a significant milestone: Over one billion surveillance cameras were installed worldwide at the end of 2021. While recent studies reveal the vast majority of the public believes that video surveillance cameras reduce crime and increase public safety, a rise in privacy regulations and cyberattacks has brought intense focus on cybersecurity in order to address concerns over privacy, data protection, and other risks. Accordingly, the industry has responded with innovative camera technologies, such as signed video and privacy masking software, as well as other cyber-hardening technologies, practices, and policies, to meet these new challenges.

With safety and security as a basic requirement, a driving force, and the cornerstone of society, we can expect the trend towards more intelligent video to continue well into the future.

ABOUT THE BOOK

The revised edition of *Intelligent Network Video* is more comprehensive in every area than the first and second editions. There is also a new chapter on cybersecurity, as well as thoroughly revised chapters on cloud and analytics. The book takes the reader on a tour through the building blocks of intelligent network video – from imaging to network cameras and video encoders, through the IT technologies of network and storage and into video management, analytics, and system design. There are also a few chapters more technical in nature, included specifically for readers interested in delving deeper into the details. These chapters also include the word "technologies" in their titles.

The purpose of the third edition is to trace the trajectory of video surveillance technology from its roots to its current state and into its potential future role in security and beyond. For the reader, it is an opportunity to explore what the latest technology has to offer, as well as to gain some insight into the direction that surveillance will take us in the years ahead.

About the author

Fredrik Nilsson is Vice President of the Americas for Axis Communications, overseeing the company's operations in North and South America and serving on the global management team. In his more than 25-year career at Axis, he has undertaken various roles in both Sweden and the United States. Since assuming responsibility for the Americas in 2003, revenues in the region have grown from $20M to $1B due in part to a focus on increasing Axis's presence in the market by opening more than 15 Axis Experience Centers throughout the region. Nilsson has also been instrumental in leading the surveillance industry shift from analog closed-circuit television to network video.

Mr. Nilsson has been an active participant in *SIA (Security Industry Association)* and was on the Board of Directors and Executive Committee for many years. A trusted industry speaker, he has delivered lectures and keynotes at many conferences, including influential shows such as *Securing New Ground, ASIS/GSX, ISC West, Expo Seguridad,* and *SACC Executive Forum.* In 2016, Mr. Nilsson received the prestigious George R. Lippert Memorial Award presented to individuals for their long-term service to SIA and the security industry, the impact of their efforts on behalf of SIA and the industry, as well as their integrity, leadership, and diplomacy in industry dealings. Mr. Nilsson was inducted into the Security Sales & Integration Industry Hall of Fame in 2017 which recognizes those who have left an indelible impression on the security industry and transformed it into the thriving business of today through qualities including leadership, innovation, achievement, humanitarianism, and integrity. Mr. Nilsson is also a past chairman of *Mission 500*, a charitable organization backed by the security industry, dedicated to helping children and communities in crisis.

An oft-quoted source for top business news outlets such as *The New York Times, USA Today, The Washington Post, NBC News, Forbes,* and *Svenska Dagbladet,* he has also contributed articles and quotes to numerous professional security, business, and IT publications both in the United States and abroad. Mr. Nilsson has made guest appearances on *CNN, CNBC, Fox News,* and *NECN.* Mr. Nilsson currently sits on the Board of Directors for MSAB, the global leader in mobile forensic technology for extraction, analysis, and management.

Prior to Axis, Mr. Nilsson served as a product manager for ABB, a global leader in power and automation technologies. A graduate of the Lund Institute of Technology, he holds a master's degree in electrical engineering and postgraduate studies in economics.

Chapter 1

The evolution of video surveillance systems

Video surveillance, also known as closed-circuit television (CCTV), is an industry that emerged in the 1970s as a viable commercial technology to deliver safety and security. In the many years since then, the industry has experienced its share of technology changes. As in any other area, the ever-increasing demands on products and solutions drive development and evolving technologies help to support them. In the video surveillance market, these demands include the following:

- Better image quality
- Streamlined installation and maintenance
- Reliable technology and function
- Better options for storing recorded video and greater capacity
- Lower cost and better return on investment
- Larger systems and better scalability
- Remote monitoring capabilities
- Integration with other systems
- Smarter products, with edge storage and built-in intelligence
- Protection against malicious intent, both physical (vandalism) and digital (cyberattack)

To meet these requirements, video surveillance has experienced several technology shifts, among them the transition from analog CCTV surveillance to fully digital, network-based video surveillance systems.

Surveillance systems started out as 100% analog and have gradually become digital, even though analog systems remain in use to some degree. Today's fully digital systems, which use network cameras and standard IT servers for video recording, have come a long way from the early analog tube cameras connected to video cassette recorders (VCRs) that required tapes to be switched manually.

The existence of several semi-digital solutions has led to some confusion in the video surveillance industry. Neither fully analog nor fully digital, these are partly digital systems that incorporate both digital and analog devices. Only systems in which video streams are continuously being transported over an IP network are truly digital and provide real scalability and flexibility. Still, some define semi-digital systems with analog cameras connected to a digital video recorder (DVR) as digital, while others reserve the term for fully digital systems with network cameras and servers. Although all these systems contain digital components, there are some very important distinctions between them.

DOI: 10.4324/9781003412205-1

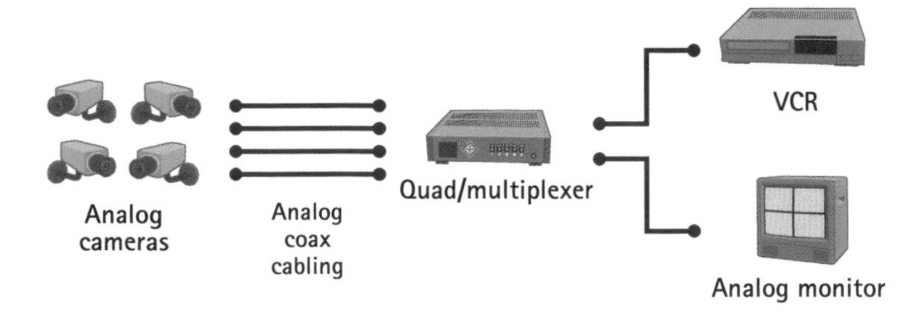

Figure 1.1 A classic analog video surveillance system.

The following sections outline the evolution of video surveillance systems. Different system configurations are explained, from fully analog to fully digital, along with the benefits of each configuration. Section 1.1 describes analog systems. Sections 1.2 and 1.3 describe semi-digital video systems. Sections 1.4 and 1.5 describe true network video systems. Only the system outlined in section 1.5 is fully digital.

1.1 VCR-BASED ANALOG CCTV SYSTEMS

The traditional analog CCTV system comprised analog cameras connected to a VCR for recording video, as shown in Figure 1.1. The system was completely analog, and the VCR used the same type of cassettes as a home VCR. Each camera was connected by its own coax cable to the VCR. The video was not compressed, and a tape's recording capacity at full frame rate was limited to a maximum of eight hours.

Eventually, time-lapse recording was incorporated into VCRs to increase tape capacity. Time-lapse enabled the recording of every second, fourth, eighth, sixteenth, or thirty-second frame. In analog systems that used time-lapse recording, these were the only recording frame rates possible and that was how the video surveillance industry came up with specifications such as 30 fps (frames per second), 15 fps, 7.5 fps, 3.75 fps, and 1.875 fps. When several cameras were used, quads became another important system component. A quad simply took inputs from four cameras and created one video signal output to show four different images on one screen. This made the system slightly more scalable but at the expense of lower resolution.

In larger systems, the multiplexer became commonplace. This device combined the video signals from several cameras into a multiplexed video signal, which made it possible to record from up to 16 cameras on one device. The multiplexer also made it possible to map selected cameras to specific viewing monitors in a control room. However, all equipment and all signals were still analog, and viewing video still required analog monitors connected to a VCR, quad, or multiplexer.

Although analog systems functioned well, the drawbacks included limitations in scalability and the need to maintain VCRs and manually change tapes. In addition, the quality of the recordings deteriorated over time. For a long time, analog cameras could only deliver black-and-white images. Today, most analog cameras support color video.

1.2 DVR-BASED ANALOG CCTV SYSTEMS

By the mid-1990s, the video surveillance industry saw its first digital revolution with the introduction of the digital video recorder (DVR). In this case, video is instead recorded to hard drives, which requires it to be digitized and compressed to store as much video as possible (Figure 1.2).

Hard disk space was limited in early DVRs, so either the recording duration had to be shorter or a lower frame rate had to be used. The disk space limitations prompted many manufacturers to develop proprietary compression algorithms. Although these algorithms worked well, users were

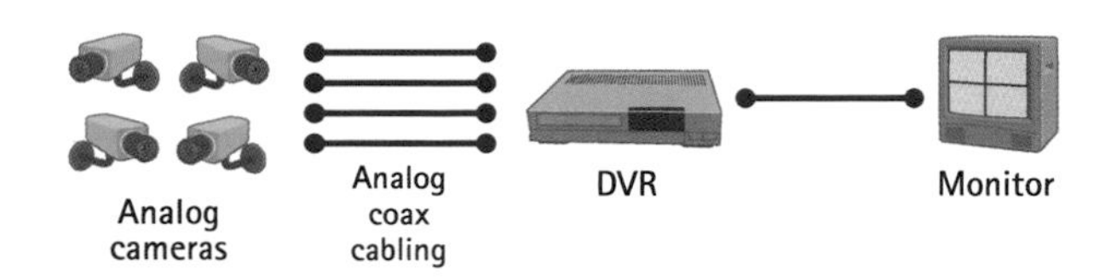

Figure 1.2 A surveillance system with analog cameras connected to a DVR, including quad or multiplexer functionality and providing digital recording.

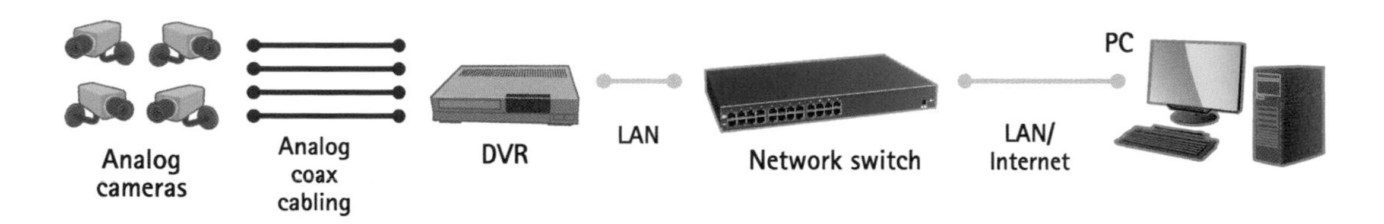

Figure 1.3 An example of how analog cameras can be networked using a network DVR for remote monitoring of live and recorded video.

tied to one manufacturer's tools when playing back the video. Over the years, the cost of disk space dropped dramatically, and standard compression algorithms became available and widely accepted. To the benefit of users, most manufacturers gave up their proprietary compression formats in favor of standards such as JPEG, MPEG-4, and H.264.

Most DVRs had multiple video inputs (typically 4, 16, or 32), which meant DVRs also included the functionality of the quad or multiplexer. DVRs replaced the multiplexer and the VCR, thereby reducing the number of components in CCTV systems.

The introduction of the DVR system brought the following major advantages:

- No tapes or tape changes
- Consistent recording quality
- Ability to search quickly through recorded video

Early DVRs used analog monitors for showing video. However, because the DVR made the video available digitally, it became possible to network and transmit the digital video through a phone modem connected to a serial port on the DVR. In later models, the modem was built into the DVR itself. Though the ability to monitor video remotely on PCs was a great benefit, the actual functionality was not very useful because the bandwidth available with phone modems was too low, often in the 10–50 kbps range (kilobits per second). That meant very low frame rates, low resolution, or highly compressed video, which made the video useless.

1.3 NETWORK DVR–BASED ANALOG CCTV SYSTEMS

DVRs were eventually equipped with an Ethernet port for network connectivity. This introduced network DVRs to the market and enabled remote live or recorded video using PCs (Figure 1.3). Some of these systems required a special Windows® client to monitor the video, whereas others used standard web browsers that made remote monitoring more flexible.

The network DVR system provides the following advantages:

- Remote monitoring of video on PC
- Remote operation of the system

Although DVRs provided great improvements over VCRs, they also had some inherent disadvantages. The DVR was burdened with many tasks, such as digitizing video from all cameras, compressing video, recording, and networking. In addition, the DVR was a black-box solution with proprietary hardware and preloaded with software that often forced the user to source spare parts from a single manufacturer, making maintenance and upgrades expensive. Cybersecurity such as virus protection was also difficult to implement. Though the DVR was often a Windows-based machine, the proprietary interface did not support virus protection. Moreover, the DVR could only offer limited scalability. Most DVRs offered 16 or 32 inputs, which made it difficult to build systems not based on multiples of 16. For example, if the system required 10 or 40 cameras, the customer would be forced to buy a 16-channel DVR or a 48-channel DVR combination.

1.4 VIDEO ENCODER–BASED NETWORK VIDEO SYSTEMS

The first step towards a networked video system based on an open platform came with the introduction of the video encoder. Encoders are typically used in existing installations where previously installed analog cameras still function well but where the DVR's shorter life span requires its replacement with a more flexible and future-proof recording solution.

A video encoder connects to analog cameras and digitizes and compresses the video. It then sends the video over an IP network through a network switch to a server, which runs video management software (VMS) for monitoring and recording, as shown in Figure 1.4. This is a true network video system because the video is sent continuously over an IP network. In essence, the tasks previously performed by the DVR are now divided: The video encoder handles digitization and compression, and the server takes cares of recording. One of the benefits of this solution is better scalability.

A video encoder–based network video system has the following advantages:

- *Non-proprietary hardware:* Standard network and server hardware can be used for video recording and management.
- *Scalability:* Cameras can be added one at a time, as encoders are available in 1-, 4-, 16-, and up to 84-channel versions.
- *Off-site recording:* Record video remotely and at central locations.
- *Future-proof:* Systems are easily expanded by incorporating network cameras.

1.4.1 NVRs and hybrid DVRs

Alternatives to the open platform – which is based on a PC with installed video management software – are different types of network video recorders (NVRs) (Figure 1.5) and hybrid DVRs. An NVR or hybrid DVR is a proprietary hardware box with pre-installed video management software for managing video from encoders or network cameras. The NVR handles only network video inputs, whereas the hybrid DVR can handle both network video and analog video inputs. Because recording and video management are available in a single unit, much like the DVR, the benefit of using an NVR or hybrid DVR is the ease of installation. However, while often easy to install, NVRs

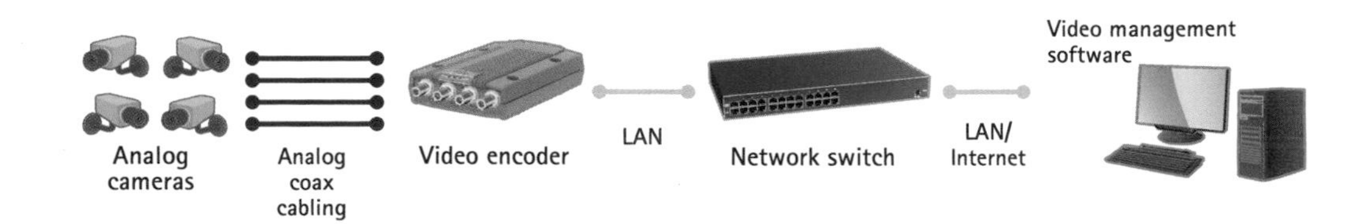

Figure 1.4 A true network video system, where video is continuously transported over an IP network. It uses a video encoder to migrate the analog security system into an open IP-based video solution.

Figure 1.5 An example of an NVR with pre-installed video management software and built-in switch.

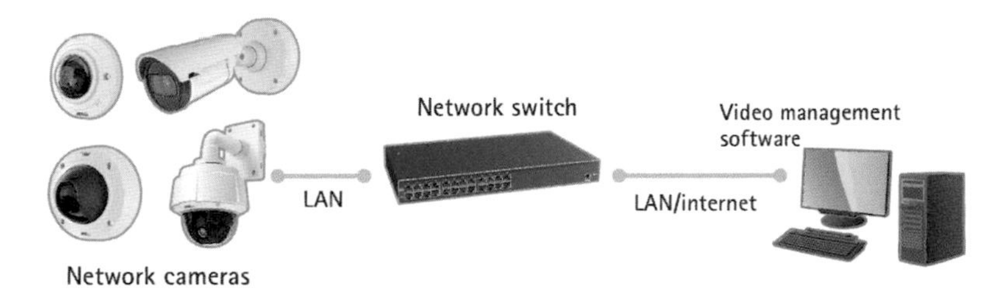

Figure 1.6 A true network video system where video streams are continuously transported over an IP network. This system takes full advantage of digital technology and provides consistent image quality from the cameras to viewers at any location.

and hybrid DVRs can be more difficult to maintain on a corporate IT network because they use proprietary platforms.

1.4.2 High-definition CCTV cameras

While other components of encoder-based systems continued to evolve, the analog cameras themselves were long found to be lacking, especially regarding resolution. An analog CCTV camera image is based on television lines (as opposed to pixels as in an IP camera), and for a long time, the maximum resolution available was 704 × 480 (NTSC). To rectify this situation, various HD-CCTV standards were developed, which provided 720p and 1080p analog resolutions. Further developments brought analog cameras with up to 4K resolution. These HD-CCTV standards also brought improvements in transmission distances and allowed all the signal traffic associated with the camera (audio, serial data, PTZ commands) to use the same coaxial cable as the camera. In many cases, the coaxial cable did not need to be replaced, as its capacity was sufficient to accommodate the increase in bandwidth.

Although it is now possible to achieve high resolutions with analog cameras, these products still offer little or none of the intelligence offered by digital network cameras. They also still require other dedicated hardware for their operation, which usually means connecting the camera to a compatible DVR. As with other analog cameras, these high-definition models can be connected to video encoders for integration into digital networks and for greater intelligence.

1.5 NETWORK CAMERA–BASED NETWORK VIDEO SYSTEMS

As its name indicates, a network camera – commonly called an IP camera – is a camera with an IP network connection. In a network camera–based video system, video is transported over an IP network through network switches and is recorded on a server running video management software (Figure 1.6). This represents a fully digital network video system.

One of the greatest benefits of a network camera is that when images are captured, they are digitized once only (inside the camera) and then remain digital throughout the system. As a result, the image quality is consistently high. This is not the case with analog cameras: Although they, too, digitize images to provide image-enhancing functions, they then convert the images back to analog. It is important to know that with every conversion from analog to digital or from digital to analog, there is some loss of video quality. Analog signals also degrade when transported over long cables and over time if stored on tape. Ideally, video should be digitized once only and remain digital throughout the system.

Another advantage of IP-based networks is that you can use the network for more than just transporting video. Besides allowing several network cameras to share the same physical cable, IP networks can carry the following:

- Power to compatible network cameras (Power over Ethernet)
- Information to and from the cameras' input and output (I/O) contacts
- Pan, tilt, and zoom (PTZ) commands
- Two-way audio

In addition, an IP network enables network cameras to be configured remotely and allows video and other data sent over the network to reach virtually any location without reductions in quality. To sum it up, IP networks provide an extremely flexible and cost-effective medium for all communication within a network video surveillance system. Network video is scalable: It can be used for any size of video surveillance system, from a single camera to thousands.

A network camera–based network video system offers many advantages:

- Access to high-resolution cameras
- Consistent image quality, regardless of distances in the network
- The same cabling can be used for video, power, PTZ, audio, digital input and output, and audio
- Access to wireless functionality
- Remote access to camera settings and focus over the network
- Access to edge intelligence and edge storage (that is, built into the camera)
- Full flexibility and scalability

Although an analog camera connected to a video encoder can be compared to a network camera, a network camera can offer many more value-adding functionalities.

The network camera is a key driver of innovation in the surveillance industry. Not only does it surpass analog video quality in many aspects, such as resolution and light sensitivity, but it also offers the advantages of built-in intelligence and storage.

Chapter 2

The components of network video

To understand the scope and potential of an integrated, fully digitized system, let us first examine the core components of a network video system: the network camera, the video encoder, the network, the server and storage, and the video management software.

One of the main benefits of network video is that its network infrastructure platform consists of commercial off-the-shelf equipment. The network hardware and software, such as cables, servers, storage devices, and operating systems, are standard IT equipment. The other components – the network camera, the video encoder, and the video management software – are unique to video surveillance and make up the cornerstones of network video solutions (see Figure 2.1). Another component is video analytics, which can be installed on the network camera or can be part of the video management software.

2.1 WHERE IS NETWORK VIDEO USED?

Network video, also known as IP-based video surveillance or IP surveillance, is a system that gives users the ability to monitor and record video and audio over an IP network, which can be a local area network (LAN), a wide area network (WAN), or the internet.

Unlike analog video systems – which use dedicated point-to-point cabling – network video uses standard IP-based networks as the backbone for transporting video and audio. In a network video system, digitized video and audio streams are sent over wired or wireless IP networks, enabling monitoring and recording from anywhere on the network.

Network video can be used in an almost unlimited number of applications. However, because the advanced functionalities of network video are highly suited to security surveillance, most applications fall into this category. The high-quality video, scalability, and built-in intelligence of network video enhances security personnel's ability to protect people, property, and assets.

Network video can also have many other purposes, such as the following:

- *Remote monitoring:* Monitor equipment, people, and places – locally and remotely. Application examples include traffic and production line monitoring.
- *Business intelligence:* Especially in retail environments. Application examples include people counting, heat mapping, and monitoring multiple retail locations.
- *Broadcasting video:* Over the internet or in broadcast production systems.
- *Healthcare*: Application examples include patient monitoring and providing a remote eye for a specialist doctor in an emergency care unit.

DOI: 10.4324/9781003412205-2

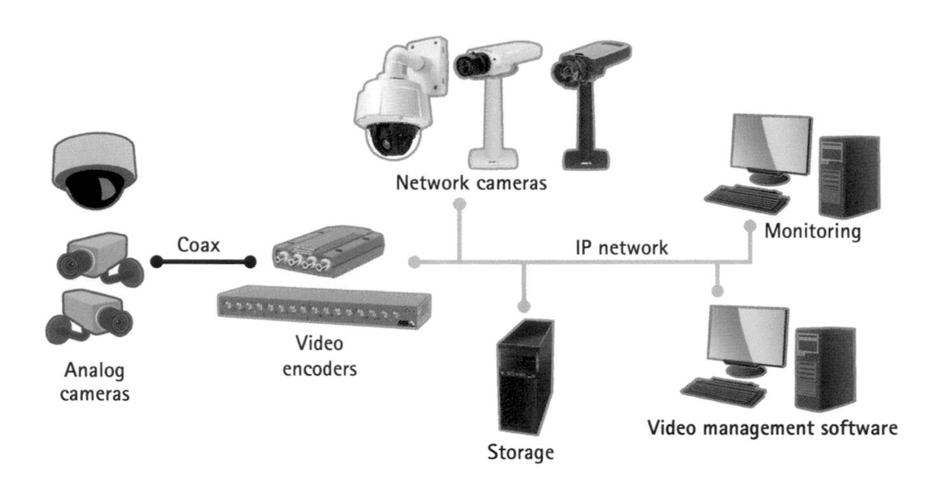

Figure 2.1 Typical network video system components, including network cameras, video encoders, and video management software. Other components, including the network, storage, and servers, are standard IT equipment.

Like most emerging technologies, network video was first deployed in a few key market segments. The primary initial driver for network video was scalability for very large installations in segments such as education, healthcare, and government. Today, network video is attractive for all market segments and system sizes.

The market segments where network video systems have been successfully installed include the following:

- *Education*: Secure and remote monitoring of school playgrounds, hallways, and classrooms at all levels of education.
- *Transportation*: Surveillance of railway stations and tracks, parking lots and garages, highways, seaports, and airports but also transportation environments, such as buses, trains, and cruise ships.
- *Banking*: Traditional surveillance in bank branches, headquarters, and ATM locations.
- *Government*: To ensure safe and secure government facilities, such as military bases, courthouses, prisons, and correctional facilities.
- *City surveillance*: Ensuring safe environments in city centers for citizens and tourists and helping police do their job more safely and efficiently.
- *Retail*: Surveillance that makes store management easier and more efficient, including business intelligence applications, such as people counting, and integration with other systems, such as point of sale (POS).
- *Healthcare*: Surveillance that provides security for both staff and patients in hospitals, on campuses, and in parking garages. Network video is also increasingly finding its way into applications such as drug diversion prevention and the monitoring of patients, where this is warranted.
- *Commercial*: Surveillance that provides safety for staff in midsize to large companies with thousands of employees and multiple locations, in office environments, at entry and exit points, warehouses, campus areas, parking lots, and garages.
- *Industrial*: Monitoring of manufacturing processes, logistics systems, warehouses, and inventory control systems.
- *Critical infrastructure*: Monitoring of public and private infrastructure vital to society's functioning, for example: power stations, cell phone towers, and water supplies.
- *Data centers*: Protecting against intruders at the perimeter and recording all movements within the buildings themselves.

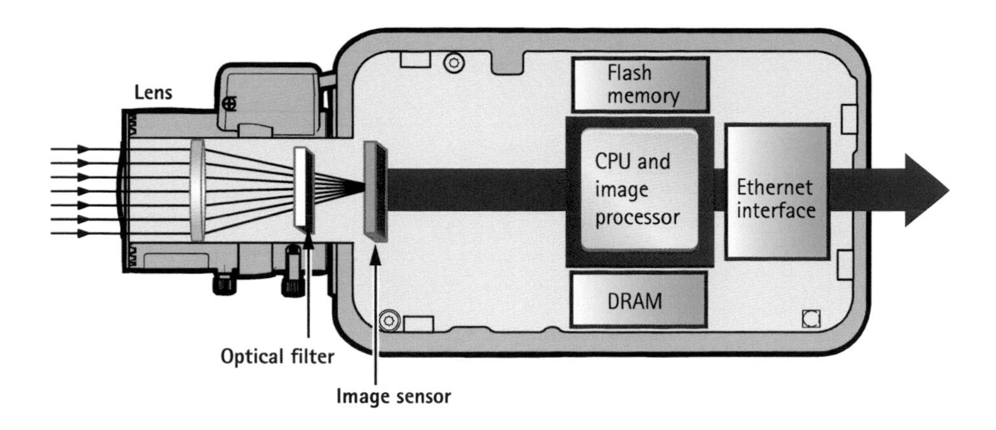

Figure 2.2 Inside a network camera. The lens exposes the image onto an image sensor connected to a processor, often an application-specific integrated circuit (ASIC) or a digital signal processor (DSP). The camera has flash memory and dynamic random-access memory (DRAM) and connects to the network through a network interface.

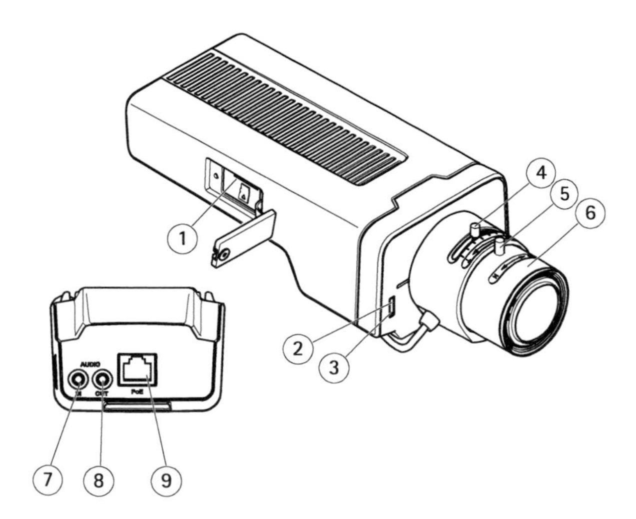

Figure 2.3 A typical network camera. Front view: (1) microSD card slot. (2) Status LED. (3) Built-in microphone. (4) Zoom puller. (5) Lock screw for focus ring. (6) Focus ring. Back panel: (7) Audio in. (8) Audio out. (9) Network connector, with Power over Ethernet (PoE).

2.2 THE NETWORK CAMERA

A network camera, also called an IP camera, can be described as a camera and computer combined into one unit. It captures and sends live images directly over an IP network, enabling authorized users to view, store, and manage video over standard IP-based networks. See Figure 2.2 and Figure 2.3.

A network camera has its own IP address and can be placed wherever there is a network connection. It has a built-in web server, email client, intelligence, storage, programmability, and much more. A web camera, or webcam, is completely different. This can be a separate camera connected to a computer through a USB port, or it can be a camera integrated into a laptop or smartphone.

2.2.1 Network cameras today

As well as outperforming their analog counterparts, network cameras also offer a number of advanced features, such as high image quality, high resolutions, onboard storage, audio, and built-in intelligence.

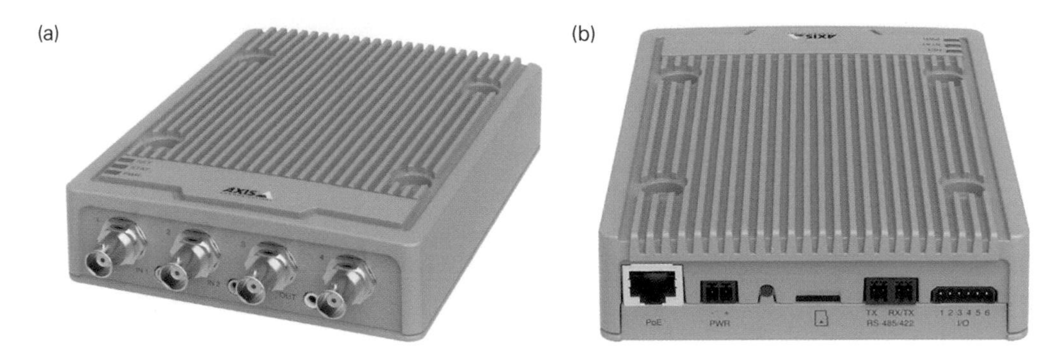

Figure 2.4 A typical video encoder. (a) Front view showing BNC connectors. (b) Rear view showing Ethernet and other connectors.

A network camera is fully bi-directional, allowing information to be sent and received. It can be an integrated part of a larger scalable system, communicating simultaneously with several applications to perform various tasks, such as detecting motion, objects, or sending multiple different video streams.

Read more about network cameras in Chapters 3 and 4. For more about the technologies these cameras use, including topics such as compression and audio, see Chapters 6 and 7.

2.3 THE VIDEO ENCODER

A video encoder integrates an existing analog camera into a network video system, bringing new functionality to the camera and eliminating the need for the associated dedicated equipment.

A video encoder provides Bayonet Neill–Concelman (BNC) (sometimes British Naval Connector) ports for connecting analog cameras and an Ethernet port for connection to a network, see Figure 2.4. The encoder digitizes the incoming analog feed so that it can be transmitted over an IP network, for viewing and recording.

The video encoder can provide a range of other functionalities, including audio, digital inputs and outputs, and PTZ (pan, tilt, zoom) control. Practically any kind of analog camera can be connected to a video encoder, including HD-CCTV cameras and specialized cameras.

Video encoders will continue to play an important role in video surveillance systems. Chapter 8 provides more information about video encoders.

2.4 THE NETWORK

IP networks range in design and size from small wireless networks to corporate networks to the internet. Just like electrical wiring and plumbing, wiring for IP networks is included already at the planning stage of most buildings.

One of the most obvious benefits of network video is its ability to use these standard IP-based networks not only for viewing and recording video but – in the case of wired networks – also for carrying power to devices that support Power over Ethernet (PoE).

Whereas bandwidth used to be a bottleneck, today's gigabit networks provide more than enough bandwidth for network video. Even cellular networks support video transmission at reasonable cost.

Network security is becoming an increasingly critical issue, especially for sensitive data, such as bank transactions, government documents, and surveillance video. Therefore, it is important to understand network security technologies, such as 802.1X, HTTPS, and firewalls. Read more about cybersecurity in Chapter 12.

See Chapter 9 for more details about network technologies, Chapter 10 for more about wired networks, and Chapter 11 for more on wireless networks.

2.5 SERVERS AND STORAGE

The server and storage components used in network video are based on standard IT equipment. Most servers running video management software (VMS) use a Windows® operating system or sometimes a Linux® or Unix™ system. Today's powerful servers make it possible to manage video from hundreds of cameras per server. Several different types of hard drive are used by these servers, including traditional spinning drives and solid-state disk (SSD) drives.

To ensure redundancy on these storage servers, RAID (redundant array of independent disks) arrays are often used. Some network video installations use network-attached storage (NAS) or storage area networks (SANs).

Using a server is required for larger installations, whereas smaller systems (up to 16 cameras) can use edge storage, such as a NAS solution or SD memory cards inside the cameras. Edge storage means that the cameras manage the recordings themselves, eliminating the need for external servers or storage.

For more information about disk types, servers, and storage, see Chapter 13.

2.6 VIDEO MANAGEMENT SOFTWARE

For very small systems, the camera's built-in web interface can provide the video management functions, as long as only one or two video streams are being viewed and no recording is required. Dedicated video management software (VMS) is needed to effectively view multiple video streams at the same time. This software is at the core of video management, monitoring, analysis, and recording (see Figure 2.5).

Hundreds of different video management applications are available, running on all types of operating systems and covering all vertical markets and scalability requirements. There are also

Figure 2.5 A screenshot of a video management software interface.

numerous integration possibilities with other systems, such as building management, access control, and industrial control. Open platform solutions often run on off-the-shelf hardware or on a purpose-built server.

Camera manufacturers publish application programming interfaces (APIs) that ensure integration and compatibility with video management software. The more open the API, the better the collection of video management software and the tighter the integration for a given network camera. Ideally, the API should be published openly and be free of charge. There are also established standards, such as ONVIF (see section 18.2.7), for interfacing to network cameras and encoders.

Some of the common features provided by video management software include the following:

- *Viewing and recording from multiple cameras*
- *Multiple recording modes:* continuous, manual, scheduled, alarm-activated
- *Multiple search functions for recorded events*
- *Camera management:* Configure multiple cameras from a single interface
- *Remote access:* browsers and mobile devices
- *Control of PTZ and panoramic cameras:* manual or automatic control
- *Configuration of I/Os:* automatic activation of external devices
- *Alarm management:* alarms, popup windows, email, and texts (SMS)
- *Audio support:* Real-time support for live or recorded audio. Support for IP speakers.

Video management systems are easily scalable, and some can scale to thousands of cameras. They also make it easier to add further functionality to the system, such as increased or external storage, firewalls, virus protection, and video analytics.

Video management systems based on open platforms can be more easily integrated with other systems, such as access control, building management, and industrial control, allowing control of multiple systems from a single interface.

For more about video management software, see Chapter 14.

Cloud-based video management means that cameras automatically record and make live or recorded video available in the cloud, giving increased scalability for small systems, easier remote system management, and increased security by recording video off-site. The technology also opens up for providing Video Surveillance as a Service (VSaaS). For more information about cloud video, see Chapter 15.

2.7 VIDEO ANALYTICS

The market for video analytics (also known as intelligent video) is driven by the need to increase the value of a video surveillance system and make it more proactive (see Figure 2.6). Video analytics make response times faster, reduce the number of operators, and can extract information for uses beyond traditional surveillance, such as people counting. Research from Sandia National Laboratories in the United States has shown that an operator is likely to miss important information after only 20 minutes in front of a monitor. Analytics can help operators handle more cameras and respond more quickly.

In a centralized system, all the analytics functionality resides in the video management system. Intelligent video algorithms consume a lot of computing power, which limits the number of cameras that can be managed by one server.

In a distributed system, the analytics are distributed to the edge and reside in the network camera. The main benefit of this is that the analysis is performed on uncompressed video in the camera,

Figure 2.6 A typical system using video analytics, which are often at the edge, that is, uploaded to the cameras or encoders. Analytics can also reside centrally, in the video management system.

which makes the results more accurate. It also makes the system fully scalable and potentially reduces bandwidth because the camera can determine – based on the video content – whether or not to send video to the server. The latest trend using deep learning at the edge has led to more and more analytics residing at the edge, providing video and metadata to the VMS and cloud.

Video analytics are necessary in systems that require fast response times or where very large numbers of cameras need to be proactively managed. For more information about video analytics, see Chapters 16 and 17.

Chapter 3

Network cameras

Network cameras, or IP cameras, offer a wide variety of features and capabilities to meet the requirements of almost any surveillance system. Today, network cameras offer improved image quality, higher resolution, and built-in storage and intelligence.

This chapter presents the components of a network camera and the different camera types. It also gives in-depth information about PTZ cameras, panoramic cameras, day-and-night cameras, as well as HDTV, and Ultra-HD network cameras. More information about camera technologies is provided in Chapter 4.

3.1 NETWORK CAMERA COMPONENTS

Like a computer, a network camera has its own network connection and its own IP address. It can be connected directly to a network wherever there is a suitable connection and can send live and recorded video and audio over the network.

The main components of a network camera (see Figure 3.1) include the following:

- *Lens*: for focusing the image on the image sensor
- *Image sensor*: either a CCD (charge-coupled device) or CMOS (complementary metal oxide semiconductor)
- *Processor*: one or more, for image processing, compression, video analysis, and networking functionality
- *Flash memory*: for storing the camera's firmware code
- *RAM*: random-access memory
- *SD card*: for local recording of video (optional)

Analog cameras use primarily two formats: PAL (the European CCTV standard) and NTSC (the American CCTV standard). In network video, both PAL and NTSC are irrelevant because network cameras use globally established resolution standards, such as HDTV 720p/1080p and 4K. Network cameras connect to Ethernet networks to transport video that is usually compressed in compliance with standards such as H.264 and H.265. This makes planning and servicing video surveillance deployments much easier. For more information about video compression formats, see Chapter 6.

Most network cameras also offer input/output (I/O) ports that enable connections to external devices such as sensors and relays. Built-in support for Power over Ethernet (PoE) eliminates the need for separate power cabling to the camera. Another time and money saver in the installation process is the ability to remotely set the field of view and focus the camera.

DOI: 10.4324/9781003412205-3

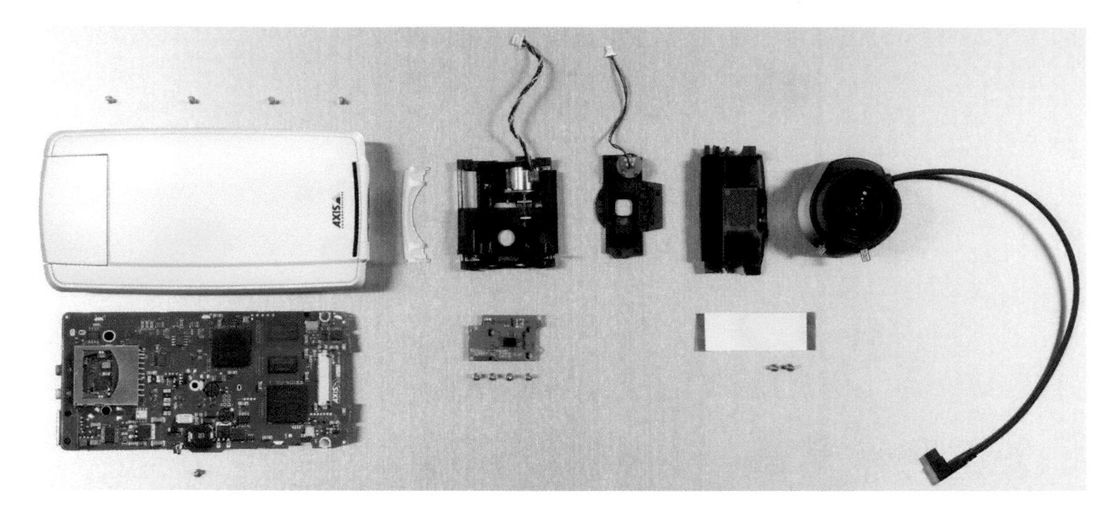

Figure 3.1 Components of a network camera.

A network camera can have many advanced features, such as built-in analytics, audio monitoring, and alarm handling. The networking functionality is essential and should include all the latest security and IP protocols. Different cameras offer different performance levels. Some cameras provide less than the full video frame rate, generally considered to be 25/30 fps, whereas others can provide 60 fps or several simultaneous video streams at full frame rate.

3.2 TYPES OF NETWORK CAMERAS

Network cameras can be grouped in several ways, but perhaps the most common grouping is according to viewing functionality. The most common types are listed here:

- *Fixed cameras:* This type of camera has a fixed viewing direction, which is set during installation. Fixed cameras can be further grouped by appearance:
    - *Fixed box camera:* Has a shape easily recognizable as a camera. The viewing direction is obvious, and the camera acts as a deterrent.
    - *Fixed dome camera:* The camera is housed behind a plastic dome, making it more discreet and more difficult to see the viewing direction.
    - *Bullet camera:* Has a slim, tubular design and is often equipped with built-in LED lighting.
- *PTZ cameras:* Cameras in which the operator can change the camera's viewing direction using, for example, a joystick or a mouse. In this context, we are talking about mechanical PTZ cameras, but many cameras also offer digital PTZ.
- *Panoramic cameras:* These use one or more sensors and wide-angle lenses to provide wide area coverage.
- *Modular cameras:* These are very small devices in which the camera sensor is usually decoupled from the main processor unit by a cable of up to 10 m (33 ft).
- *Thermal cameras:* These generate images based on the heat emitted by all objects.

Other ways of classifying network cameras are:

- By the intended installation environment, which can be further specified by the following:
    - *Indoor use only or for both indoor and outdoor use.* This applies generally to the ambient levels of moisture and dust, as well as to the prevailing extremes of temperature.
    - *By the level of protection* – many cameras are provided as vandal-resistant, and some use an anti-ligature design. For more information about enclosures, see Chapter 18.

- By *resolution*, with some camera models offering very high resolutions.
- Cameras for particular environments:
  - *Onboard cameras:* For use in transportation installations that subject cameras to extreme levels of dust, vibration, shocks, and temperature fluctuations.
  - *Body-worn cameras:* These are designed specifically to be worn by law enforcement officers and security guards but can also be useful in healthcare and retail environments to provide live and/or recorded video to augment the overall video surveillance system.
  - *Explosion-protected cameras:* These are housed in special casings that reduce the risk of explosion in potentially combustible areas.

3.3 FIXED BOX CAMERAS

A fixed box camera is easily recognizable as a camera and has a fixed viewing direction, making it the best choice in situations where the camera should be very noticeable. A fixed box camera may come with a fixed, varifocal, or motorized zoom lens. Most fixed box cameras support exchangeable lenses, which is convenient and cost-effective because the lens can be swapped if requirements change. Fixed box cameras can be installed in protective enclosures and can also be mounted on a pan-tilt motor for greater viewing flexibility. Figure 3.2 shows an example of a fixed box camera.

3.4 FIXED DOME CAMERAS

Fixed dome cameras consist of a fixed camera pre-installed in a small dome-shaped housing. The camera is easily adjusted to point in any direction. This camera's main benefit lies in its discreet, unobtrusive design. In addition, persons in the camera's field of view find it difficult to see the direction in which the camera is pointing. A fixed dome camera is also more resistant to tampering than a fixed camera.

Fixed domes can be grouped in several ways, including the following:

- *Number and type of sensors, which include:*
  - *Single-sensor fixed domes*
  - *Multisensor fixed domes*
  - *Multi-directional, multisensor domes*
- *Design and appearance*, including models for use in transportation vehicles, models with anti-ligature designs, corner-mount designs, and models with built-in LED lighting

One of the limitations of some fixed dome cameras is that they do not support exchangeable lenses, as the space inside the dome housing is limited. However, a varifocal lens is often available, which enables adjustment of the camera's field of view, manually or remotely if the lens is motorized.

Figure 3.2 A fixed box camera.

Figure 3.3 A fixed dome camera.

Figure 3.4 An example of a bullet camera.

Fixed dome cameras are designed with different types and levels of protection, such as vandal and dust resistance, and IP66 and NEMA 4X ratings for outdoor installations. These cameras are usually mounted on a wall, ceiling, or pole. Figure 3.3 shows an example of a fixed dome camera.

3.5 BULLET CAMERAS

Bullet cameras are like fixed box cameras in that they are a visible deterrent due to their shape. However, bullet cameras usually have a more compact design, and their shape is cylindrical, hence, the name. Many bullet cameras are equipped with built-in LED lighting to provide images even in dark conditions. Different resolutions and fields of view are supported by different models, and most can be installed outdoors, with no extra housing required. Figure 3.4 shows an example of a bullet camera.

3.6 PAN, TILT, ZOOM (PTZ) CAMERAS

A PTZ camera is designed with electrical motors that allow the camera to move in three dimensions and as such is often used in situations where an operator is present and can control it. By panning, tilting, and zooming, the camera can move to show another part of the scene being monitored. For example, an operator can follow a person through a PTZ camera that they control manually with a joystick. A PTZ camera's optical zoom typically ranges between 3× and 40×. Figure 3.5 shows an example of a PTZ camera.

Figure 3.5 A PTZ camera.

Figure 3.6 A PTZ camera that lacks full panning capability but which is still useful in many surveillance scenarios and generally comes at a lower cost.

Through their pan, tilt, and zoom functionalities, PTZ network cameras can cover a wide area. How much a PTZ camera can move differs between types and models. The most advanced models offer a full 360° pan angle and usually a tilt angle of 180°, but sometimes even more, so they can see "above the horizon".

Figure 3.7 A wall-mounted PTZ camera.

Figure 3.8 A PTZ camera in a drop-ceiling mount. When installed, only the trim ring and the dome-covered lens assembly are visible below the ceiling.

PTZ cameras are ideal in discreet installations thanks to their design. Some mounting options allow the camera to blend in and for part of the camera to be installed in a ceiling. These mounts are known as drop-ceiling mounts, recessed mounts, or flush mounts (see Figure 3.8). The dome (clear or smoked) makes it difficult for people to identify the camera's viewing angle. A high-end PTZ network camera also has the mechanical robustness that is needed to run in guard tour mode, in which the camera moves continuously between presets, allowing one PTZ network camera to cover an area where several fixed cameras would otherwise be needed. The main drawback is that only one location can be monitored at any given time, leaving the other positions unmonitored.

The PTZ network camera gets all its PTZ commands over the IP network, removing the need for extra cabling. Also, most PTZ network cameras can be powered using Power over Ethernet (PoE) instead of a dedicated power source. Both these factors help lower the total installation cost, and it is more cost-efficient to install PTZ cameras than it was in the past.

PTZ cameras provide several unique benefits, and network PTZs have increased the capabilities of the cameras by increasing resolution, providing better image quality, as well as local storage and built-in analytics.

Another advantage of some PTZ cameras is the use of high-resolution (HDTV) sensors. Because these sensors use progressive scan, there is less motion blur in the images they produce, which is particularly beneficial for moving cameras, as movement adds blur to the images if traditional interlace scanning technology is used. For more information about interlaced and progressive scanning, see Chapter 4.

3.6.1 Pan, tilt, roll, and zoom (PTRZ)

Pan, tilt, roll (sometimes rotate), and zoom (PTRZ) is a feature included in some fixed dome cameras. This allows the installer to set the camera's view during installation, by remotely maneuvering the camera to the desired viewing direction and zoom level. This is achieved by using the same type of motors as in a true PTZ camera, with the addition of an extra motor that adjusts the rotation of the lens. However, this feature is only intended for setting the initial view during installation and is not designed for continuous operation.

3.6.2 Presets and guard tours

Many PTZ cameras use preset positions, sometimes 100 or more (Figure 3.9). Once the preset positions have been set in the camera, it is quick and easy for the operator to go from one position to the next. Many PTZ cameras also have built-in guard tours (Figure 3.10). A guard tour automatically moves the camera from one preset position to the next, in a predetermined or random order. The viewing time between one position and the next is configurable. Different guard tours can be active at different times of the day.

3.6.3 E-Flip

Imagine a scenario in which a PTZ camera is mounted in a ceiling in a store and is tracking a person suspected of shoplifting. If the person passes directly below the camera, a camera with E-flip electronically and automatically rotates the images 180° so that the person and the rest of the image remain the right side up. Without E-flip, the whole image would be upside-down when the person passes underneath the camera.

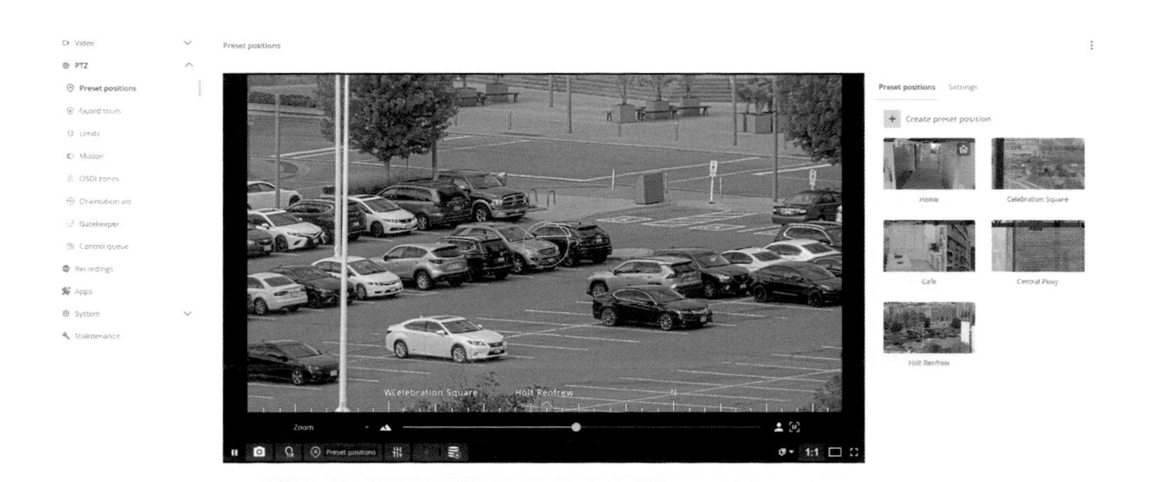

Figure 3.9 A typical interface for preset positions. Some cameras allow the creation of hundreds of preset positions.

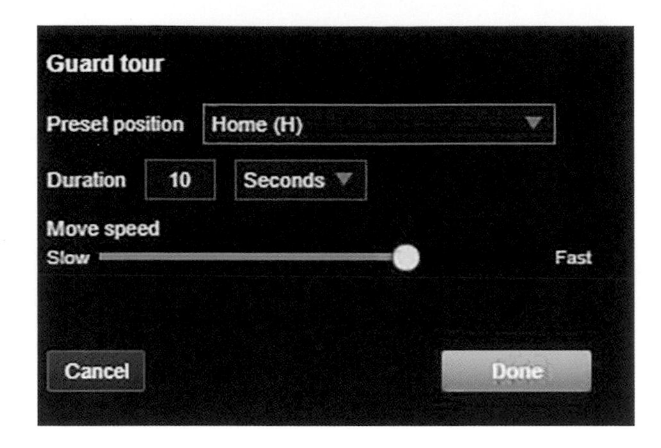

Figure 3.10 Example of a dialog for guard tour setup.

Figure 3.11 A camera with auto-flip can deliver 360° panning even if it has a mechanical stop. The camera pans 180°, tilts to its starting position, and then continues the panning operation.

3.6.4 Auto-Flip

The slip ring is the mechanical component that allows the camera to do full 360° panning. Some PTZ cameras do not have slip rings because of cost or size limitations, and these cameras have a mechanical stop to prevent 360° panning. Some cameras can bypass the mechanical stop by automatically freezing the image for a second while panning 180° and simultaneously tilting the camera block to the previous position. The camera then continues the smooth panning that was started by the operator. These movements are illustrated by Figure 3.11.

3.6.5 PTZ performance

The mechanical performance levels of PTZ cameras can differ greatly. Datasheets often specify the maximum performance in degrees per second. If the stated performance is more than 360° per second, this means the camera can complete a full circle in less than one second. Normally, this is considered adequate for a high-performance PTZ camera. Some PTZ cameras can move even faster – up to around 450° per second. Although speed is important, the camera's movements also need to be controllable at very low speeds. Therefore, some PTZ cameras specify the slowest speed at which they can be controlled. For example, a speed of 0.05° per second means it will take two hours to complete a full 360° rotation.

3.6.6 Joystick control

Joysticks make it very easy to control a PTZ camera. USB joysticks, which connect to the PC used for video monitoring, are commonly used in network video applications. Most professional joysticks also come with buttons that can be used for presets. Figure 3.12 shows an example of a PTZ joystick.

3.7 MODULAR CAMERAS

Modular cameras are designed to blend into the environment and be virtually impossible to discover. Analog versions have existed for a long time and have often been used together with network video encoders (see Chapter 8 for more about encoders). To get all the benefits of network video while making the camera as discreet as possible, the sensor unit has a high-resolution sensor that is often decoupled from the central unit that contains the processor. The cable between the senor unit and central unit is typically less than 10 m (33 ft). In many cases, you can combine different central and sensor units in several different ways. Figure 3.13 shows an example of a modular camera.

One scenario for these cameras is to install a pinhole camera at eye level, for example, flush-mounted in an ATM to provide discreet surveillance. There are also height-strip cameras for placement at

Figure 3.12 A pan, tilt, and zoom joystick.

Figure 3.13 A modular camera. A cable connects the sensor unit to the central unit (also known as the main unit).

store entrances. These cameras offer general overview surveillance or close-up shots for identification. Because they are so small and very hard to detect, modular cameras are subject to very few tampering attempts.

3.8 THERMAL CAMERAS

Thermal cameras create images based on the heat that every object radiates. Figure 3.14 shows an example of a thermal camera. Images are generally produced in black-and-white but can be artificially colored to make it easier to distinguish different shades. A thermal camera can see what a standard camera that works in the visible and near-infrared range cannot (see Figure 3.15). The contrast in thermal images is higher when there are greater temperature differences in a scene.

Thermal cameras are ideal for detecting people, objects, and incidents in deep shadow, complete darkness, or in other challenging conditions, such as smoke and dust. Because thermal images do not enable reliable identification, thermal cameras are used primarily to detect suspicious activities. They complement and support conventional network cameras in a surveillance installation by providing perimeter or area protection, discreet surveillance, and security in dangerous or off-limits areas such as tunnels, railway tracks, and bridges.

Figure 3.14 An outdoor thermal camera.

Figure 3.15 The same scene as seen by a standard camera (left) and a thermal camera (right).

A thermal camera can also be designed as a thermometric camera, which can trigger alarms if the temperature rises above or falls below a set threshold. These cameras are ideal for remote monitoring of power-generating facilities, fire hazard areas, or industrial processes involving self-igniting materials. For detailed information about thermal cameras, see Chapter 5.

3.9 PANORAMIC CAMERAS

A panoramic camera can be thought of as a nonmechanical PTZ camera. Depending on the type, the camera delivers 180–360° surveillance, providing full overview images as well as close-up images but with no moving parts. This is achieved by taking advantage of the high resolution provided by megapixel sensors and wide-angle lenses. These cameras can also provide multiple video streams from different fields of view. Panoramic cameras often look like fixed dome cameras (see Figure 3.16).

Wide-angle lenses and hi-res sensors offer multiple viewing modes, such as 360° overview, 180° panorama, quad views (simulating four different cameras), and view areas with support for digital PTZ functionality.

A panoramic camera allows the operator to zoom in on any part of a scene without mechanical moving parts, which eliminates wear and tear – the pan, tilt, and zoom movements are all digital. Zooming in on a new area is immediate, whereas a mechanical PTZ camera can take up to one second for the same operation. Because a panoramic camera's viewing angle is not visible, it is ideal for discreet installations. The disadvantage is that the image detail is nowhere close to that of a real PTZ camera when zooming in at more than 4× or 5×. The higher the resolution of the sensor, the better the zoom quality, but this comes at the price of decreased light sensitivity.

Panoramic cameras provide their functionality in different ways and can be grouped as follows:

- *Single-sensor cameras:* Providing a 360° view or a fish-eye view, these cameras use image dewarping to present natural-looking images.
- *Multisensor 180° cameras:* These cameras stitch the images together to achieve a high-quality panoramic view and single-camera behavior for the operator. The sensors are fixed direction and not adjustable.
- *Multisensor, multi-directional cameras:* The sensors in these cameras can be freely moved and rotated to provide the desired coverage, which can be overviews or zoomed in and detailed parts of the scene.
- *Multi-directional cameras with PTZ:* These are combination products that include several adjustable sensors in a ring and a PTZ camera mounted in the center of the ring. This allows an operator to quickly move in on and zoom to a particular area seen in the panoramic view.

Note that even though a panoramic network camera may include multiple sensors, in some cases, only a single IP address is required for the whole product to connect to the network. This also

Figure 3.16 A panoramic camera using a single sensor and a fish-eye lens.

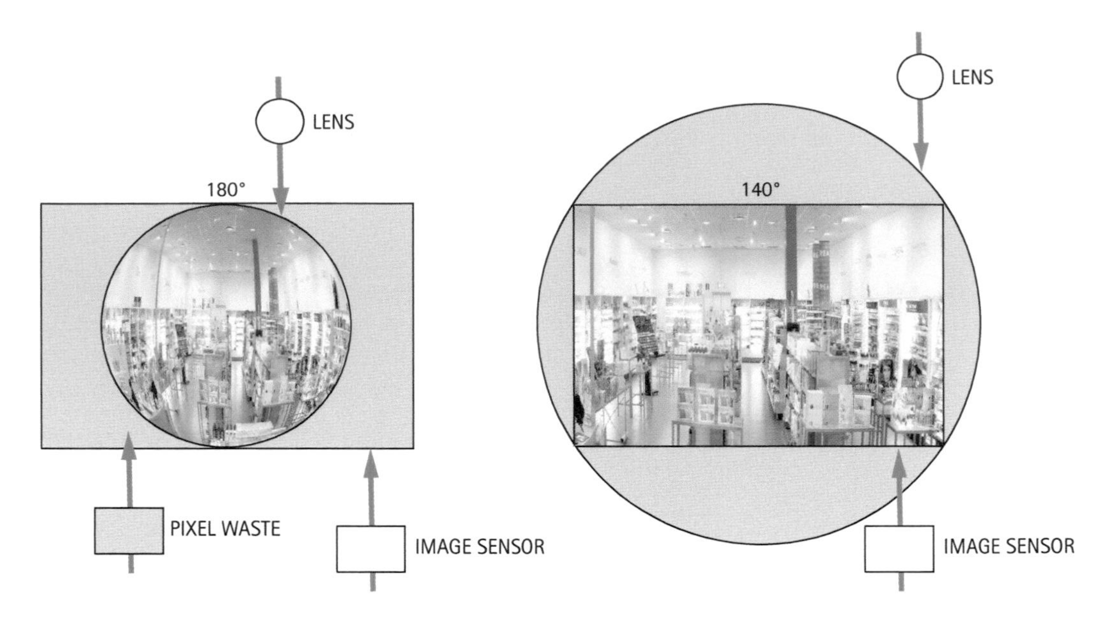

Figure 3.17 (a) With 180° field of view, only part of the sensor is used to make the image. (b) With 140° field of view, all the sensor's pixels are used to make the image.

means that any software used with the camera, for example, video management software (VMS) or an analytics application, also only requires a single license. However, if a VMS charges per video stream, then you will need one license per stream.

3.9.1 Selecting the right viewing angle

Depending on the type of lens used, a panoramic camera can have a viewing angle of up to 180°. A wide viewing angle, however, comes with some challenges, one of these being that the image is greatly distorted by the fish-eye effect of the lens. Therefore, the image needs to be de-warped to make it viewable. De-warping requires a lot of processing power in the camera or viewing station. Another problem is that only a very limited part of the image sensor can be used, which reduces resolution and thereby image quality (see Figure 3.17).

The images in Figure 3.18 illustrate the difference between different fields of view.

3.9.2 Cameras with wide viewing angles

Cameras with extra wide viewing angles, typically 120–150°, are technically in-between what are traditionally referred to as panoramic cameras and regular cameras. They offer surveillance for large areas (see an example in Figure 3.19). In this case, the horizontal and vertical fields of view are combined with a highly effective megapixel sensor to provide a sharp image with good detail, especially in the corners. When this kind of camera is housed in a vandal-resistant casing, it is ideal for effective surveillance in public transport terminals, stores and malls, hotels, banks, and school areas.

The camera in Figure 3.19 supports digital PTZ, which can be used to simulate a varifocal lens to adjust the camera's field of view remotely. The camera's digital PTZ and multiview streaming abilities allow different areas of a scene to be cropped out from the full view. This gives the camera the ability to act as several virtual cameras and helps minimize the bitrate and storage needs. The camera can send multiple simultaneous streams at different compression rates. For example, one stream can be used for viewing and another for recording.

3.9.3 180° panoramic cameras

A 180° panoramic camera provides panoramic views in a high-resolution image. The camera can be mounted on a wall or ceiling, and the example in Figure 3.20 shows an image from a wall-mounted,

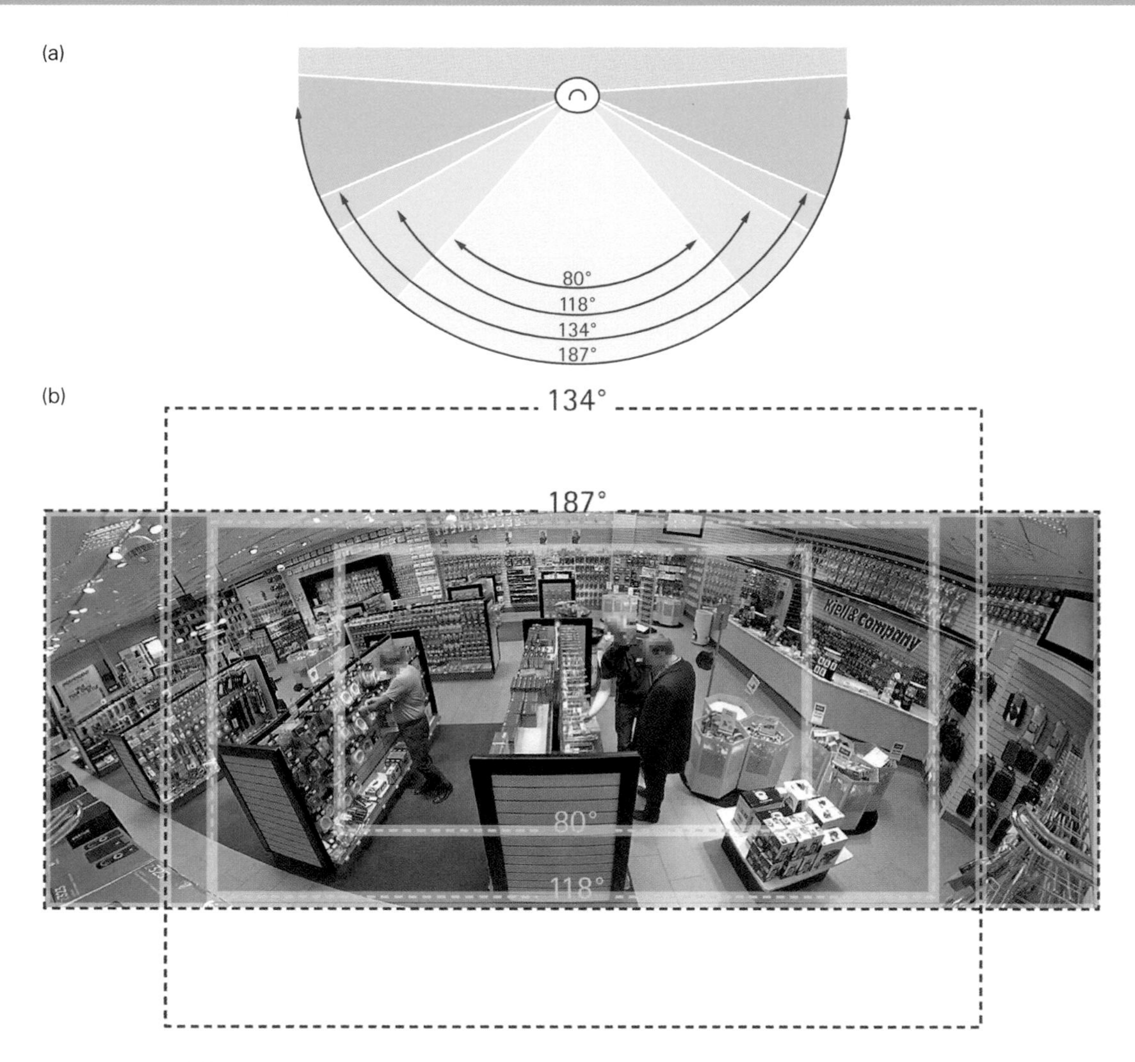

Figure 3.18 Different viewing angles have different fields of view. (a) The diagram shows the different viewing angles. (b) The image shows the resulting field of view for different viewing angles, where the 187° view is a panoramic view.

Figure 3.19 A corner view from a panoramic camera with 134° viewing angle.

Figure 3.20 Example of a panorama view.

fixed mini-dome that also offers digital PTZ and multiview streaming with de-warped views. It is ideal for public areas where event monitoring such as motion detection is required.

3.9.4 360° Panoramic cameras

When mounted in a ceiling, a 360° panoramic camera provides a variety of views. Figure 3.21 shows images for the same scene from a ceiling-mounted panoramic camera. Different aspects of the scene can be viewed from different perspectives.

This kind of camera can also offer quad view, which is useful, for example, when the camera is positioned at the intersection of two aisles. The operator can digitally pan, tilt, and zoom into areas of interest in four individually cropped-out and de-warped view areas (see Figure 3.22.)

3.9.5 Multisensor panoramic cameras

Multisensor systems are gaining ground in the surveillance industry. These systems use multiple sensors paired with individual lenses to create an omnidirectional system that provides panoramic images with no distortion. Figure 3.23 shows an example of a multisensor panoramic camera.

There are other advantages with multisensor cameras. The installation process is more efficient, as a greater area can be covered by fewer units, as well as the fact that these cameras often use a single network connection (just one port on a switch). Also, as a VMS may count a multisensor camera as a single camera, this may mean fewer licenses are required.

As a multisensor camera typically includes three or four individual cameras combined into one unit, this makes them more expensive than wide-angle panoramic cameras with a single sensor. Also, images from multisensor cameras usually overlap and are often stitched together to make a complete image. Figure 3.24 shows an example of a stitched image from a multisensor camera.

3.9.6 Comparing and combining panoramic and PTZ cameras

A panoramic camera offers a wider field of view than a mechanical PTZ camera, and there is no wear and tear caused by moving parts. The PTZ camera delivers much better image detail, especially for objects far away. The typical application for a panoramic camera is for situational awareness – to detect *whether* a specific activity is occurring. Identifying *what* is happening normally requires multiple fixed cameras and one or more PTZ cameras. The images in Figure 3.25 illustrate the difference in field of view between panoramic and PTZ cameras.

Sometimes, the choice between overview and detail is very difficult to make. Perhaps a security manager wants the best of both worlds without having to spend money or time setting up several different cameras to monitor a single large area. The solution is a combination that includes a multisensor camera and a PTZ camera. All the integrated cameras in this combination product can be

(a)

(b)

Figure 3.21 (a) An overview image. (b) The results from the same scene after de-warping.

powered and controlled through the same network cable. This type of camera is sometimes referred to as an omni-camera. It has a permanent panoramic overview as well as the ability to focus on an area and follow a subject once activity has been detected. This is an ideal camera for many situations and areas, such as city squares, stadiums, parking lots, airports, and logistic centers. Figure 3.26 shows an example of this type of combination camera.

3.10 ONBOARD CAMERAS

Onboard cameras are specially designed for discreet and efficient surveillance in trains, subway cars, buses, and emergency vehicles. The environment in onboard surveillance requires a particularly rugged design. Cameras need to be dust- and water-protected, and they need to withstand tough conditions, such as vibrations, shocks, bumps, and temperature fluctuations. The modest appearance, which is usually based on a fixed mini-dome camera, is often combined with an active tampering alarm that helps detect and prevent tampering attempts, such as redirection and defocusing. Figure 3.27 shows an example of an onboard camera in its intended environment.

Figure 3.22 Example of a quad view from a ceiling-mounted panoramic camera.

Figure 3.23 A multisensor camera.

Figure 3.24 Example of a stitched image from a multisensor camera.

Figure 3.25 (a) PTZ cameras can cover larger areas and can see details far away, but they can only see one section a time. (b) Cameras with wide and panoramic viewing angles give an overview of the whole area.

Figure 3.26 A multisensor and PTZ camera combination. The multisensor camera has four sensors and lenses for monitoring a larger area. It also controls the integrated PTZ camera. This solution allows the operator to keep the big picture and to zoom in on details at the same time.

Figure 3.27 An onboard camera mounted in the ceiling of a bus.

3.11 BODY-WORN CAMERAS

Body-worn cameras are designed for use by personnel in various professions, for example, by police officers in their interactions with the public. Common characteristics of body-worn cameras are their low weight, robust design, water-resistance, and ease-of-use. Some camera models also support the attachment of a mini sensor (also a camera), for example, for mounting on a helmet, if a particular situation should require this.

A body-worn camera is battery-powered and usually comes with a docking station that provides both battery charging and uploading of recorded material to the central system. The docking station may have a system controller built into it, that is, the intelligence that handles the recording upload. Alternatively, the system controller may be available as a separate component, with support for connecting multiple docking stations.

Figure 3.28 Body-worn cameras, as carried by, for example, patrolling police officers and security guards.

3.12 EXPLOSION-PROTECTED CAMERAS

When a camera is installed in a potentially explosive environment, its housing must meet very specific safety standards. An explosive atmosphere is defined as an area where flammable substances, such as liquids, vapors, gases, or combustible dusts, are likely to occur and mix with air in such proportions that excessive heat or sparks might make them explode. Examples include gas stations, oil platforms and refineries, chemical processing plants, printing industries, gas pipelines and distribution centers, grain handling and storage facilities, aircraft refueling and hangars, and hospital operating theaters.

It is the explosive environment that must be protected from potential igniters from the camera and other equipment. In other words, the camera must be explosion-protected, explosion-proof, or flame-proof. These terms are used synonymously and mean that the camera will contain any explosion originating within its own housing and will not generate conditions that could ignite vapors, gases, dust, or fibers in the surrounding air. It does *not* mean that the camera itself will withstand an exterior explosion. For more on explosion-protected cameras, see Chapter 18.

Figure 3.29 An example of a docking station for body-worn cameras.

Figure 3.30 Example of a pan, tilt, and zoom camera designed and certified for use in potentially explosive atmospheres and harsh environmental conditions. Because the housing is made of stainless steel, it is suitable for offshore, onshore, marine, and heavy industrial environments.

3.13 BI-SPECTRAL PTZ CAMERAS

This type of camera combines visual and thermal video streams in a single PTZ camera. As the thermal part of the camera also has zoom capability, this is used for long-range detection. The optical camera can then be used for identification and PTZ capabilities. These cameras are often used in sensitive installations where early detection is important. The camera only requires a single IP address.

3.14 BEST PRACTICES

In rapidly growing markets, new products are constantly being introduced by new vendors. The network video market is no exception, and there are currently hundreds of different brands available. Because network cameras include much more functionality than analog cameras, choosing the right camera becomes not just more important but also more difficult.

Selecting a camera can involve these considerations:

- *Camera type:* Should it be a fixed camera or a PTZ camera? Perhaps a panoramic camera is more suitable?
- *Indoors or outdoors:* In an outdoor environment, anything but an auto-iris lens is unsuitable, and a protective housing is mandatory.
- *Lighting conditions:* What is the lighting in the scene? Is a day-and-night camera necessary? Does it need backup IR lights? Is a low-light camera that provides colors at night needed? Perhaps a thermal camera is more suitable?
- *Resolution:* Higher resolution provides better image detail but has a few disadvantages. What is the optimal resolution for the application?
- *Analytics*: Are there requirements for built-in analytics, such as video motion detection, object detection, or people counting?

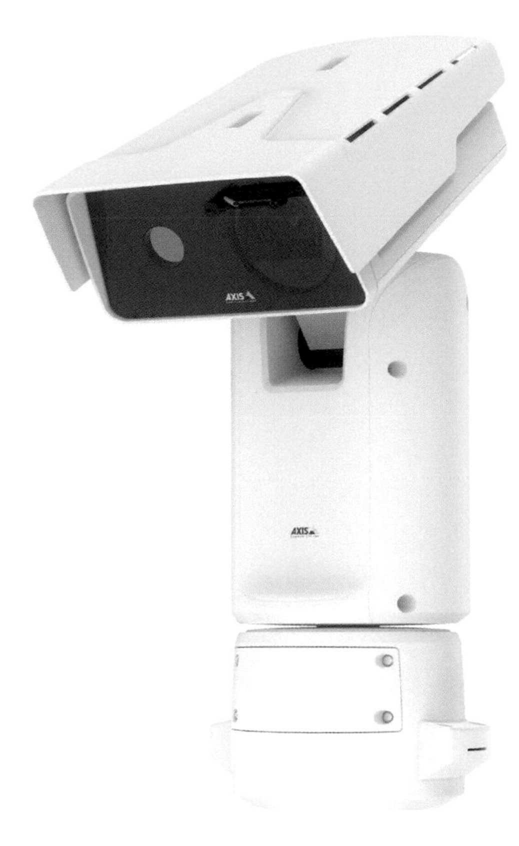

Figure 3.31 An example of a bi-spectral PTZ camera.

- *Network functionality*: Does the selected camera have all the appropriate networking and security protocols required in today's enterprise networking?
- *Vendor:* What is the brand promise? With a plethora of brands on the market, will the selected brand be on the market years later? Is the quality high enough? What is the warranty, and can it be extended?

CHAPTER 4

Camera technologies

Cameras are complex mechanical and electrical devices that are based on many different technologies. To make the most of a video surveillance system, it is important to understand these technologies.

Perhaps the most important factor for any camera is image quality. This is especially true in video surveillance, where lives and property may be at stake. But what is the definition of image quality, and how can image quality be measured and guaranteed?

There are at least three definitions of image quality:

1. How aesthetically pleasing is the image to the eye?
2. How well does the image reflect reality?
3. How well does the image match the purpose? For example, the purpose may be to get a good overview of events, or it may be to enable the identification of a person or object.

Although one image might not be able to meet all three criteria, most would agree that in a surveillance context, it is important that the image is clear, sharp, correctly exposed, and that it delivers the critical information. In a surveillance context, how pleasing an image is to the eye is of little importance. In video surveillance, you need contrast (the ability to discern between features, background, and subject) and sharpness even when there is a lot of movement or when the lighting is poor.

How well a network camera performs depends on many factors. To be able to predict and judge image quality, you need to understand the processes and elements that produce the image.

This chapter provides a discussion about light, lenses, and image sensors and their relation to network cameras, scanning techniques (deinterlacing), image processing (including wide dynamic range), and resolution. Pan-tilt-zoom (PTZ) functions are also discussed. Chapter 6 discusses compression, which also greatly impacts the quality of recorded and streamed video images.

4.1 LIGHT

Light plays a major role in the quality of an image. This section discusses the properties of light and how it affects image quality. It also discusses illuminance and what a camera's lux measurement means. Lastly, it explains how cameras can take advantage of near-infrared light to produce good-quality black-and-white images in low-light environments.

4.1.1 Characteristics of light

Visible light comes in different forms, from different directions, and with different color hues – all of which affect image quality.

DOI: 10.4324/9781003412205-4

The following are some common forms of light in a scene:

- *Direct light:* This comes from a point source or small bright object (such as sunlight or a spotlight) and creates sharp contrasts with highlights and shadows.
- *Diffuse light*: Light from a source that is so much larger than the subject that it illuminates the subject from several directions (for example, gray sky, an illuminated screen, a diffuser, or light bouncing off a ceiling). Diffuse light reduces contrast, which affects the brightness of colors and the level of detail that can be captured.
- *Specular reflection:* This is light from one direction that hits a smooth surface (such as water, glass, or metal) and is reflected in another direction. Specular reflections within an image can present problems and reduce visibility. A polarizing filter in front of a camera lens can sometimes reduce these reflections.

The direction of the light sources in relation to the subject is also vital, as it determines how much detail can be obtained from an image. The following are the main light directions:

- *Frontal light:* Light that comes from behind the camera. The scene in front of the camera is well-illuminated, and this is the ideal lighting situation.
- *Sidelight:* Light illuminating the scene from the side. This may create great architectural effects, but it also produces shadows.
- *Backlight:* Light comes from the scene itself or from behind the scene and straight into the camera lens. This is difficult to handle. Unwanted silhouettes of objects may appear, and details and color can be lost.

To manage difficult lighting situations, try to avoid backlight, or try adding some artificial light sources. In indoor installations, where backlight or reflections are often present (for example, light from windows), add some frontal lighting. Use diffusers or reflectors to create good illumination. For a discussion on how cameras handle backlight situations and scenes with complex or high-contrast lighting conditions, see section 4.5.

4.1.2 Illuminance

As a rule, the more light on the subject, the better the image. Too little light makes focusing difficult, and the image will be noisy or dark. Too much light will make the image overexposed. Also, different forms and different directions of light affect image quality in different ways.

The level of illuminance required to produce a good-quality image depends on the camera and its light sensitivity. In other words, the darker the scene, the more sensitive to light the camera needs to be. Light sensitivity is often specified in lux (lx), which is a measure of illuminance, and the specified value corresponds to a level of illuminance at which the camera produces an acceptable image. The lower the lux specification, the more light-sensitive the camera.

At least 200 lux is normally needed to illuminate an object so that the camera can produce a good-quality image. Although a camera may specify that images are possible at 0.1 lux, the image quality might not be so good. The 0.1 lux specification means only that the camera can capture an image at 0.1 lux – it says nothing about the image quality.

4.1.2.1 Definition of lux

Lux (also known as lumen per square meter or meter-candle) is the amount of light falling onto a surface per square meter. In the lux scale, one lux is equal to the amount of light falling on 1 square meter (1 m^2) at a distance of 1 meter from a candle. Correspondingly, 10 lux is the amount of light measured at 1 m from 10 candles. Foot-candle is another unit for illuminance. 1 foot-candle is equal to 10.7 lux.

Different light conditions offer different illuminance. Surfaces in direct sunlight receive 100,000 lux, whereas surfaces in full moonlight receive 0.1 lux. Many natural scenes have complex illumination,

Figure 4.1 (a) Frontal sunlight: details and colors emerge. (b) Backlight: Details and colors are lost, as the camera is placed on the opposite side of the gas station. (c) Problem 1: Windows can create reflections and poor exposure. (d) Blinds might help: Reflections are gone but image is underexposed. (e) Solution to problem 1: additional frontal lighting; details appear, and exposure is correct. (f) Problem 2: Partial backlight renders vending machine text unreadable. (g) Solution to problem 2: additional frontal lighting (text on machine more visible).

Figure 4.2 (a) Trees and telephone lines where the foreground (at ground level) is illuminated by 5 lux. (b) A corridor where the floor is illuminated by about 150 lux. (c) A shopping mall where the floor is illuminated by about 500 lux. (d) A building on a sunny morning where the building is illuminated by about 4,000 lux.

Table 4.1 Examples of levels of illuminance

Illuminance	Example
0.00005 lux	Starlight
0.0001 lux	Moonless overcast night sky
0.01 lux	Quarter moon
0.1 lux	Full moon on a clear night
10 lux	Candle at 30 cm (1 ft)
50 lux	Family living room
150 lux	Office
400 lux	Sunrise or sunset
1,000 lux	Shopping mall
4,000 lux	Sunlight, morning
32,000 lux	Sunlight, midday (min)
100,000 lux	Sunlight, midday (max)

with both shadows and highlights that give different lux readings in different parts of a scene. The light also shifts in both intensity and direction during the day. Therefore, it is important to keep in mind that a single lux reading cannot indicate the light conditions for a scene as a whole, and it says nothing about the direction of the light. Illuminance is measured using a lux meter (see Figure 4.3), and a lux measurement is always specified for a particular surface.

Figure 4.2a, b, c, and d show examples of environments with a lux reading for a specific area of a scene.

Figure 4.3 A lux meter is a tool that measures the illuminance of objects and can be used to better understand lighting and its effect on image quality.

Remember that lux meters and cameras do not collect the same light information. Illuminance or lux refers to how an object in a scene is lit (incident light), not how the light is collected by the camera. A lux meter measures visible light only and does not consider the amount of light reflected by an object. The camera, on the other hand, does record reflected light, so the actual amount of light captured may be lower or higher than the lux reading. Also, glossy objects reflect more light than dull objects, and weather conditions affect lighting and reflection as well. While snow intensifies the reflected light, rain absorbs much of the reflected light.

4.1.2.2 Lux rating of network cameras

Many manufacturers specify the minimum level of illumination their cameras need to produce acceptable images. These specifications can be used to compare cameras from the same manufacturer, but they are not as useful when comparing cameras from different manufacturers (Figure 4.4). This is because different manufacturers use different methods and have different criteria for what makes an acceptable image. The light sensitivity measurements of day-and-night cameras make matters even more complex. Day-and-night cameras usually have very low lux values because of their sensitivity to near-IR light. Because lux is only defined for visible light, it is not totally correct to use lux to express IR-sensitivity, yet doing so is common practice.

4.1.3 Color temperature

Many types of light, such as sunlight and incandescent lamps, can be described in terms of their color temperature (Figure 4.5). The color temperature is measured in degrees Kelvin (K), and the scale is based on the radiation of all heated objects. The first visible light radiating from a heated object is red. As the temperature of the heated object rises, the radiating color becomes bluer. Red has a lower color temperature than blue, which, ironically, are the opposite to the colors often used to indicate hot or cold water. Near dawn, sunlight has a low color temperature (redder), whereas during the day, it has a higher temperature (yellower, more neutral colors).

At midday, sunlight has a color temperature of about 5,500 Kelvin, which is also about the temperature of the surface of the sun. A tungsten light bulb has a color temperature of about 3,000 Kelvin. Some light sources, such as fluorescent lamps that are gas-discharge lamps rather than heated filaments, are further from the approximation of a radiating heated object. Such light sources cannot be described as accurately in terms of degrees Kelvin. Instead, the closest color temperature is used.

LED (light-emitting diodes) illuminators are highly efficient light sources with advanced control methods for extremely low energy consumption. LEDs are popular complements to network cameras and can provide the required lighting in demanding scenes. LEDs can be operated using Power over Ethernet (PoE), their light can be easily directed using optics, and they have a very long life. The color temperature for white-light LEDs is about the same as for daylight.

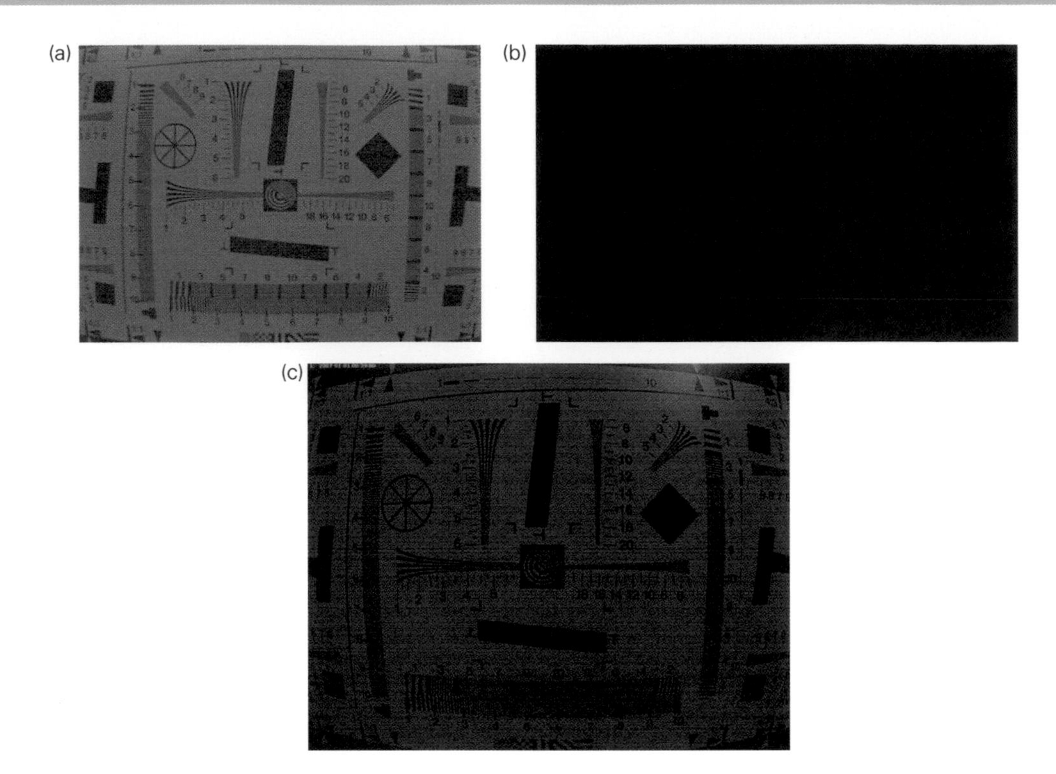

Figure 4.4 These images are from three similarly priced cameras from different vendors and were captured at 1 lux using the cameras' default settings in day mode. Cameras (a) and (b) are set to produce an image at 1 lux, while camera (c) is set to produce an image at 0.5 lux. Cameras (a) and (b) are supposed to have the same light sensitivity, and camera (c) is supposed to be the most light-sensitive. However, at 1 lux, the image from camera (a) clearly has the most brightness and contrast.

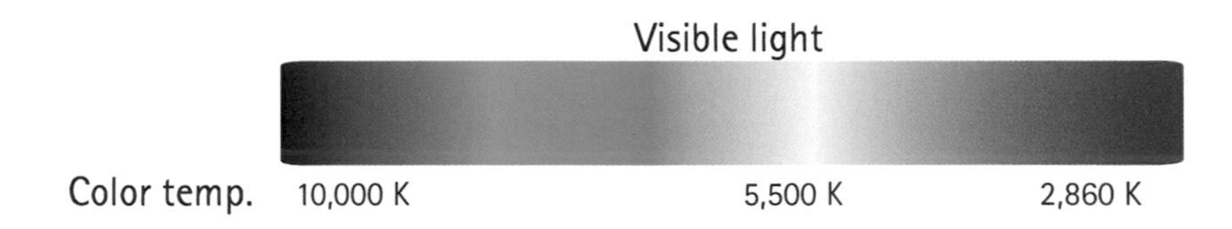

Figure 4.5 The color temperature of visible light (from highest to lowest).

Scenes illuminated by light sources with different colors (or spectral distribution) look different to a camera. Human vision, however, has a remarkable way of coping with such changes so that colored objects appear to maintain their color. This is sometimes referred to as color constancy. For a camera to do the same, it must adapt to the local illumination. This process is sometimes referred to as white balancing, which is discussed in section 4.1.4. In its simplest form, it uses a known object (usually gray) and makes color adjustments to an image so that a gray color (and all other colors) in a scene appears as perceived by human vision. Figure 4.6 shows scenes illuminated by different light sources.

4.1.4 White balance

Most modern cameras have an automatic white balance system that ensures the image has the right color balance (Figure 4.7). It is also usually possible to adjust the white balance manually or to select from presets tailored to indoor and outdoor environments.

To achieve the right color, neutral colors (black, gray, white) in a scene should stay neutral in the resulting image, regardless of the illumination. With automatic white balance, the camera uses two or three different gain factors to amplify the red, green, and blue signals.

(a)

(b)

(c)

(d)

Figure 4.6 Scenes illuminated by different light sources.

(a) Indoor office lights. Standard light bulbs (3,000 K) are more reddish than daylight and create a brown or yellow tone in the image. To compensate, the camera uses white balance techniques.
(b) Industrial long-tube fluorescent lights are designed to offer unobtrusive light. However, the images will appear with a green and dull tone. Images need white balance compensation.
(c) Some artificial electrical light may offer something similar to daylight.
(d) Sunlight at midday: Colors differ depending on the time of day.

4.1.5 Invisible light

As discussed previously, the color (or spectral distribution) changes when the temperature of the light source changes. The colors are still visible in the range from just below 3,000 to 10,000 Kelvin, so we can see them. For cooler or hotter objects, the bulk of the radiation is generated within the invisible wavelength bands.

Outside the visible range of light, we find infrared (IR) and ultraviolet (UV) light, which cannot be seen or detected by the human eye. However, most camera sensors can detect some of the near-infrared (NIR) light, from 700 nanometers up to about 1,000 nanometers. If this light is not filtered out, it can distort the color of resulting images (Figure 4.8).

Therefore, a color camera is fitted with an IR-cut filter, which is a piece of glass placed between the lens and the image sensor. This IR blocking filters out the near-IR light and delivers the same color interpretations that the human eye produces. UV light, on the other hand, does not affect a surveillance camera, as the camera's sensor is not sensitive to this light.

(a)

(b)

(c)

Figure 4.7 (a) An image with a reddish tint. (b) An image with a greenish tint. (c) An image with white balance applied.

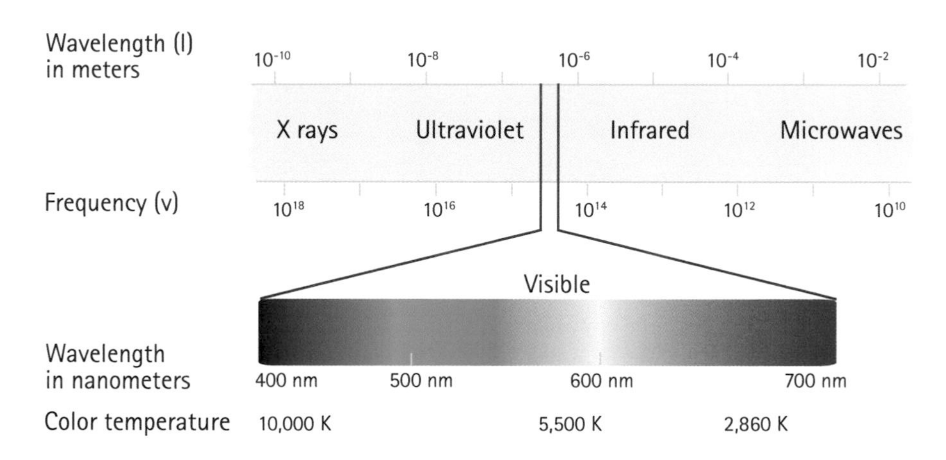

Figure 4.8 The wavelengths of light. Near-infrared light spans the range 700–1,400 nanometers.

Some vendors market a camera as day-and-night by turning it into a black-and-white camera by decreasing the chrominance (color) sensitivity of the camera, which somewhat improves the camera's light sensitivity. However, this is not true day-and-night functionality, which is only achieved with a mechanically removable IR-cut filter. When the filter is removed, the image sensor can pick up light in the near-infrared (NIR) wavelength range. This increases the camera's light sensitivity to 0.001 lux or even lower. In most true day-and-night cameras, the IR-cut filter is removed automatically when the light hitting the sensor falls below a certain level, for example, below 1 lux (Figure 4.9).

A day-and-night camera delivers color images during the day. As the light dims, the camera switches to night mode and delivers video in black and white to reduce noise (graininess) and to provide clear, high-quality images (Figure 4.10).

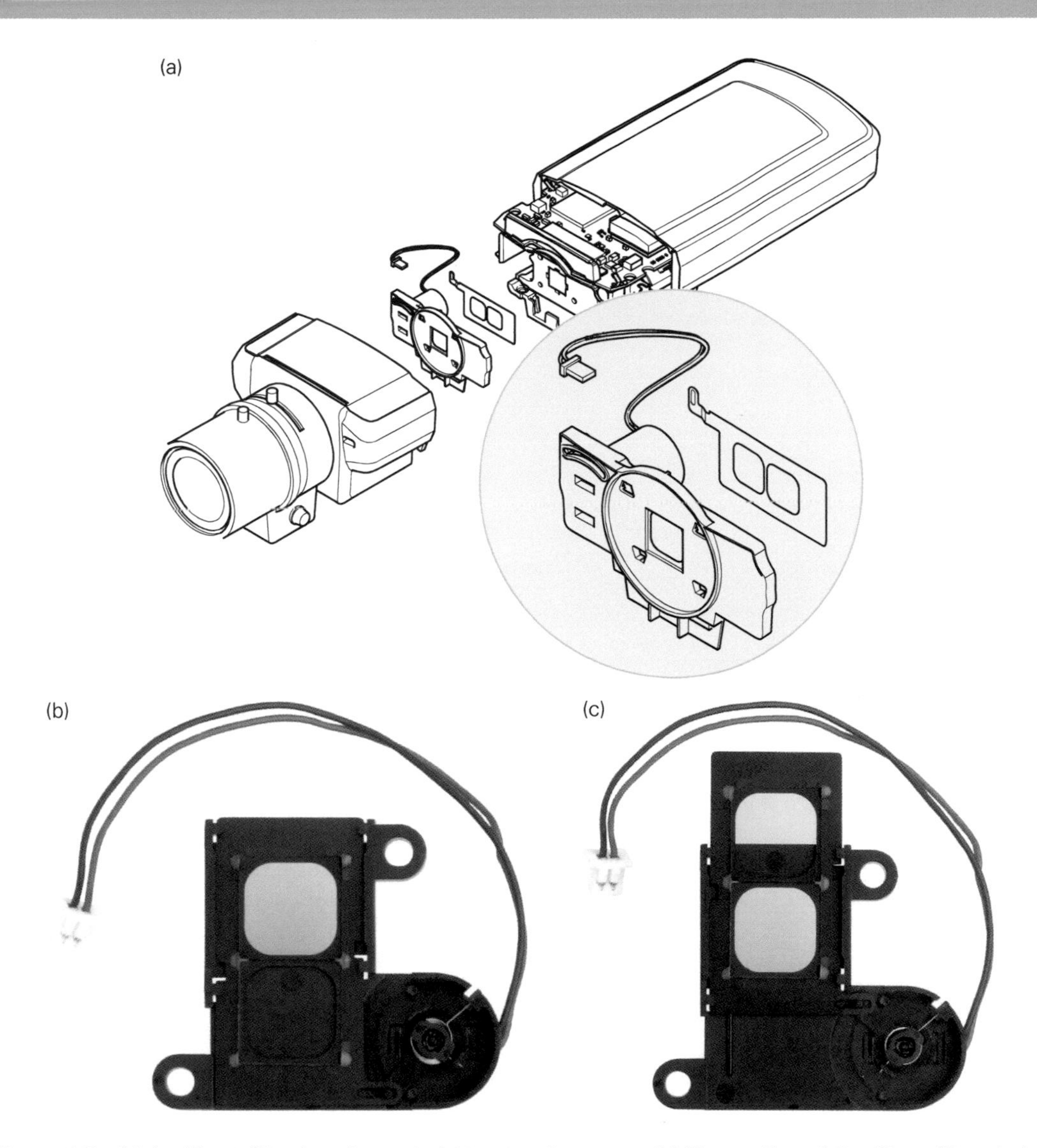

Figure 4.9 (a) An IR-cut filter in a day-and-night network camera. (b) The position of the IR-cut filter during daytime. (c) The position of the IR-cut filter during nighttime.

Cameras that can use near-infrared light are sometimes marketed as IR-sensitive cameras, but this does not mean that these cameras produce heat-sensitive infrared images. Such images require true infrared cameras that are specialized in detecting the long-wave infrared (LWIR) light (heat) that radiates from all living and inanimate objects. In infrared images, warmer objects, such as people and animals, typically stand out from cooler backgrounds. True infrared cameras are called thermal cameras, which are covered in more detail in Chapter 5.

4.1.6 IR illuminators

IR illuminators should be used if the goal is 24/7 surveillance in areas with low light. Many different types of IR illuminators are available. Some are external and stand-alone, some can be mounted on cameras or their mounts, and some cameras come with built-in IR illuminators (see the images in Figure 4.11). Some IR illuminators produce light at a low wavelength, about 850 nm, which shows a faint red hue visible to the human eye if the room is completely dark. Totally covert IR illuminators with higher wavelengths of about 950 nm also are available, although these higher wavelengths do not have the same reach as at 850 nm. Illuminators also come fixed at different angles, for

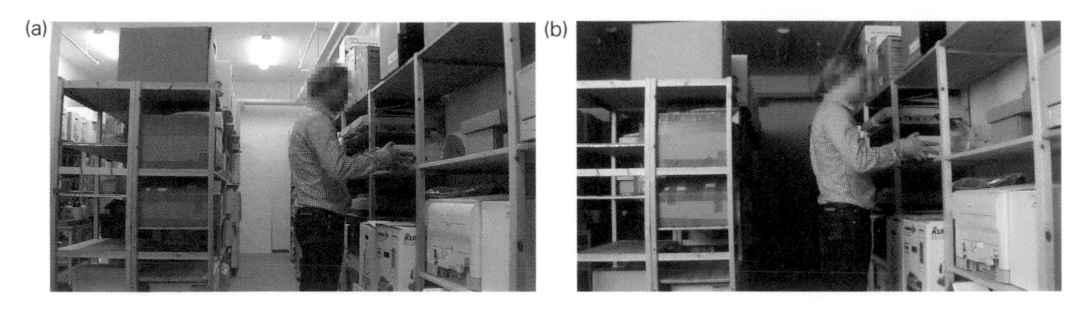

Figure 4.10 Images captured by a day-and-night camera. (a) Captured in day mode. (b) Captured in night mode.

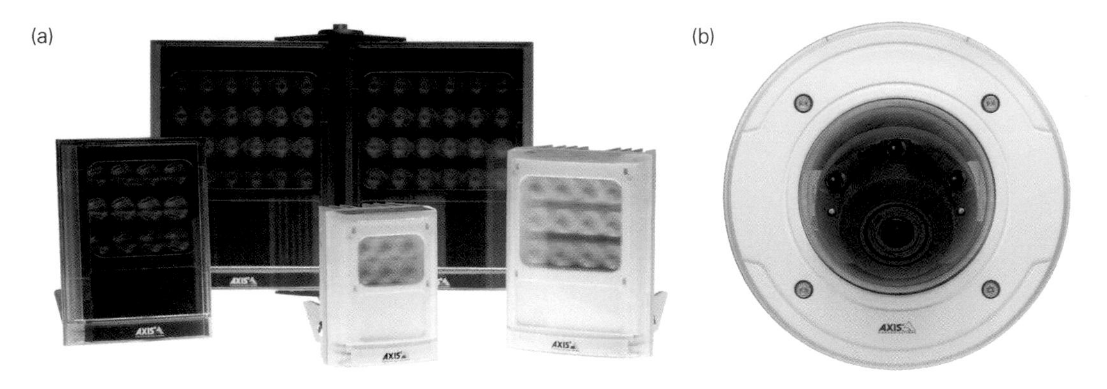

Figure 4.11 (a) IR illuminators. (b) A camera with built-in IR illuminators.

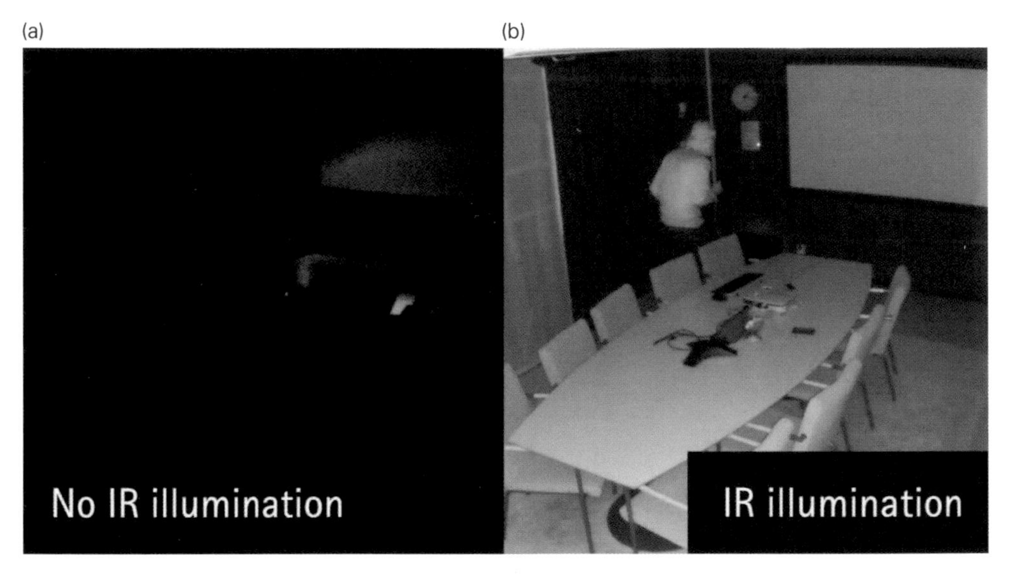

Figure 4.12 Image of a scene taken by a camera with built-in IR illuminators. (a) IR illumination is off. (b) IR illumination is on.

example, 20, 30, or 60 degrees. A wide-angle illuminator normally equals a shorter reach. Most IR illuminators today use LEDs (light-emitting diodes), which is a cost-efficient solution where illuminators last for five to ten years. New LED technology has made it possible to produce efficient solutions with just a few LEDs mounted on the camera. The lifespan of these LEDs is very long, often longer than the camera's own lifespan. Modern built-in LEDs also adjust the angle of the IR light as the camera's viewing angle changes. Figure 4.12 shows snapshots of a scene with and without IR illuminators.

4.1.7 Day-and-night applications

Day-and-night or IR-sensitive cameras are useful in certain environments or situations that restrict the use of artificial light. These include low-light video surveillance applications, where light conditions are less than optimal, covert surveillance situations, and discreet installations, for example, in residential areas where bright lights could disturb residents at night. An IR illuminator that provides near-infrared light can also be used together with a day-and-night camera to further enhance the camera's ability to produce high-quality video in low-light or nighttime conditions.

Network cameras today are not only very light-sensitive but are also very good at producing color images at night. In some low-light scenarios, these cameras are an even better option than day-and-night cameras.

4.2 LENSES

The first camera component to capture the light is the lens, which, therefore, has a large impact on image quality. A camera lens normally consists of an assembly of lenses. Camera lens functions that affect image quality include the following:

- *Defining the field of view*, that is, defining how much of a scene and the level of detail that the camera shall capture.
- *Controlling the amount of light* that passes through to the image sensor so that the image is correctly exposed.
- *Focusing* by adjusting either elements within the lens assembly or the distance between the lens assembly and the image sensor.

This section discusses different types of lenses, as well as characteristics such as field of view, iris, f-number, focusing, mounts, and lens quality.

4.2.1 Lens types

There are three main types of lenses:

- *Fixed lens*
- *Varifocal lens*
- *Zoom lens*

A fixed lens (see Figure 4.13) has only one field of view because the focal length is fixed. There is a choice between normal, telephoto, or wide-angle view. To choose the correct lens, you need to know

Figure 4.13 A fixed lens.

Figure 4.14 A varifocal lens.

beforehand exactly which focal length is needed. A 4 mm focal length is common for fixed network camera lenses.

A varifocal lens (see Figure 4.14) offers a range of focal lengths and, therefore, different fields of view. The field of view can be adjusted, which also means the lens must be refocused. Varifocal lenses for network cameras often have focal lengths that range from 3.0 to 8 mm.

Like varifocal lenses, zoom lenses allow the user to select different fields of view. But a zoom lens also has autofocus, which means there is no need to refocus when changing the field of view. Focus is maintained within a specified focal length range, for example, 6–48 mm. Lens adjustments can be either manual or motorized and remote controlled. The zoom capacity of a lens refers to the ratio between the longest and shortest focal lengths. For example, a lens with 3× zoom has a maximum focal length that is three times longer than its shortest focal length.

4.2.1.1 IR-coated lenses

IR-coated lenses are treated with special materials to provide IR correction for color and infrared light. The coating also has antireflective qualities that improve the properties of the infrared light that passes through the lens and reaches the sensor.

4.2.2 Lens quality

A light phenomenon called diffraction limits a lens' resolving power, which describes the smallest size of details that can be imaged. The bigger the lens, or the larger the aperture, the better the resolving power. All lens designs are the result of compromises and are optimized for a particular task. The quality of the lens material, the coatings on a lens, and how the lens assembly is designed all influence the resolution power a lens can provide.

Lenses are made of glass or plastic. Although glass lenses are generally better, there is no guarantee that this is always the case. The properties and design of a lens are more important. No lens is perfect, and all lenses create some form of aberration or image defects because of their limitations. Types of aberrations include the following:

- *Spherical aberration:* Light passing through the lens' edges is focused on a different distance than light striking near the center.
- *Astigmatism:* Off-axis points are blurred in the radial or tangential direction. Focusing can reduce one at the expense of the other, but it cannot bring both into focus at the same time.

- *Distortion (pincushion and barrel):* The image of a square object has sides that curve in or out.
- *Chromatic aberration:* The position of sharp focus varies with the wavelength.
- *Lateral color:* The magnification varies with wavelength.

4.2.3 Lens mount standards

There are three main mount standards used for interchangeable network camera lenses: CS-mount, C-mount, and S-mount.

CS-mount and C-mount look the same, and both have a 1 inch thread and a pitch of 32 lines per inch. The screw mount size is defined as 1–32 UN 2A by the unified thread standard established by the American National Standards Institute (ANSI). CS-mount is an update to the C-mount standard that allowed for reduced manufacturing cost and sensor size. CS-mount is much more common today than C-mount. Manufacturers sometimes make the thread of their cameras' lens mount slightly smaller or larger than the standard to ensure perfect alignment between the lens and the sensor.

CS-mounts differ from C-mounts in the flange focal distance (FFD), which is the distance from the mounting flange to the sensor when the lens is fitted on the camera.

- *CS-mount:* The flange focal distance is 12.5 mm (≈ 0.49 in ≈ ½ in).
- *C-mount:* The flange focal distance is 17.526 mm (0.69 in ≈ 11/16 in).

Using the wrong type of lens many make it impossible to focus the camera. A C/CS adapter ring – essentially a 5 mm spacer – can be used to convert a C-mount lens to a CS-mount lens (Figure 4.15).

S-mount (Figure 4.16) is also called M12-mount and is common in small cameras, such as modular cameras and fixed mini-domes. It is called M12 because it has a nominal outer thread diameter of 12 mm and 0.5 mm pitch (M12×0.5). This follows the metric screw thread standard established by the International Organization for Standardization (ISO). Because S-mount lenses are often mounted directly on the circuit board, they are sometimes called board lenses.

Thermal cameras often use other lens mounts than C- or CS-mount. The first documented thermal lens standard is TA-lens. The letters "TA" stand for Thermal A, where "A" stands for the first documented standard. The lens mount is an M34×0.5 screw mount. This is large enough for sensors with a diameter up to at least 13 mm, which means that it works for many popular LWIR sensors.

Fixed dome cameras often use a type of board lens that has a round high-precision surface with a diameter of 14 or 19 mm. Aligning this type of lens is a challenge, and therefore it is best mounted with a custom-made holder. These lenses are not interchangeable.

PTZ cameras use modules – known as camera blocks or block lenses – that consist of a sensor and lens assembly with an auto-iris, as well several mechanical parts for autofocusing and motorized zooming. PTZ camera lenses are not interchangeable.

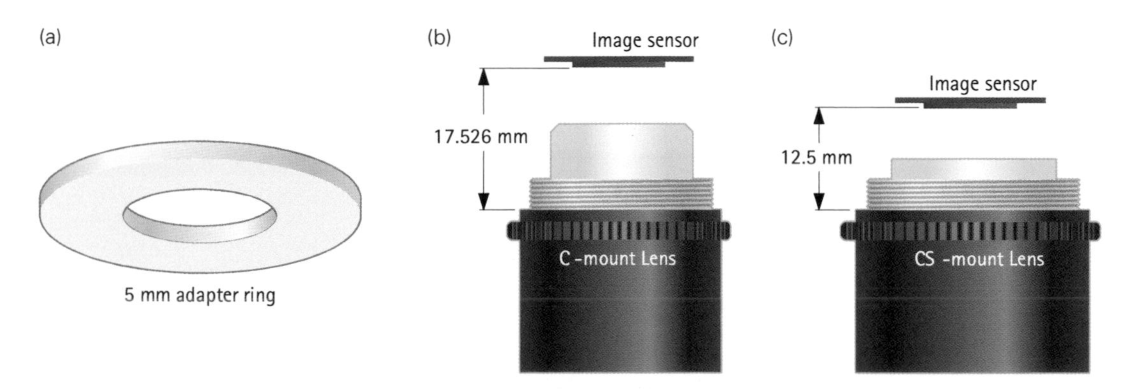

Figure 4.15 (a) C/CS adapter ring. (b) C-mount lens. (c) CS-mount lens.

Figure 4.16 An S-mount lens.

Figure 4.17 An example of a camera fitted with an EF lens.

Some high megapixel cameras (over 10 megapixels) use high-end lenses from professional photography, such as Canon EF lenses.

4.2.4 Field of view (focal length)

It is important to know what field of view (FoV) the camera must cover because it determines the amount and level of information that can be captured in an image. Field of view is divided into several subareas where the most common in network video are the following (also see Figure 4.18):

- *Normal view* means that the image has the same field of view as the human eye.
- *Telephoto view* is a magnification of a narrower field of view, generally providing finer details than the human eye can deliver.
- *Wide-angle view* is a larger field of view with fewer details than in normal view.

The focal length of the lens and the size of its image sensor determine a camera's field of view. Focal length is defined as the distance between the entry lens (or a specific point in a complex lens assembly) and the point where all the light rays converge at a point, typically the camera's image sensor. The longer the focal length, the narrower the field of view. To achieve a wide field of view, the focal length should be shorter.

(a)

(b)

(c)

Figure 4.18 (a) Wide-angle view. (b) Normal view. (c) Telephoto field of view.

To understand terms such as field of view, focal lengths, and so on, it helps to compare with a traditional camera that uses 35 mm film. The human eye has a fixed focal length that is equivalent to a lens with a focal length of 50 mm on a classic 35 mm film camera. Traditional lenses with focal lengths ranging from 35 to 70 mm are, therefore, considered "normal" or "standard" lenses (Figure 4.19a).

Telephoto lenses (Figure 4.19b) for traditional cameras have focal lengths of more than 70 mm. A telephoto lens is used when the surveillance object is either small or far from the camera. A telephoto lens delivers good detail for long-distance viewing. Telephoto images usually have a slight geometrical distortion, which appears as curvatures at the edges. Generally, a telephoto lens has less light-gathering capability and, therefore, requires good lighting to produce a good-quality image.

Wide-angle lenses (Figure 4.19c) have focal lengths of less than 35 mm. The advantages include a wide field of view, good depth of field, and decent low-light performance. The downside is that this type of lens produces geometrical distortion. Lenses with a focal length of less than 20 mm create what is often called a fish-eye effect. Wide-angle lenses are not often used for long-distance viewing.

A camera's field of view is determined by both the lens' focal length and the size of its image sensor. So the focal length of the camera's lens is irrelevant unless you also know the size of the image sensor, which can vary. The focal length of a network camera lens is generally shorter than the focal length of its counterpart among classic photographic camera lenses. This is because the size of an image sensor (the digital equivalent to a 35 mm film) is much smaller than the size of a frame on a 35 mm film.

The typical sizes of image sensors are ¼, ⅓, ½, and ⅔ inch. There are two ways to determine the field of view for a given lens. The easy way is to put it in relation to a traditional 35 mm camera. The size of the image sensor has a given conversion factor. Multiplying the conversion factor with the focal

Figure 4.19 (a) Normal lens with standard focal length. (b) Telephoto lens with long focal length). (c) Wide-angle lens with short focal length.

Table 4.2 The field of view for a given network camera

Sensor	= factor	× network camera lens	= traditional camera's focal length
(inch)	(inch)	(mm)	(mm)
1/4	9	3–8	27–72
1/3	7	3–8	21–56
1/2	5	3–8	15–40

length of the network camera lens gives what the focal length of an equivalent 35-mm camera lens would be. See Table 4.2.

The second way is to calculate it using the formula in Figure 4.20. Given the focal length of the lens and the distance to the scene or subject from the camera position, this formula helps determine the height and width of the scene that will be captured.

Calculation in feet:

When using a camera with a ¼ inch CCD sensor and a 4 mm lens (f), how wide (W) does an object have to be to be visible at 10 feet (D)? A ¼ inch sensor is 3.6 mm wide (w).

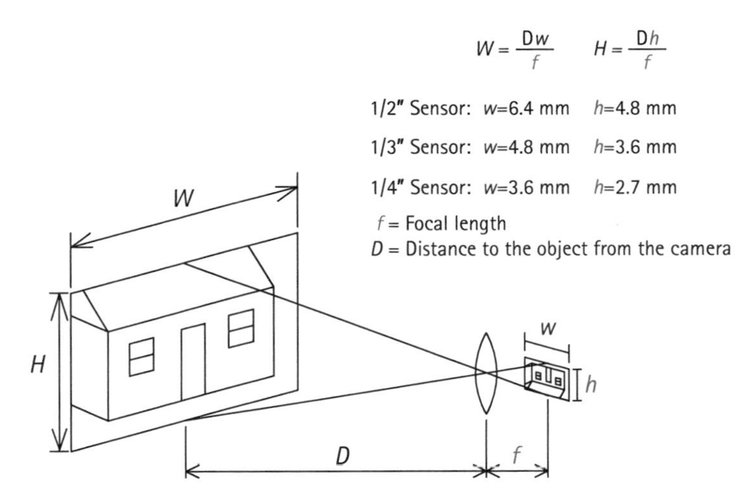

Figure 4.20 In the formula, (w) and (h) correspond to the physical width and height of the image sensor.

Calculation in meters:

When using a camera with a ¼ inch CCD sensor and a 4 mm lens (f), how wide (W) does an object have to be to be visible at 3 m (D)? A ¼ inch sensor is 3.6 mm wide (w).

$$w = D \times \frac{w}{f} = 3m \times \frac{3.6mm}{4mm} = 2.7m$$

The fastest way to calculate the field of view and which lens is required is to use an online lens calculator (Figure 4.22) or a rotating lens calculator (Figure 4.21).

Some manufacturers make their lens calculators available online. These tools make calculations quick and convenient, but a common drawback is that they ignore the potential geometrical distortions of lenses.

4.2.5 Matching lens and sensor

Camera lenses are manufactured to match the sizes of image sensors. Selecting a suitable lens for the camera is of fundamental importance (Figure 4.23). A lens made for a ½ inch sensor works with ½-, ⅓-, and ¼-inch sensors but will not work with a ⅔ inch sensor.

If using a lens made for a smaller sensor than the one in the camera, the image will have black corners. Using a lens made for a larger sensor than the one in the camera means that some information will be lost, as it falls outside the sensor. Therefore, the field of view will be smaller than the lens is capable of. The image will appear to have a telephoto effect, making everything look zoomed in. This is known as a focal length multiplier greater than 1. Assuming that the resolution is the same, the smaller the image sensor for a given lens, the more a scene is magnified. See Figure 4.24.

4.2.6 Aperture (iris diameter)

A lens' ability to allow light to pass through it is measured by its aperture, or iris diameter. The bigger the lens aperture, the more light that can pass through it. In low-light environments, it is generally better to have a lens with a large aperture.

To get a good-quality image, the amount of light passing through a lens must be optimized. If too little light passes, the image will be dark. If too much light passes, the image will be overexposed.

The amount of light captured by an image sensor is controlled by two elements that work together:

- *Aperture*: The size of the lens' iris opening.
- *Exposure*: The length of time an image sensor is exposed to light.

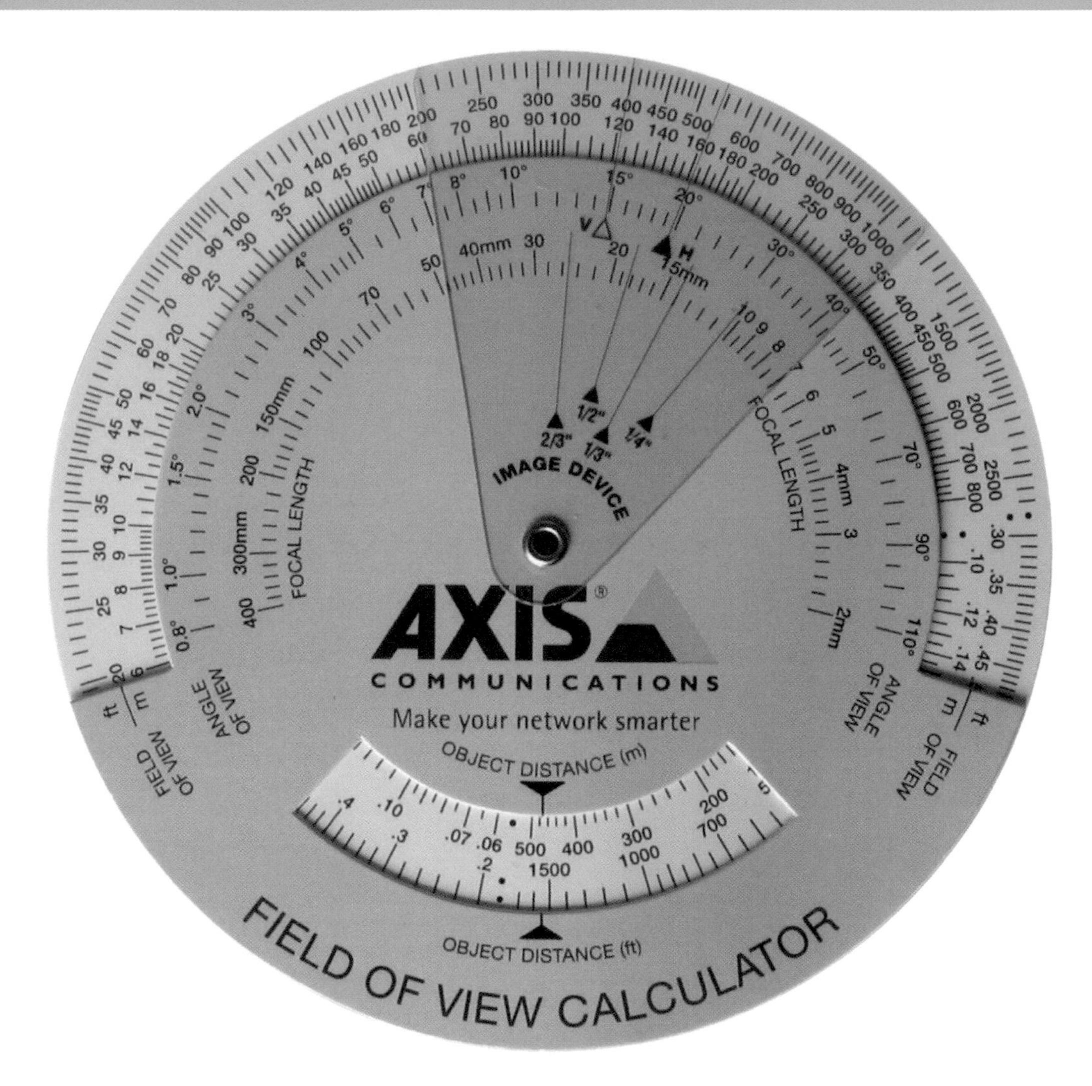

Figure 4.21 A rotating lens calculator is a good tool for quickly calculating the lens required or the field of view.

The longer the exposure time, the more light an image sensor receives. Well-lit scenes need shorter exposure times, whereas darker scenes need longer exposures. A certain exposure level can be achieved using either a large iris opening and a short exposure or a small iris opening and a long exposure. It is important to know that increasing the exposure time also increases motion blur. Increasing the iris opening reduces the depth of field.

Gain is another parameter that affects how much light the sensor is exposed to. Gain is a signal amplifier – that is, it tunes the sensitivity of the image sensor. But increasing the gain also increases the noise level.

In traditional cameras, it is the shutter that controls the exposure time. However, most network cameras do not use mechanical shutters. Instead, the image sensor acts as an electronic shutter. It switches on and off to collect and discharge electrical charges. There are two types of electronic shutters. Many network video cameras, especially those with CMOS sensors, use a rolling shutter. Unlike a global shutter, which exposes all the pixels at the same time in a single snapshot, the rolling shutter captures the image by scanning the frame line by line. In other words, not all parts of the image are captured at the same time. Instead, each line is exposed in a slightly different time

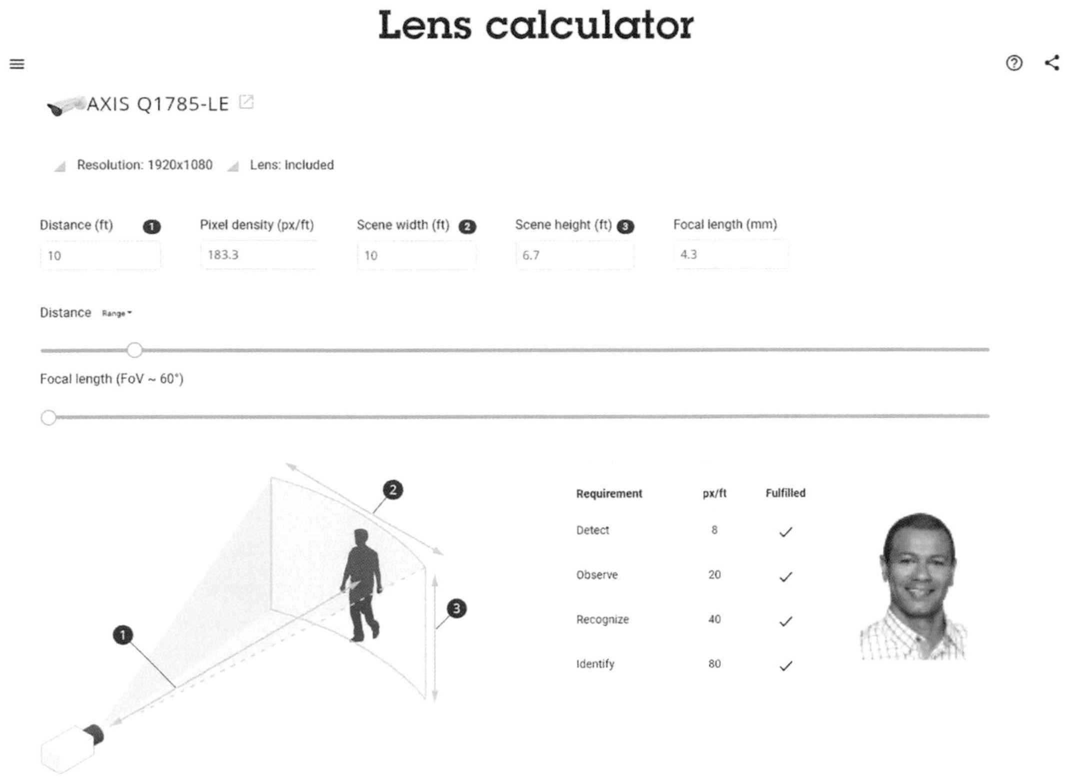

Lens calculator

AXIS Q1785-LE

Resolution: 1920x1080 Lens: Included

Distance (ft) 1	Pixel density (px/ft)	Scene width (ft) 2	Scene height (ft) 3	Focal length (mm)
10	183.3	10	6.7	4.3

Distance Range

Focal length (FoV ~ 60°)

Requirement	px/ft	Fulfilled
Detect	8	✓
Observe	20	✓
Recognize	40	✓
Identify	80	✓

Figure 4.22 An example of an online lens calculator.

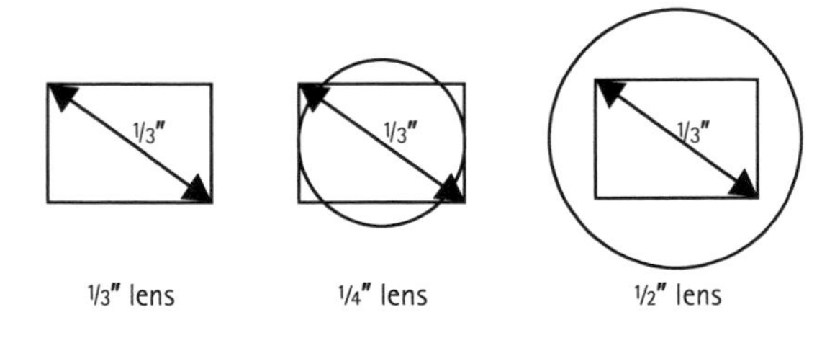

Figure 4.23 Examples of different lenses paired with a ½ inch sensor.

window. Therefore, if the camera shakes or vibrates, each exposed line is moved slightly in relation to the other lines, making the image warp or wobble. Rolling shutters may cause fast-moving subjects to also appear distorted (see Figure 4.25).

Rolling shutter distortion induced by vibrations can be avoided with stabilization techniques. Optic stabilization instantaneously compensates for the motion. With electronic stabilization, the rolling shutter must first scan at least one line before the digital processing to stabilize the image can begin. Even so, this method works very well, and the technology is improving rapidly. For more information about image stabilization, see section 4.5.8.

4.2.7 Types of iris control

The ability to control a camera's iris opening plays a central role in image quality. An iris is used to maintain the optimum light level hitting the image sensor so that images are properly exposed. The iris can also be used to control the depth of field, which is explained in more detail in section 4.2.9.

(a) 

(b) 

Figure 4.24 The field of view for the same lens and resolution but with different sensor sizes. (a) Using a ¼ inch sensor. (b) Using a ¼ inch sensor. Because the sensor in (b) is less than ⅓ inch, the image covers a smaller part of the scene. It corresponds to the parts of image (a) contained within the red border. In other words, the result of using a lens that is too big for the sensor is a magnification of a smaller area.

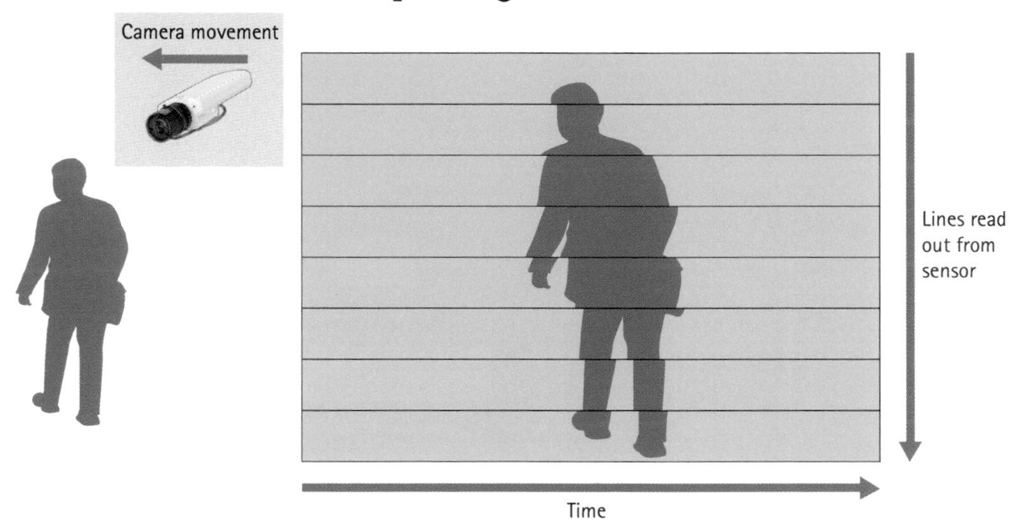

Figure 4.25 The principle of rolling shutter distortion. A shaking camera causes misalignment between the scanned lines.

Iris control can be fixed or adjustable, and adjustable iris lenses can be manual or automatic. Automatic iris lenses can further be classified as either auto-iris or P-Iris lenses.

With fixed iris lenses, the iris opening cannot be adjusted and is fixed at a certain f-number. The camera can compensate for changes in light level by adjusting the exposure time or by using gain.

With manual iris lenses, the iris can be adjusted by turning a ring on the lens to open or close the iris. This makes them less suitable for outdoor surveillance or in other environments with changing light conditions.

There are two types of auto-iris lenses: DC-iris and video iris. Both use a galvanometer to automatically adjust the iris opening in response to changes in light levels. Both use an analog signal to control the iris opening, but they differ in the location of the circuitry that converts the analog signal into control signals. In a DC-iris lens, the circuit resides inside the camera. In a video iris, the circuit is inside the lens.

In bright situations, a camera with an auto-iris lens can be affected by diffraction and blurring when the iris opening becomes too small. This problem is especially common in hi-res cameras because the pixels in those image sensors are smaller than the pixels in lower resolution cameras. Therefore, the image quality is dependent on using the right aperture. To optimize image quality, a camera needs to have control over the iris opening. The problem with an auto-iris lens is that this control cannot be made available to the camera or the user.

The P-Iris was designed to address the shortcomings of an auto-iris lens while retaining the benefits of an automatically controlled iris. A P-Iris lens has a motor that allows the position of the iris opening to be precisely controlled, as well as specialized software for optimized image quality.

In bright situations, a camera with a DC-iris lens and a megapixel sensor with small pixels typically produces blurry images because the aperture becomes too small. This type of image blur is called diffraction. The P-Iris limits the closing of the iris to prevent diffraction.

The P-Iris provides improvements in contrast, clarity, resolution, and depth of field. Having a good depth of field (where objects at different distances are in focus simultaneously) is crucial when monitoring areas with a lot of depth, such as long corridors, server halls, or parking lots. See section 4.2.9 for more on depth of field.

4.2.8 F-number (F-stop)

In low-light situations, particularly indoors, a vital factor to look for in a network camera is the ability of the lens to collect light. The f-number is a measure of this light-gathering ability. It is the ratio (N) of the lens' focal length (f) to the diameter of the aperture (D).

$$N = \frac{f}{D}$$

F-numbers can be expressed in several ways. The two most common are f/D and F*D*. The slash indicates division as explained in the formula earlier. An f-number of $f/4$, or F4, means that the aperture is equal to the focal length (f) divided by 4. So if a camera has an 8 mm lens, light must pass through an aperture 2 mm in diameter.

The smaller the f-number, the better the lens' light-gathering ability – that is, more light can pass through the lens to the image sensor. A small f-number is achieved through a large aperture relative to the focal length (see Figure 4.26 and Figure 4.28).

The difference in the amount of light between two sequential f-numbers is called an f-stop. Traditional camera lenses usually use a standardized sequence (F1.0, F1.4, F2.0, F2.8, F4.0, F5.6, F8.0, and so on), where each f-stop represents a halving of the light intensity compared to the previous stop. This is illustrated in Table 4.3, which shows the amount of light relative to F5.6. This means that an F1.0 lens lets in 32 times more light than an F5.6 lens. In other words, it is 32 times more light-sensitive.

In low-light situations, a smaller f-number generally produces better image quality. However, some sensors may not be able to take advantage of a lower f-number in low-light situations because of their design. And a choosing a lens with a higher f-number means an increase of the depth of field, which is explained in section 4.2.7.

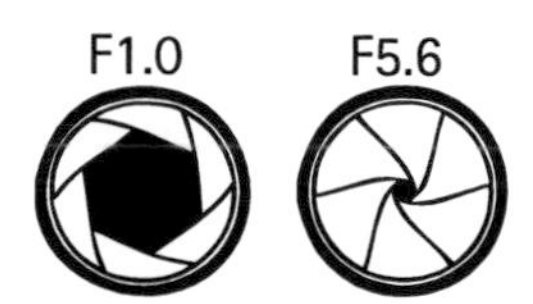

Figure 4.26 A smaller f-number means a larger aperture, while a larger f-number means a smaller aperture.

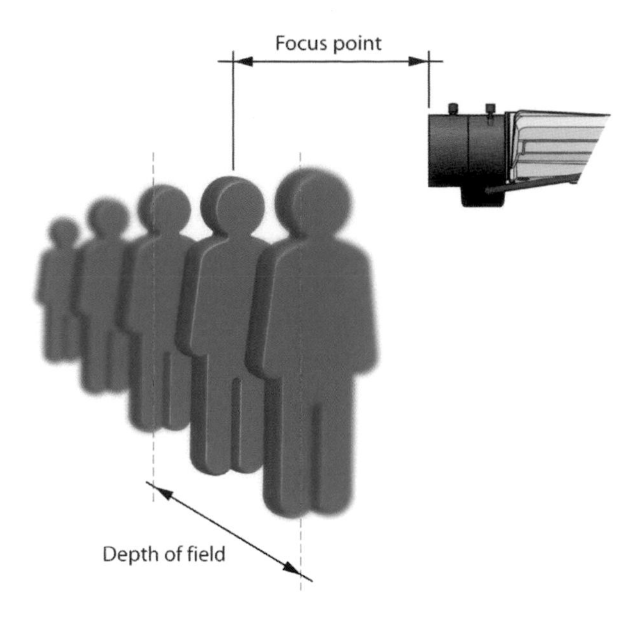

Figure 4.27 In this line of five people, the camera focuses on the second person. The faces of the person behind and the person in front can also be identified – they are within the depth of field.

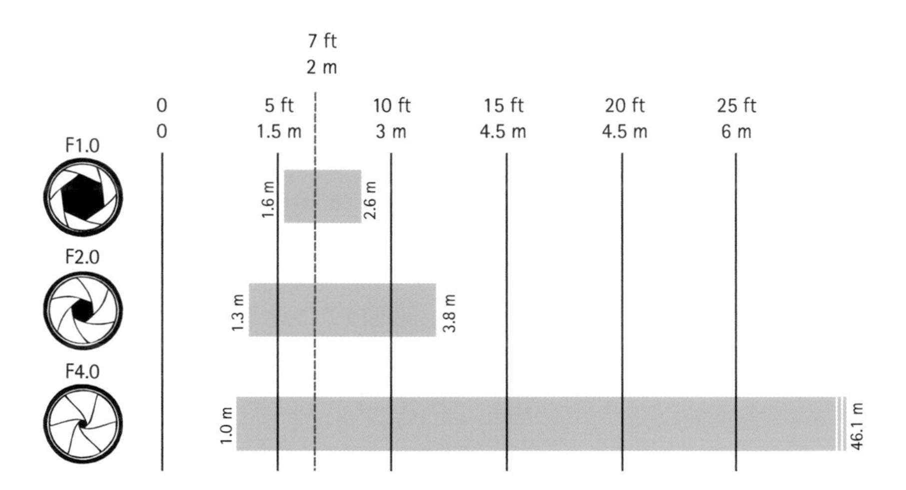

Figure 4.28 Iris opening in relation to depth of field. Depending on the pixel size, very small iris openings may blur an image due to diffraction.

Table 4.3 Amount of light relative to f-stop F5.6

F-stop	F1.0	F1.4	F2.0	F2.8	F4.0	F5.6
Relative light level	32	16	8	4	2	1

Although network cameras with DC-iris lenses have an f-number range, often only the maximum light-gathering end of the range is specified. A lens with a lower f-number is usually also more expensive than a lens with a higher f-number.

4.2.9 Depth of field

A lens can only focus on a single point, but there is an area that stretches in front of and behind the focal plane that still appears sharp. This volume of space is known as the depth of field (see

Table 4.4 How focal length, aperture, and camera-to-object distance affect the depth of field

	Short depth of field	*Long depth of field*
Focal length	Long (telephoto lens)	Short (wide-angle lens)
Aperture	Large (small f-number)	Small (big f-number)
Distance to object	Short	Long

Figure 4.27). The depth of field is not a fixed distance or volume of space. It changes in size and is usually described as either short (shallow) or long (deep). In an image with short depth of field, only a small zone appears sharp. In an image with long depth of field, a larger zone appears sharp.

Depth of field is important in many video surveillance situations. For example, when monitoring a parking lot, the camera needs to be able to deliver sharp images of cars and their license plates at several distances, say, 20, 30, and 50 meters (60, 90, and 150 feet).

Depth of field is affected by several factors. The ones most often mentioned and that can be controlled by the user are as follows:

- *Focal length:* The longer the focal length, the shallower the depth of field.
- *Aperture (iris diameter):* The larger the aperture, the shallower the depth of field.
- *Camera-to-object distance:* The shorter the distance, the shallower the depth of field.

The depth of field becomes shorter as the aperture increases. Thus, the smaller the aperture the better. Table 4.4 gives a summary of the effect that focal length, aperture, and camera-to-object distance have on depth of field.

Note that unlike aperture and focal length, camera-to-object distance is not set by physical camera parameters but is simply an effect of how close the camera is to the object. If the camera's position is limited by surrounding elements, such as poles or buildings, and the range at which objects need to be in focus is non-negotiable, as is often the case in video surveillance situations, what is left to play with is the lens' focal length and aperture.

Figure 4.28 gives an example of the depth of field for different f-numbers at a focal distance of 2 m (7 ft). A large f-number (smaller aperture) enables objects to be in focus over a longer range.

The depth of field changes gradually, not abruptly from sharp to unsharp. Everything immediately in front of or behind the focusing plane gradually loses sharpness. To our eyes, the points that are projected within the depth of field still appear sharp (see Figure 4.27). The *circle of confusion* is a term used to define how much a point needs to be blurred to be perceived as unsharp. When our eyes can see the circle of confusion, those points are no longer acceptably sharp and are outside the depth of field.

In low-light situations, there is a trade-off between achieving good depth of field and reducing the blur from moving subjects. Getting a good depth of field and a properly exposed image in low light require a small aperture and a long exposure time. However, the long exposure time increases motion blur. A smaller aperture generally increases the sharpness because it is less prone to optical errors. But if the aperture is too small, diffractions can appear in the image. With round apertures, the diffractions produce a pattern of circles called an *Airy disc* or *Airy pattern* (after George Biddell Airy). So in an image of a round object, the object would be surrounded by fainter rings of color (see Figure 4.29). Similar blurring effects occur with other aperture shapes, but the geometrics of the aperture would render a pattern other than a circular one.

Getting the optimal result is a matter of weighing the environment against the image requirements and priorities. See Table 4.5.

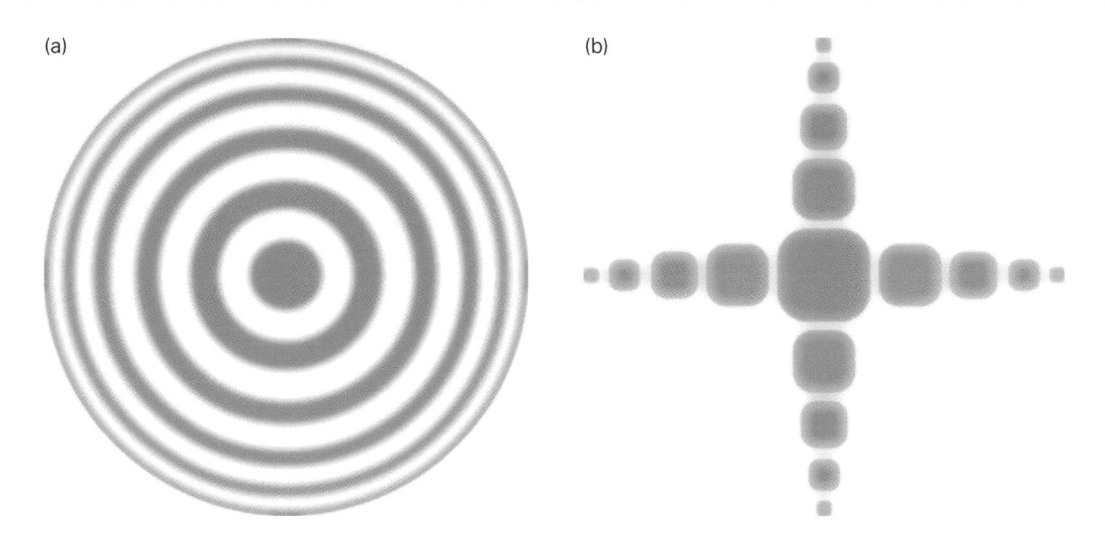

Figure 4.29 (a) Circular apertures that are too small produce a diffraction pattern of concentric rings. (b) Square apertures produce a cross pattern.

Table 4.5 What is achievable under various lighting conditions

	Sunshine	Low-light setting 1	Low-light setting 2
Exposure time	Short	Long	Short
Aperture	Small (large f-number)	Small (large f-number)	Large (small f-number)
Gain or sensitivity	Limited	Limited	Full
Image priority	All	Depth of field (DoF)	Movement
Result	Good DoF, clear image, frozen action[a]	Good DoF, clear image, blurry movement[a]	Limited DoF, noise, frozen action

a A very small iris may reduce sharpness.

The following are some guidelines of what is achievable under various lighting conditions:

- *Sunshine:* With lots of light, the camera uses a small iris opening and a short exposure time. This gives maximum image quality with good depth of field and a clear image without motion blur.
- *Overcast:* If depth of field is a priority, increase the exposure time to reduce the aperture but noting that motion blur will increase. Reduce the exposure time for clearer images of moving objects. Note also that the depth of field is compromised with a larger aperture.
- *Evening and night:* In low-light situations, the camera adjusts the gain of the image sensor to deliver good images but note that noise levels increase and appear as grainy effects in the image.

4.2.10 Focusing

Focusing a network camera often requires making very fine adjustments to the lens. With some auto-iris network cameras, there might be problems setting the focus in very bright or very dark conditions.

With less sophisticated auto-iris lenses, it is best to adjust the focus in low-light or by using a darkening filter, such as a neutral density (ND) filter or even a welder's glass. The filter fools the camera into thinking it is darker than it really is. In low-light conditions, the iris opens automatically,

Figure 4.30 Using the focus puller, the installer can make a rough focus adjustment before leaving the site.

which shortens the depth of field and helps the user focus more precisely. The more sophisticated auto-iris lenses used in many modern IP cameras allow you to open the iris through the camera's user interface. In those cases, a darkening filter is not needed.

Given a greater depth of field, it may be easier to set the focus in bright conditions. But as the light decreases and the iris diameter increases, the depth of field gets shorter, and the image becomes blurry.

For optimal performance, it is best to follow the manufacturer's instructions when adjusting the camera lens.

A camera with autofocus automatically adjusts the lens mechanically so that the image is focused. This is a requirement in pan-tilt cameras where the camera direction is constantly changing. Some fixed cameras may also have autofocus, but because focusing a camera normally only needs to be done once, during installation, it is often difficult to justify the extra cost. Sometimes these cameras have a focus window, which sets up focus positions that allow easier autofocusing.

If the camera has remote focus, there is no need to worry about the time-consuming task of manually adjusting the focus on-site. The installer can use the focus puller for approximate focusing on-site (Figure 4.30). Exact adjustment of the focus can then be done remotely from a computer. Some cameras include full remote-focus capabilities, eliminating the need for any manual adjustment on-site. Using, for example, a browser, the user can see the live view and check that the focus is correct. See Figure 4.31. With a built-in pixel counter, users can also make sure that they get enough pixels across the subject to be able to identify it, whether it is a person's face or a license plate (for more information about resolution requirements, see Chapter 18).

4.2.11 Lenses for high-resolution cameras

Cameras with megapixel, HTDV, or Ultra-HD (4K, 8K) resolution place higher demands on lenses. Therefore, replacing a lens on a high-resolution camera needs careful consideration. The main reason is that the pixels on high-resolution sensors are much smaller than the pixels on sensors with lower resolutions and, therefore, require a higher-quality lens.

It is not only the quality of the transparent material that matters but also the construction of the aperture (the iris). When light passes through the aperture, it spreads out. This effect is called diffraction, and it cannot be avoided, although it can be limited somewhat through careful construction of the aperture. Using a P-Iris lens (see section 4.2.7) optimizes the aperture in high-resolution cameras.

Figure 4.31 Through the camera's user interface, the user can adjust the focus from any location. During the remote focusing process, the camera's iris opens so that the back focus can reset, then the focus is fine-tuned.

It is also critical to use the correct lens resolution. A ⅓ inch lens made for a 1080p camera is often not suitable for a 4K camera. This is because such a lens can resolve details with reasonable contrast only up to a certain number of lines per millimeter. It is best to match the lens resolution to the camera resolution to fully exploit the camera's capability and to avoid the aliasing effects. The appropriate lens for a high-resolution camera over 5 megapixels can be expensive. Read more about aliasing in section 4.5.7.

4.3 IMAGE SENSORS

When a network camera captures an image, light passes through the lens and falls on the image sensor. The sensor consists of photosensitive diodes, called photosites, which register the amount of light that falls on them. The photosites convert the received amount of light into a corresponding number of electrons. The stronger the light, the more electrons are generated. The buildup of electrons in each photosite is converted into a voltage. An analog-to-digital (A/D) converter converts the voltage into digital data as picture elements, more commonly known as pixels. Once an image is formed, it is sent for processing in a stage that determines, among other things, the colors of each individual pixel that make up an image. Image processing is discussed in section 4.5.

4.3.1 Color filtering

Image sensors register the amount of light from bright to dark but no color information. Since these sensors are "color blind", a filter in front of the sensor allows the sensor to assign color tones to each pixel. Two common color registration methods are RGB (red, green, and blue) and CMYG (cyan, magenta, yellow, and green).

The RGB method takes advantage of the three primary colors, red (R), green (G), and blue (B), much like the human eye does. When mixed in different combinations, red, green, and blue produce most of the colors that humans see. By applying red, green, and blue color filters in a pattern over the pixels of an image sensor, each pixel can register the brightness of one of the three colors of light. For example, a pixel with a red filter over it records red light while blocking all other colors.

COLOR AND HUMAN VISION

The human eye has three different color receptors: red, green, and blue. A camera is built to imitate the human visual system. This means that the camera translates light into colors that can be based not just on one light wavelength but a combination of wavelengths.

When the wavelength of light changes, the color changes. Blue light has a shorter wavelength than red light. However, a combination of wavelengths or a single wavelength can be interpreted by the human visual system as the same color because the system is not able to tell the difference. Hence, yellow light, for example, could be either a single wavelength or a combination of several (for example, green and red). Thus, there are certain different combinations of wavelengths that look the same.

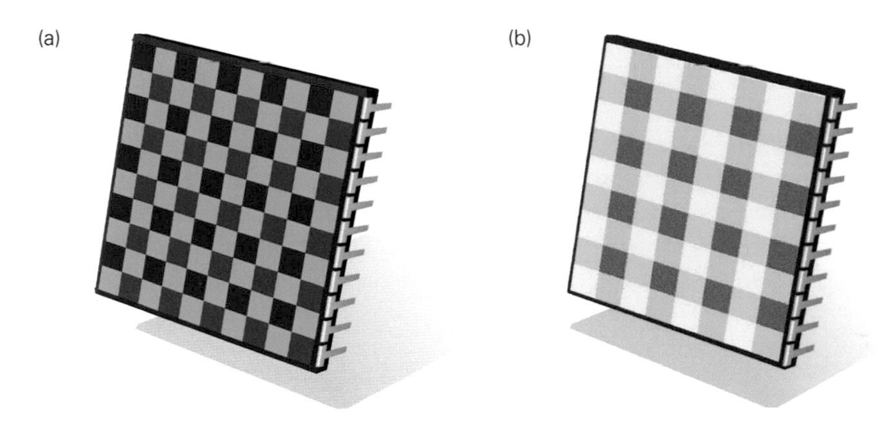

Figure 4.32 (a) Bayer array color filter placed over an image sensor. (b) CMYG (cyan, magenta, yellow, green) color filter array.

The Bayer array, which has alternating rows of red-green and green-blue filters, is the most common RGB color filter (see Figure 4.32).

Since the human eye is more sensitive to green than to the other two colors, the Bayer array has twice as many green color filters. This also means that with the Bayer array, the human eye can detect more detail than if the filter used the three colors in equal measures.

Another way to filter or register color is to use the complementary colors – cyan (C), magenta (M), and yellow (Y). Complementary color filters on sensors are often combined with green (G) filters to form a CMYG color array (see Figure 4.32). The CMYG system generally offers higher pixel signals due to its broader spectral band pass. However, the signals must then be converted to RGB since this is used in the final image, and the conversion implies more processing and added noise. The result is a reduction of the initial gain in signal-to-noise ratio, and the CMYG system is often not as good at presenting colors accurately.

The CMYG color array is often used in interlaced, CCD (charge-coupled device) image sensors, whereas the RGB system is primarily used in progressive scan image sensors. The following section presents CCD and CMOS (complementary metal-oxide semiconductor) image sensor technologies. Section 4.4 gives more information about interlaced and progressive scan.

4.3.2 CMOS and CCD technologies

Figure 4.33 shows CCD and CMOS image sensors. CCD sensors are produced using a technology developed specifically for the camera industry, whereas CMOS sensors are based on standard technology that is already extensively used in products such as PC memory chips. Modern CMOS sensors use a more specialized technology, and the quality of the sensors has increased rapidly in recent years.

(a)

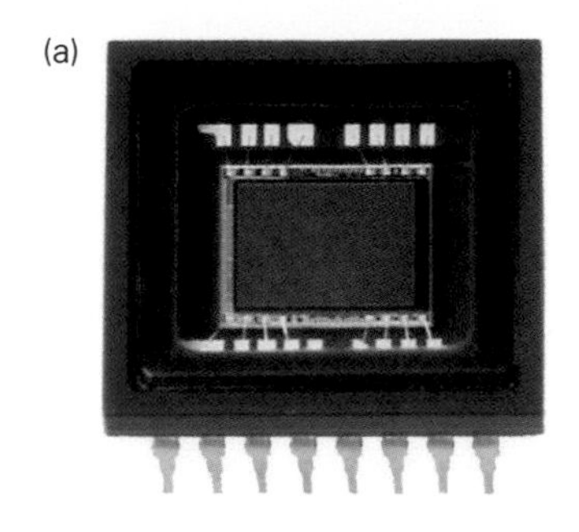

(b)

Figure 4.33 (a) CCD sensor. (b) CMOS sensor.

In a CCD sensor, every pixel's charge is transferred through a very limited number of output nodes (often only one) in which the charges are converted to voltage levels, buffered, and sent from the chip as an analog signal. The signal is then amplified and converted into a set of numbers using an analog-to-digital (A/D) converter outside the sensor. All the pixels in the sensor can be devoted to capturing light, and the uniformity of the output (a key factor in image quality) is high. The CCD sensor has no built-in amplifiers or A/D converters, so tasks are performed outside it. CCD sensors are not very common in network cameras today due to the rapid improvements and advantages of CMOS sensors.

4.3.2.1 CMOS technology

Unlike CCD sensors, CMOS chips contain all the logic needed to produce an image. They incorporate amplifiers and A/D converters, which lowers the total cost of the camera. Every CMOS pixel contains conversion electronics. Compared to CCD sensors, CMOS sensors have better integration possibilities and more functions. However, this addition of circuitry inside the chip can lead to a risk of structured noise, such as stripes and other patterns. CMOS sensors also have a faster readout, lower power consumption, higher noise immunity, and a smaller system size. Megapixel, HDTV and 4K CMOS sensors are more widely available and less expensive than megapixel CCD sensors.

It is possible to read individual pixels from a CMOS sensor. This allows "windowing", which means that parts of the sensor area can be read out, instead of the entire sensor area at once. This enables the delivery of a higher frame rate from a limited part of the sensor and allows digital PTZ functions to be used. It is also possible to achieve multiview streaming, which lets one camera act as several virtual cameras. By allowing several cropped view areas to be streamed simultaneously from the sensor, one camera can simulate the views and tasks that would otherwise require multiple cameras.

CMOS sensors have come to dominate the video surveillance camera market thanks to their improved performance, higher resolutions, and lower cost. Much of this development was driven by the demands of the consumer industry.

4.3.3 More about image sensors

Apart from size, resolution, and sensor type, there are several other characteristics that differentiate sensors, such as the following:

- *Pixel size*
- *Light sensitivity of a pixel (including fill factor)*
- *Maximum signal-to-noise ratio*
- *Dynamic range*
- *Fixed-pattern noise*

Both the size of the image sensor and the size of the individual pixels affect image quality. Most network cameras use ¼ inch or ⅓ inch image sensors no larger than 4.8 × 3.6 mm. A larger image sensor can contain many more pixels than a smaller sensor and delivers higher-resolution images and greater detail. When comparing image sensors, always pay attention to the pixel count and pixel size. Two similar-sized image sensors that differ in pixel count and pixel size will likely produce different resolutions and have different levels of light sensitivity. See section 4.3.4 for more information on high-resolution sensors.

The bigger the pixels in an image sensor, the better they are at storing electrons generated from their exposure to light. Generally, a larger pixel entails a larger maximum signal-to-noise ratio, resulting in images that are less noisy in highlights. Also, a larger pixel often means a higher fill factor and, therefore, higher light-sensitivity. The fill factor is the ratio of the area devoted to light gathering compared with the total area, which includes the area devoted to circuitry within a pixel. A sensor with pixels completely devoted to light gathering has 100% fill factor. Each pixel on a CMOS sensor has circuitry, so the sensor's fill factor will be less.

An image sensor also affects the dynamic range, that is, the range from the maximum useful level to the lowest noise level. To be able to capture both dark and bright objects in the same scene without too much noise, the camera's image sensor must have a high dynamic range (HDR), which is also known as wide dynamic range (WDR). Many natural scenes have a wide range of brightness levels, which is sometimes difficult for a camera to handle. Typical examples are indoor scenes with a bright window or an outdoor scene with a dark foreground and a bright sky. There are various techniques that enable a camera to go beyond the limited dynamic range of a typical sensor. In these, the individual pixels are exposed or treated differently to reduce noise. Without such techniques, a higher gain on dark pixels would also amplify noise and give a lower-quality image.

The quality of an image sensor is also determined by how much fixed-pattern noise it has. Fixed-pattern noise has a pattern that does not change over time. It is caused by non-uniformity of the pixels on an image sensor and by electrons from heat generated by the sensor. It is mostly noticeable during long exposures.

4.3.4 High-resolution sensors

Megapixel and HDTV network cameras deliver high-resolution video images, providing great detail and the possibility to identify people and objects, a key requirement in video surveillance applications. Megapixel sensors are fundamental to HDTV, Ultra-HD, megapixel, and multi-megapixel cameras, for their extremely detailed images and for multiview streaming.

Megapixel CMOS sensors are more widely available and generally less expensive than megapixel CCD sensors, although there are costly CMOS sensors too.

It is difficult to make a fast megapixel CCD sensor, which, of course, is a disadvantage, as it adds to the challenges of building a multi-megapixel camera based on CCD technology.

Assuming that the sensor technology and characteristics are equal, the sensor size should increase along with the increase in pixel quantity to have the same light sensitivity. Still, many sensors in multi-megapixel cameras (5 megapixels or more) are about the same size as 1080p sensors. This means that the size of each pixel in a multi-megapixel sensor is smaller than the pixels in a 1080p sensor. Because its pixels are smaller, a multi-megapixel sensor is typically less light-sensitive per pixel than a 1080p sensor. However, size is not everything – quality also matters. As technology evolves rapidly, a modern small pixel may be better than a dated large pixel, and thus, a smaller sensor may well be more light-sensitive than a larger one. Again, testing and evaluating performance is the only way to know for sure that the sensor and other camera components are matched in such a way that they meet expectations.

Megapixel sensors can be used innovatively in panoramic cameras to provide high-quality, full-overview images or close-up images using instant zoom with no moving camera parts. For more details on this subject, see section 3.9 on panoramic cameras.

4.4 IMAGE SCANNING TECHNIQUES

Interlaced scanning and progressive scanning are the two techniques currently available for reading and displaying information produced by image sensors. Interlaced scanning is mainly used in CCDs. Progressive scanning is used in both CCD and CMOS sensors. Network cameras can use either technique, while most analog cameras are based on standards that can only use the interlaced scanning technique to transfer and display images.

4.4.1 Interlaced scanning

Interlaced scanning was originally introduced to improve the image quality of a video signal without consuming additional bandwidth. The method soon became universal in traditional, analog television sets. To put it simply, the technique splits each frame into two fields. The scanning starts at the top-left corner and sweeps all the way to the bottom-right corner, skipping every second row on the way. This reduces the signal bandwidth by a factor of two, allowing for a higher refresh rate, less flicker, and better portrayal of motion.

There are some downsides to interlaced video, including the following:

- *Motion artifacts:* If objects are moving fast enough, they will be in different positions when each individual field is captured. This may cause motion artifacts.
- *Interline twitter:* A shimmering effect shows up when the subject being filmed contains very fine vertical details that approach the horizontal resolution of the video format. This is why TV presenters usually avoid wearing striped clothing.

All analog video formats and some modern HDTV (high-definition television) formats are interlaced. The artifacts or distortions created through the interlacing technique are not very noticeable on an interlaced monitor. However, when interlaced video is shown on progressive scan monitors such as computer monitors, which scan the lines of an image consecutively, these artifacts become noticeable. The artifacts, which can be seen as “tearing”, are caused by the slight delay between odd and even line refreshes, as only half the lines keep up with a moving image while the other half waits to be refreshed. This is especially noticeable when the video is paused to view a single frame.

4.4.2 Deinterlacing techniques

To show interlaced video on computer screens and reduce unwanted tearing effects, different deinterlacing techniques can be used. The problem with deinterlacing is that two image fragments, captured at different times, must be combined into an image suitable for simultaneous viewing.

There are several ways to limit the effects of interlacing, including the following:

- *Line doubling*
- *Blending*
- *Bob deinterlacing*
- *Discarding*
- *Motion adaptive deinterlacing*

Line doubling removes every other field (odd or even) and doubles the lines of each remaining field (consisting of only even or odd lines) by simple-line doubling or, even better, by interpolation. This results in the videos having effectively half the vertical resolution scaled to the full size. Although this prevents accidental blending of pixels from different fields (called the comb effect), it causes noticeable reduction in picture quality and less smooth video (see Figure 4.34b, d).

Blending mixes consecutive fields and displays two fields as one image. The advantage of this technique is that all fields are present, but the comb effect may be visible because the two fields, captured at slightly different times, are simply merged. The blending operation may be done differently, giving more or less loss in the vertical resolution of moving objects. This is often combined

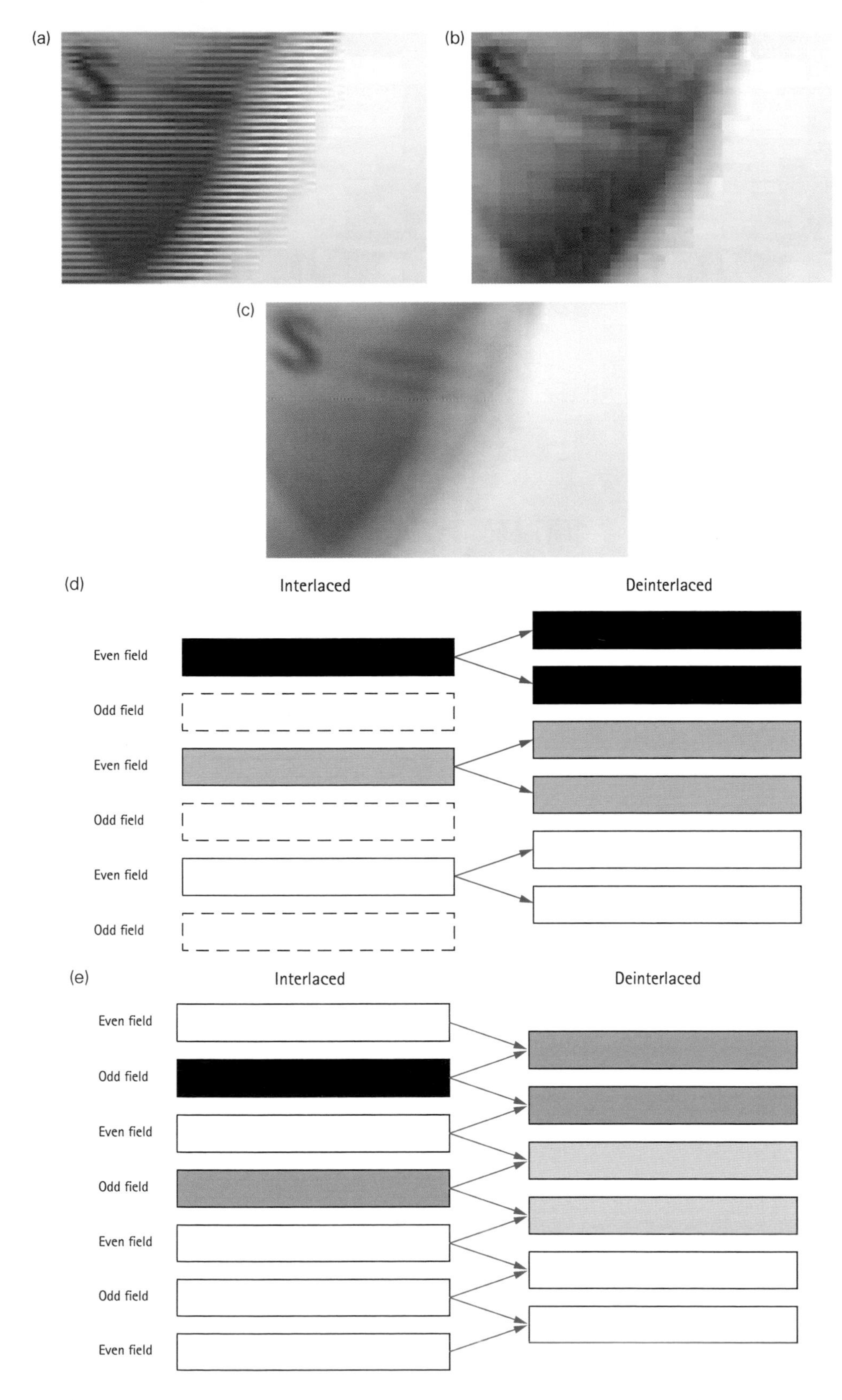

Figure 4.34 (a) Original interlaced image shown on a progressive scan monitor. (b) Image deinterlaced with line doubling. (c) Image deinterlaced with blending. Line doubling technique. Blending technique. (d) Line doubling technique. (e) Blending technique.

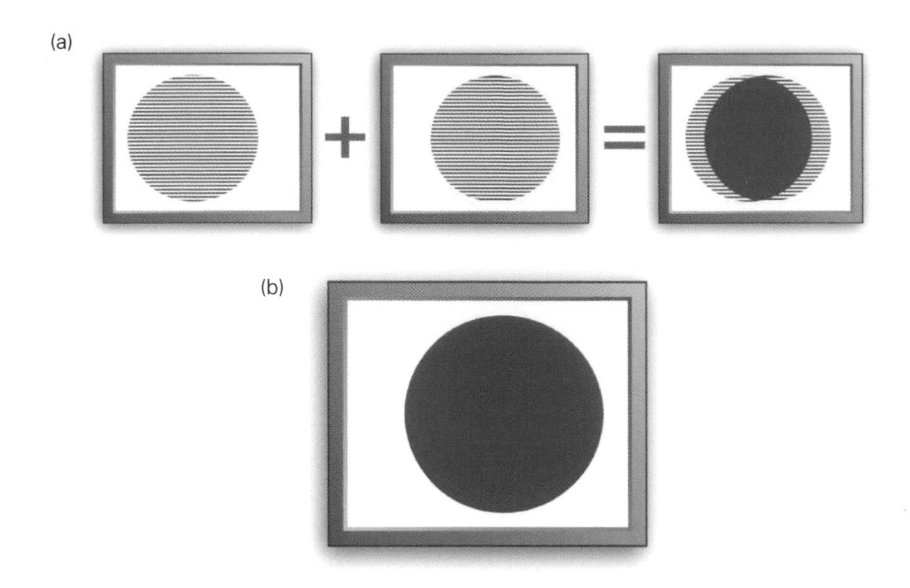

Figure 4.35 (a) An interlaced scan image shown on a progressive (computer) monitor. Odd lines (left). Even lines, 17/20 milliseconds (NTSC/PAL) later (center). Freeze frame on moving dot using interlaced scan (right). (b) Freeze frame on moving dot using progressive scan.

with a vertical resize so that the output has no numerical loss in vertical resolution. The problem with this technique is that there is a loss in quality because the image has been downsized and then upsized. The loss in detail makes the image look softer, and the blending creates the ghosting artifacts (see Figure 4.34c, e).

In bob deinterlacing the fields are "bobbed" up and down, whereby each field is made into a full frame. Each odd frame is pushed down by half a line, and each even frame is pushed up by half a line. This method requires knowing which field (odd or even) should be displayed first. The total number of frames for the video is doubled, and to play the video at the right speed, the frame rate is doubled. This technique leads to some line flickering and requires more computer power for playback.

Discarding is a trivial method that removes interlacing artifacts by scaling down the video horizontally, throwing out every other field and only keeping half the information. The video size is smaller, and only half of the temporal resolution is kept, but no artifacts are visible.

Motion adaptive deinterlacing is a more advanced technique. It incorporates the previously described technique of blending or averaging, together with a calculation for motion. If there is no motion, the two consecutive fields are combined to form a complete frame, with full resolution and sharpness. When motion is detected, vertical resolution is compromised as before. In pixel-based motion adaptive deinterlacing, the non-moving parts of an image are shown in full resolution while pixels that cause artifacts in the moving parts of the image are discarded. Lost data at the edges of moving objects can then be reconstructed using multi-direction diagonal interpolation. This technique requires a lot of processing power.

Interlaced scanning has served the world of analog camera, television, and VHS video well for many years and is still the most suitable technique for some applications. However, with the development of TVs and displays using LCD, plasma, and LED technology, progressive scan is now dominant. (See Figure 4.35.)

4.4.3 Progressive scanning

The progressive scanning technique captures, transmits, and displays all the lines in the image in a single frame. Scanning is done line by line, from top to bottom. In other words, captured images

Figure 4.36 Comparison between progressive, interlaced, and 2CIF-based scanning techniques. In these examples, the cameras are using the same lens, and the car has the same speed (20 km/h, 15 mph). (a) Progressive scan is used in network cameras, full size 640 × 480. (b) Interlaced scan, mostly used in analog CCTV cameras, full size 704 × 576. (c) 2CIF-based scan as used in DVRs, full size 704 × 240 (PAL).

are not split into separate fields as in interlaced scanning, so there is essentially no flickering effect. Therefore, moving objects are better presented on computer screens using progressive scan. In a video surveillance context, progressive scanning can be critical to identify the details of a moving subject, such as a person running away.

When a camera captures a moving object, the sharpness of a still image depends on the technology used. Compare the JPEG images in Figure 4.36, which were captured by three different cameras using progressive scan, 4CIF interlaced scan, and 2CIF, respectively. For more information on resolution, see section 4.6.

Note the following in the images in Figure 4.36:

- All image systems produce a clear image of the background.
- In the interlaced scan images (b), jagged edges result from motion.
- In the 2CIF sample (c), motion blur is caused by lack of resolution.
- Only progressive scan (a) makes it possible to identify the driver.

In progressive scan, single frames from a video sequence can be printed with almost photographic quality. This is very beneficial and can even be crucial if the material is to be used as evidence in a court of law.

4.5 IMAGE PROCESSING

The lens and sensor are key camera components and determine much of the quality of an image. The quality can be further improved by the image processor, which is built into the network camera. The image processor can use various techniques to process the image, including the following:

- *Control of exposure time, aperture, and gain*
- *Backlight compensation and wide dynamic range (WDR)*
- *Bayer demosaicing* that converts raw data into a color image
- *Noise reduction*
- *Color processing*, such as saturation and white balance
- *Image enhancement*, such as sharpening and contrast

4.5.1 Exposure

Thanks to the human eye's ability to adapt to different lighting conditions, we can see and navigate in both very dark and very bright environments. A camera, too, must be able to cope with changes in brightness, so it needs to find the correct exposure.

A camera has three basic controls to achieve the appropriate exposure and, thus the ideal image quality:

- *Exposure time:* how long the sensor is exposed to incoming light
- *Aperture or f-stop:* the size or the iris opening that lets light through to the sensor
- *Gain:* image level amplifier that can make the images look brighter

Exposure time, aperture, and gain are often set automatically by the camera, but many cameras also allow manual adjustment. By adjusting one or more of these settings, an image of a scene will appear relatively unchanged even when the lighting at the scene changes.

The images in Figure 4.37 are taken with different illumination. Thanks to auto-exposure, the images appear relatively similar. The images are taken in the same room, but in the top image, the figure is standing on top of a cabinet, and in the bottom image, the figure is standing inside the same cabinet, so the lighting is different.

Note that increasing the gain of the image not only increases the image luminance but also the noise. Adjusting the exposure time or using a smaller f-stop (larger iris opening) is, therefore, preferable. However, sometimes the iris is fully open, and the exposure time required is longer than the time between two frames. For example, if the frame rate is 30 fps and the exposure time is 1/15 second, the exposure time is twice as long as the time between each frame. In this case, a decision must be made – lower the frame rate or increase the gain? The answer depends on the application's requirements and priorities.

(a)

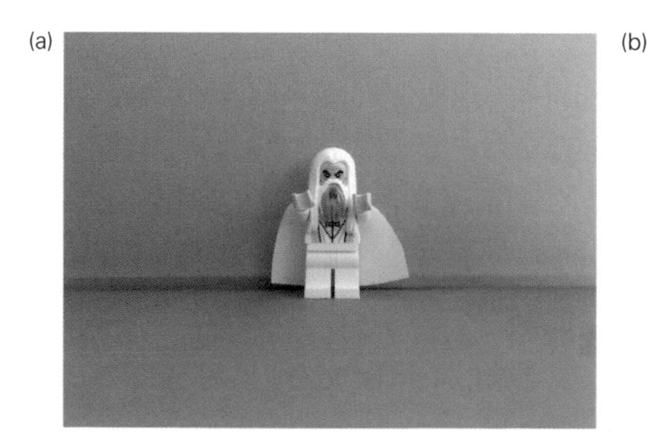

(b)

Figure 4.37 Even though they have different illuminance, auto-exposure can make two images of the same scene appear much the same. (a) The figure is illuminated by afternoon sunlight coming from the right through a window about 50 cm (19 11/16 in) away. (b) The figure is illuminated only by indirect light.

(a)

(b)

Figure 4.38 (a) Strong backlight, without backlight compensation. (b) With backlight compensation applied, limited areas of high illumination can be ignored.

4.5.2 Backlight compensation

Although a camera's automatic exposure controller tries to get the brightness of the image to appear as the human eye would see the scene, it can be easily fooled. Strong backlight is difficult for a camera to handle (Figure 4.38).

In the case with strong backlight, such as in Figure 4.38a, the camera believes that the scene is very bright. So the camera makes the iris opening smaller or reduces the exposure time, which results in a dark image. A feature called backlight compensation helps prevent this from happening. It strives to ignore limited areas of high illumination as if they were not present. The resulting image (Figure 4.38b) makes it possible to identify all the objects in the scene. The bright areas, however, are overexposed. Without backlight compensation, the image would be too dark, and color and details would be lost in the foreground. Such a situation also might be dealt with by increasing the dynamic range of the camera, as discussed in the next section.

In addition to dealing with limited areas of high illumination, a network camera's automatic exposure controller must decide what area of an image to base the exposure value on. For example, the foreground (such as a bus, a city square, or railway tracks) can hold more essential information than the background that often takes up a part of the upper half of an image (such as the sky or a line of trees). The less important areas of a scene should not determine the overall exposure. The camera designer's solution should be to divide the image into sub-images and assign different weights for the exposure algorithm (4.1.1). In advanced network cameras, the user can select which of the predefined areas should be more correctly exposed. The position of the window or exposure area can be set to center, left, right, top, or bottom.

4.5.3 Wide dynamic range (WDR)

Assigning different weights for different parts of an image (see Figure 4.39) is one way to determine exposure values. But there are many surveillance situations that present lighting challenges that cannot be solved by changing the exposure time or creating exposure zones.

An overcast day with few shadows has a low dynamic range. That is, there are no deep blacks and no extreme highlights. But on sunny days with very distinct shadows, there could be a big difference between the brightest and darkest areas. This is called a wide dynamic range (WDR), also known as high dynamic range (HDR). Some dynamic ranges extend further than the camera or the human eye can perceive. In these situations, the camera must decide which area to expose properly.

Typical situations include the following:

- Entrances with daylight outside and a darker environment inside. This is very common in retail, banking, and office environments.

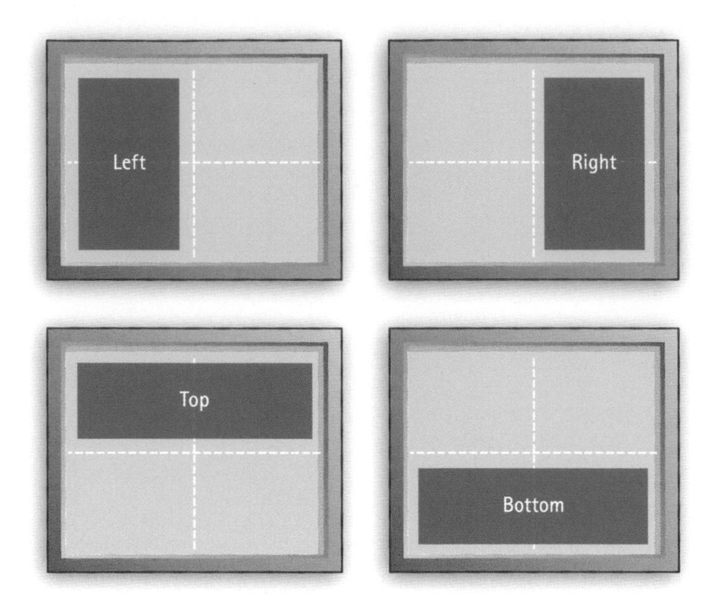

Figure 4.39 Assigning different weights to sections of an image to better determine which area should be more correctly exposed.

- Vehicles entering a parking garage or tunnel, also with daylight outside and low-light indoors.
- In city surveillance, transportation, perimeter surveillance, and other outdoor applications where some parts of the scene are in direct sunlight and others are in deep shadow.
- Vehicles with strong headlights, heading directly towards the camera.
- Environments with lots of reflected light, such as the glass in office and shopping malls.

Wide dynamic range (WDR) imaging is a method for restructuring images with full scene content so that people, objects, vehicles, and events can be reliably identified in situations where there is a wide range of lighting conditions. Without WDR imaging, the camera would produce an image where objects in the dark area of the scene would barely be visible (Figure 4.40).

Both images are lacking information from the full scene. A good WDR surveillance camera can capture both these extremes in a single image (see Figure 4.41), that is, it can clearly show details in the well-lit entrance as well as in the dark shadows inside the parking garage.

4.5.3.1 Measuring dynamic range

Not all situations need a WDR camera. Using a regular camera that is configured to avoid clipping the highlights, the sample images in Figure 4.42 show how image noise increases as the amount of light decreases. These images and Table 4.6 together give an indication of when WDR technology should be used.

The lower the image quality and contrast, the less useful the image, for example, in forensic investigations. Looking at the images in Figure 4.42, ask yourself which image is good enough to fulfill its purpose. In scenes with a good balance between shade and lighting, there is a better chance the images will be of sufficient quality to meet requirements. If the scene is closest to type (a) in Table 4.6 (sunlight/shadow), you can expect a standard camera to deliver images with a quality equal to image 6, which is good enough for identification and many forensic purposes. As the lighting conditions deteriorate, you will need to lower your expectations on image quality accordingly and tolerate a higher level of noise. If the scene is closest to type (b) (window illumination), a standard camera might deliver an image quality equal to image 9, which provides little more than detection level. In conditions like type (c) (dark indoor scene), you can expect noise levels equal to

(a)

(b)

Figure 4.40 (a) Image is underexposed. (b) Image is overexposed.

Figure 4.41 The same scene, but here, the image has been captured using WDR technology (capture WDR). The areas that were underexposed are brighter and the areas that were overexposed have been levelled.

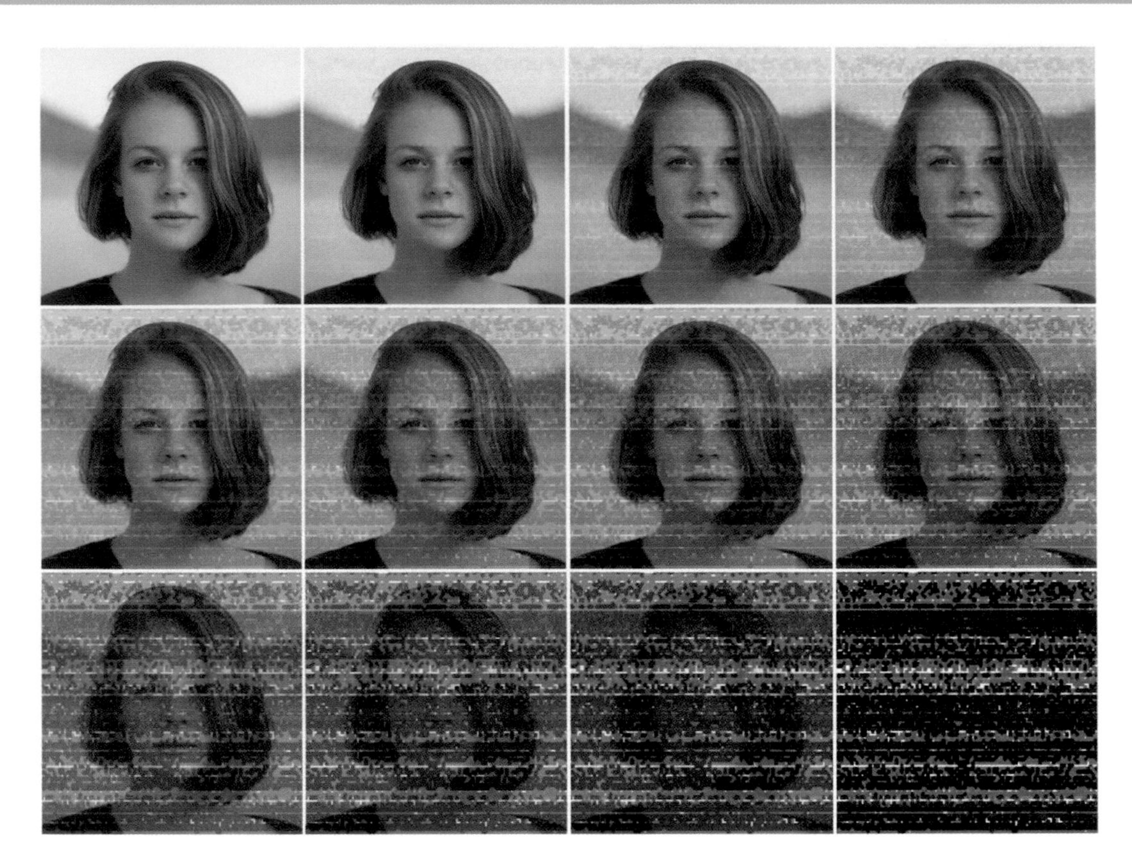

Figure 4.42 Images of subject taken by a standard camera with default settings. As the light drops, the noise increases.

Table 4.6 Type of scene and its typical illumination ratio

Type of scene	Example	Illumination ratio
(a) Sunlight/shadow	A sunlit train station with part of the platform shielded from the sun	1:20
(b) Window illumination indoor	A lobby with large windows	1:200
(c) Dark indoor scene with opening	A typical warehouse scene with doors or openings to the outside sunlight	1:2000

or greater than in image 12. Remember that this is only an indication since standard cameras obviously vary in quality.

As dynamic range varies so much, the dynamic range capability of a camera is often presented in the logarithmic unit of decibels (dB). It expresses the ratio of radiance between the brightest and the dimmest object that can be captured by the camera. Note that this is not the same as the illumination ratio used in Table 4.6. If the radiance ratio is 1000:1, the dB value is 60 dB. The value is calculated as the logarithm (in this case, 3) of the ratio times 20, as when calculating the amplitude ratio of voltage.

$$Ratio_{dB} = 10 \times log_{10}\left(V_1^2 / V_0^2 \right) = 20 \times log_{10}\left(V_1 / V_0 \right)$$

$$Ratio_{dB} = 20 \times log_{10}\left(1000/1 \right) = 20 \times 3 = 60\ dB$$

The dimmest detectable level can be defined as the noise lower threshold of the sensor pixel. Any signal below this level is drowned in noise. With this definition, a good image sensor can reach a dynamic range of about 70 dB. Through techniques such as multi-frame exposures, some sensors can increase the upper detection limit. This extends the dynamic range to more than 100 dB, but this is not necessarily always the best WDR solution.

Although some modern surveillance cameras use sensors with extended dynamic range that allow them to better handle difficult scenes, the dB value cannot fully describe the WDR capacity. The decibel unit is merely an approximation of the dynamical capabilities of the camera. As is the case with lux ratings, manufacturers may use their own measuring methods, which makes datasheet comparisons unreliable. To fully understand a camera's capabilities, it is best to test them on-site.

4.5.3.2 Types of WDR

Each type of WDR is designed to address different types of situations and use cases. WDR types include the following:

- *Contrast WDR*
- *Capture WDR*
- *Forensic WDR*

Contrast WDR (also known as WDR-dynamic contrast) allows the camera sensor to capture an image with a higher bit depth than what the camera ultimately can send out. The bit depth is an internal property that translates to dynamic range. Contrast WDR performs advanced tone mapping, where some brightness levels are dropped to decrease the bit depth to a format that a computer screen can handle. Tone mapping looks at both the darkest and the brightest parts. The result is an image that has more details at both ends.

There are two types of tone mapping. With global tone mapping, all pixels are handled in the same way, which means that the same levels are removed everywhere in the image. With local tone mapping, individual decisions are made for different regions in the image to decide which levels to remove. Local tone mapping requires much more processor power but gives a better result.

Capture WDR (also known as WDR-dynamic capture) captures several images with different exposure levels for each image. These pictures are then combined in a composite, where both the brightest and the darkest parts are kept, resulting in an image with more clarity and sharpness. However, this image has a higher bit depth than a computer screen can handle. Therefore, tone mapping needs to be applied just as with contrast WDR. Capturing several images in the time span normally used for one image requires an extremely fast and sensitive sensor. Still, the output is much better than with contrast WDR.

Forensic WDR (also known as WDR-forensic capture) applies advanced algorithms that optimize image quality specifically for forensic purposes. The algorithms lower noise levels and increase the image signal to display every detail in the best possible way. The method also includes the ability to seamlessly transition between WDR and low-light mode. The resulting images often look different from what we have come to expect from television broadcasts and still-image photography, but this WDR technology makes it easier to detect and identify critical details in a scene.

Image results from a WDR camera will differ depending on aspects such as the complexity of the scene and the amount of movement. As with any video surveillance situation, the most important questions are as follows: What do you want to see? How do you want to present the captured image?

WDR technologies typically use different exposures for different objects within a scene and employ ways to display the results. However, the screen displaying the video may also have limited dynamic range.

(a)

(b)

Figure 4.43 A parking garage where the camera is placed inside the garage. (a) Image taken without WDR imaging. (b) Image taken with capture WDR.

(a) (b)

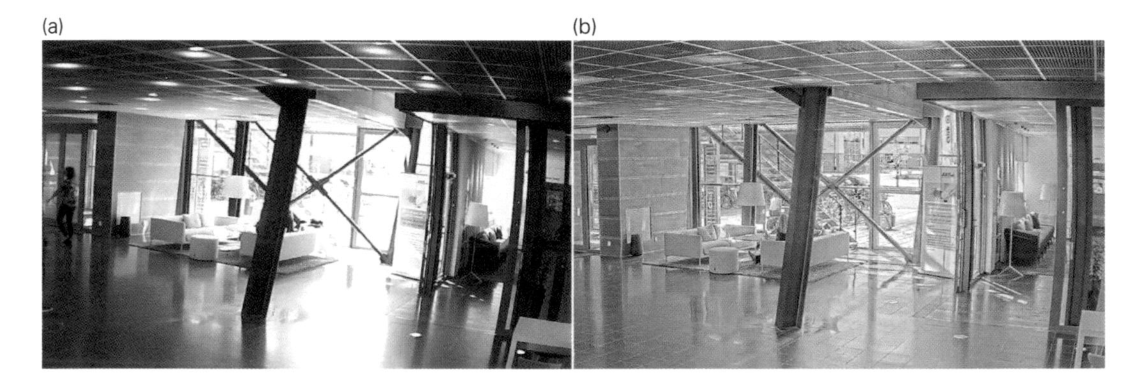

Figure 4.44 A scene illustrating how forensic WDR levels the shadows and highlights, creating a high-contrast image suitable for forensic investigations. (a) Without WDR, the sunlight creates strong backlight that wipes out details. (b) With forensic WDR, it is possible to make out much more detail.

Although WDR technologies solve some problems, it sometimes introduces others. Here are some examples:

- Noise can be very different in different regions of an image. In particular, dark regions may contain very visible levels of noise.
- Pixels between two different exposure regions may show visible artifacts. This can be seen in images with high dynamic content and multiple different levels of lighting at the same time.
- Different exposure regions may have been allocated a dynamic range that is too low, making every part of the image bad.
- Colors may look bleak.

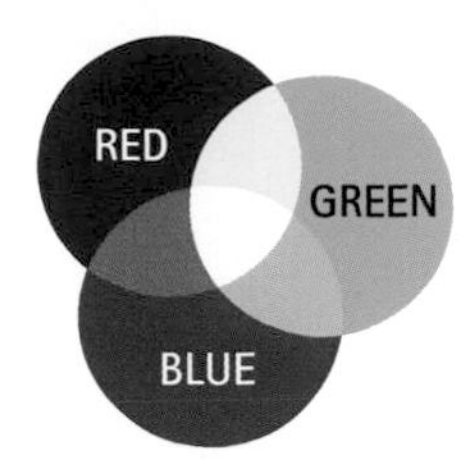

Figure 4.45 Mixing equal amounts of red, green, and blue light results in white.

4.5.4 Bayer demosaicing

After the sensor captures the raw image, the camera processes it to produce a high-quality color image.

In a process called demosaicing, an algorithm is used to translate the raw data from the image sensor into full-color information for each pixel. Because each pixel only records the illumination behind one of the color filters, it needs the values from the other filters – interpolated from neighboring pixels – to calculate the actual color of a pixel. For example, if a pixel with a red filter registers a bright value and the neighboring green-and-blue pixels are equally bright, then the camera's processor determines that the color of the pixel with the red filter must be white (Figure 4.45).

4.5.5 Noise

Because no sensor is perfect, all cameras exhibit uncertainty about the pixel values produced. This is known as random noise, and it means that even if the scene and the illumination are the same, the same pixel will not give the exact same value each time it is read out. Random noise sometimes appears as banding, with entire rows being affected. Every sensor also has a few bad, nonfunctional pixels.

A typical sensor also has differences between individual pixels. This means that adjacent pixels exposed to exactly the same light do not respond exactly the same way. This appears as fixed-pattern noise that does not change over time. Some fixed-pattern noise changes with temperature and exposure time and is, therefore, more pronounced in hot environments or when the exposure time increases at night.

Reducing noise is a key task in a video surveillance camera. Part of the noise is generated within the camera, but part of it is actually due to the nature of light itself, so it affects all cameras. This noise is mostly visible in bright daylight images, for example, in blue skies. Noise is lower in cameras that use sensors with larger pixels, which can collect more light in each pixel.

An image taken in low light might appear grainy or have specks of color because the noise is amplified along with the signal (Figure 4.46). This is true for both random noise and fixed-pattern noise. In a video camera, fixed-pattern noise can be recognized by the constant position of the specks, but it cannot be distinguished from random noise in a still-image camera.

Noise can be reduced by various filtering techniques that replace flawed pixels with new values calculated from neighboring pixels. Most cameras include one or more filters in their processing. Most filters, however, have visible side effects that appear as reduced resolution, motion blur, or other artifacts.

Modern surveillance cameras with powerful processors are well equipped to analyze and reduce noise levels, even in high resolution and at full–frame rate. Some techniques for minimizing noise are listed here:

- *Spatial processing* analyzes a single frame to find pixels that are very different in color or intensity from their surrounding pixels.
- *Temporal processing* compares consecutive frames to find artifacts that are not static over time and which can be regarded as potential noise.

Figure 4.46 Part of the image is shown with noise, which appears as specks of colors that distort the image.

4.5.6 Sharpening and contrast

Digital sharpening and contrast enhancement (Figure 4.47) are tools for fine-tuning digital images. Digital sharpening changes the image edge's contrast, not the resolution.

- *Sharpening* increases the local contrast by lightening the light pixels and darkening the dark pixels at the edges. But be aware that sharpening can also amplify noise.
- *Contrast enhancement* affects not only the edges but all the pixels in the image equally. Contrast enhancement changes how the original pixel values are mapped or converted to values used by a display.

Fog, haze, smoke, and similar conditions have a special effect on images that can be corrected by contrast enhancement. Whole image areas lose their visibility, and objects lose their contrast. Defogging automatically detects the foggy elements and tweaks them digitally to deliver a clearer image (see Figure 4.48). An advanced contrast enhancement algorithm analyzes the image pixel by pixel. So the amount of tweaking varies throughout the scene and only affects the dim areas, not the areas that are already bright and clear. In some cases, defogging can make the difference between whether a vehicle is identified or not.

4.5.7 Aliasing

Sometimes a subject contains finer details than the size of the pixels in an image sensor. The sensor cannot detect these details because the pixel resolution is too low. To see fine details, the image scale must be increased with a telephoto lens, or a sensor with a higher resolution must be used.

Repeated unresolved patterns, such as a herringbone pattern on a jacket, can cause problems for some cameras due to an effect called *aliasing*. The effect is due to poor resolution and appears as distortion in the form of unwanted larger patterns that can even be colored in the image. The two images in Figure 4.49 are representations of the same pattern on a jacket as seen with two different image sensor resolutions. The image to the right has a pronounced aliasing effect.

Cameras translate continuous grades of tone and colors to points on a regular sampling grid. When details are finer than the sampling frequency, a good camera averages the details to avoid aliasing. This discards irresolvable details, but if too many details are filtered out, the image may appear soft.

4.5.8 Image stabilization

In outdoor installations, zoom factors above 20× can prove impractical if there are vibrations and motion caused by traffic or wind. Vibrations in a video can be reduced if the camera is equipped with an image stabilizer, which is especially useful in windy environments such as ports or areas

Figure 4.47 Sharpening versus contrast. (a) Image before sharpening and changing of contrast levels. (b) Contrast reduced – the darker areas become more visible, and the lighter areas become less bright. (c) Sharpening applied.

Figure 4.48 Images of a scene in foggy weather. (a) Defogging off. (b) Defogging on.

prone to heavy vibrations such as highways or bridges. Even in office buildings, there is normally some level of vibration.

There are several types of image stabilization. Mechanical image stabilization or optical image stabilization changes the path to the sensor by moving the sensor or the lens. Image stabilization can also be done by the camera's processor. This is called electronic image stabilization (EIS) and is achieved through a combination of motion estimation and image shift.

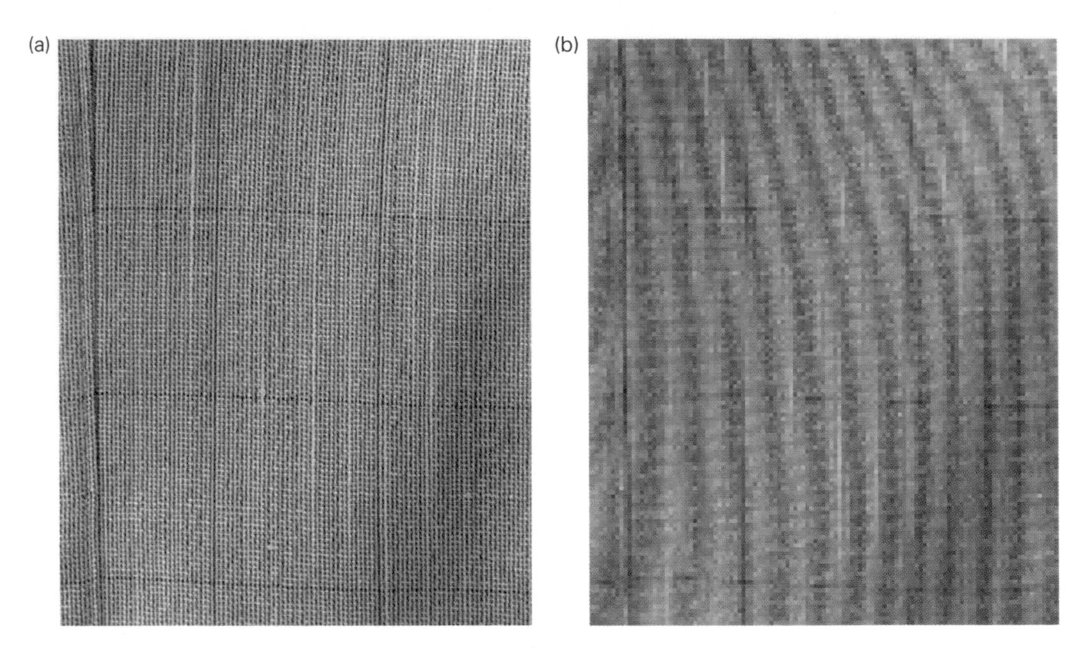

Figure 4.49 The same pattern on a jacket as seen with two different image sensor resolutions. Note the pronounced aliasing effect in the image to the right.

Blurring caused by very rapid movements occurring during the camera's exposure time can be reduced by decreasing the exposure time. It also is possible to use special lenses that compensate for motion during the exposure.

EIS also reduces the file size of the compressed image and lowers the video stream's bitrate. In addition to the obvious benefit of getting more useful video (see Figure 4.50), using EIS saves valuable storage space. Historically, EIS has primarily been available in PTZ cameras but can also be found in some high-end fixed cameras.

4.5.9 Privacy masking

In many cases, a network camera can be used to monitor very large areas, parts of which the operator should not be looking at. For example, a PTZ camera installed outside a stadium in a city center might be able to zoom in on nearby apartments. In such instances, privacy masking is important because it allows some areas of a scene to be blocked for viewing and recording (Figure 4.51).

Cameras with digital PTZ also have privacy mask functionality. But in cameras with mechanical PTZ capabilities, the privacy mask not only follows the size and position of the object but it also shifts the angle and grows and shrinks with the object.

4.6 RESOLUTION

Resolution in an analog or digital world is similar, but there are some significant differences in how it is defined. In a digital system, an image consists of square pixels.

4.6.1 NTSC and PAL resolutions

In North America and Japan, the NTSC (National Television System Committee) standard is the predominant analog video standard, whereas in Europe, the PAL (Phase Alternating Line) standard is used. Both standards were established by the television industry.

NTSC has a resolution of 480 lines and uses a refresh rate of 60 interlaced fields per second (or 30 full frames per second). A new naming convention, which defines the number of lines, scanning type, and refresh rate for this standard is 480i60, where "i" stands for interlaced scanning.

Figure 4.50 (a) Snapshot of video with electronic image stabilization (EIS). (b) Without EIS.

Figure 4.51 Example of privacy masking. Note the gray boxes at the lower edge of the image.

PAL has a resolution of 576 lines and uses a refresh rate of 50 interlaced fields per second (or 25 full frames per second). The new naming convention for this standard is 576i50. The total amount of information per second is the same in both standards.

Common Intermediate Format (CIF) is a standardized system for converting the NTSC and PAL standards to digital format. When analog video is digitized, the maximum number of pixels that can be created is based on the number of TV lines available for digitization. The maximum size of the digitized image is 720 × 480 pixels (NTSC) or 720 × 576 pixels (PAL). This resolution is also known as D1.

4CIF has 704 × 480 pixels (NTSC) or 704 × 576 pixels (PAL). 2CIF has 704 × 240 pixels (NTSC) or 704 × 288 pixels (PAL), which means that the number of horizontal lines is divided by two. In most cases, each horizontal line is shown twice when shown on a monitor. It is an effort to maintain correct ratios in the image and is called line doubling. It is also a way to cope with motion artifacts caused by interlaced scan. Section 4.4.2 discusses deinterlacing, and Figure 4.52 shows a comparison of the different NTSC and PAL image resolutions.

Sometimes, a quarter of a CIF image is used. This resolution is called Quarter CIF or QCIF for short.

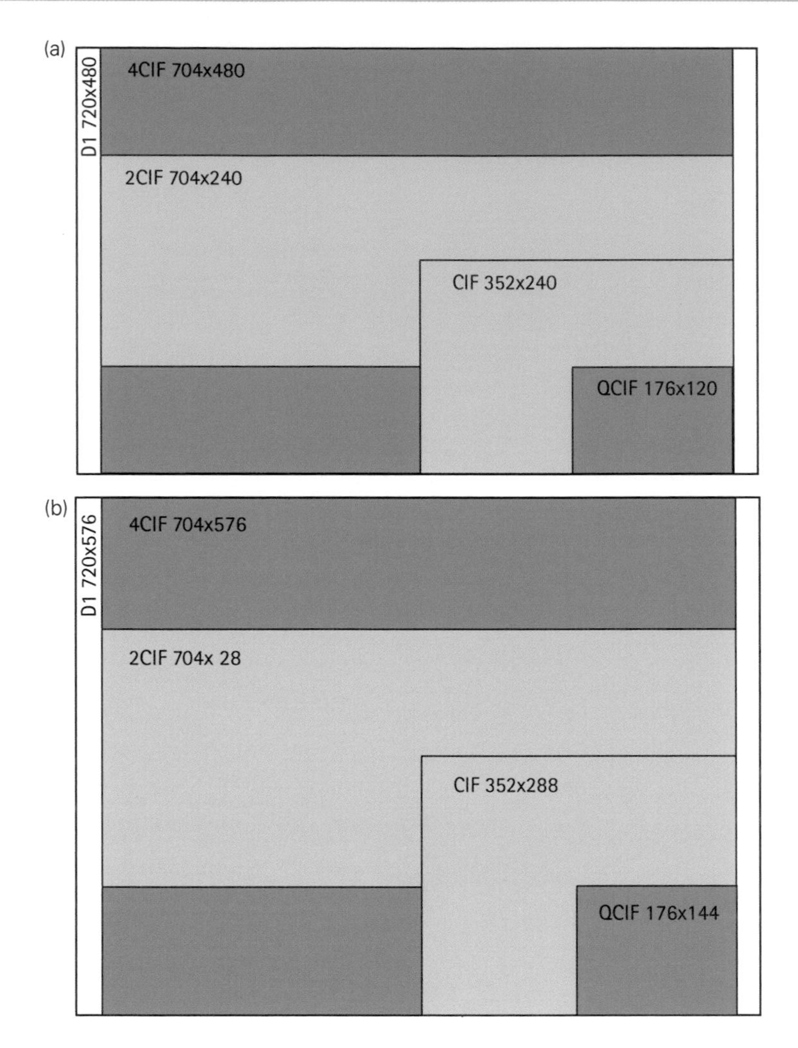

Figure 4.52 (a) NTSC image resolutions. (b) PAL image resolutions.

Since the introduction of network cameras, it has been possible to design surveillance systems that are fully digital. This renders the limitations of NTSC and PAL irrelevant. Several new resolutions stemming from the computer and digital television industry have been introduced. They are worldwide standards and give better flexibility.

4.6.2 Megapixel resolutions

A network camera with megapixel resolution uses a megapixel sensor to deliver an image containing one million pixels or more. The more pixels a sensor has, the greater the potential it has for capturing finer details and producing a higher-quality image. Megapixel network cameras are used in scenarios where details are critical, such as when people and objects need to be identified, when a larger area needs to be monitored, or when the scene needs to be divided into multiple view areas.

Table 4.7 and Figure 4.53 summarize some megapixel formats.

In the video surveillance industry, best practices have emerged regarding the number of pixels required for certain situations. The DORI standard (Detection, Observation, Recognition, Identification) specifies the following levels of resolution for different purposes:

- *Detection:* 25 px/m (8 px/ft) allows for reliable determination of whether a person or vehicle is present.
- *Observation:* 63 px/m (20 px/ft) provides characteristic details of an individual, such as distinctive clothing.
- *Recognition:* 125 px/m (40 px/ft) determines with a high degree of certainty an individual that has been seen before.
- *Identification:* 250 px/m (80 px/ft) enables the identity of an individual beyond a reasonable doubt.

For an overview image, the general opinion is that 70–100 pixels are enough to represent 1 m (20–30 pixels per foot) of a scene. For situations that require detailed images, such as face identification, the requirements can rise to 500 pixels per meter (150 pixels per foot). To identify people passing through an area 2 m wide and 2 m high (6.56 × 6.56 ft), the camera needs a resolution of at least 1 megapixel (1000 × 1000 pixels).

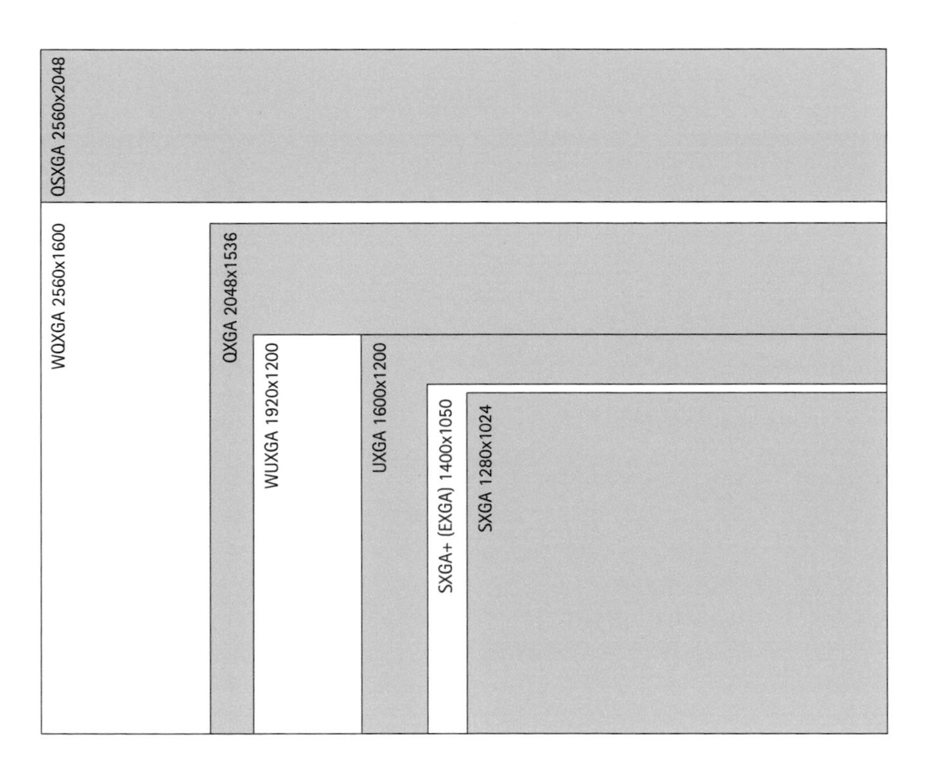

Figure 4.53 Megapixel image resolutions.

Table 4.7 Megapixel formats

Display format	Number of megapixels	Pixels
SXGA	1.3	1280 × 1024
SXGA+ (EXGA)	1.4	1400 × 1050
UXGA	1.9	1600 × 1200
WUXGA	2.3	1920 × 1200
QXGA	3.1	2048 × 1536
WQXGA	4.1	2560 × 1600
QSXGA	5.2	2560 × 2048

Figure 4.54 A megapixel camera view versus an analog camera view.

Figure 4.55 The images from an analog camera (left) are not as sharp as those from a megapixel camera (right).

Megapixel resolution is one area in which network cameras excel over analog cameras (Figure 4.54). The maximum resolution a regular analog camera can deliver after digitizing the video signal is D1, which is 720 × 480 pixels (NTSC) or 720 × 576 pixels (PAL). There are, however, also HD-CCTV analog cameras available, which do provide higher resolution. See section 1.4.2 for more information.

Figure 4.56 The gray lines show the view from an analog camera, while the total image shows the megapixel camera view. Given that the number of pixels per area is the same, one 1080p (2 megapixel) camera covers a much larger scene than four 4CIF cameras.

The D1 resolution corresponds to a maximum of 414,720 pixels, or 0.4 megapixels. By comparison, a common megapixel format is 1280 × 1024 pixels, giving a resolution of 1.3 megapixels. This is more than three times higher than the resolution that analog CCTV cameras can give.

4.7 HIGH-RESOLUTION CAMERAS

In the early years of megapixel network cameras, one of the most highlighted benefits was their ability to provide video at much higher resolution than analog cameras. Now that the market has matured, there is very little demand for cameras below 720p (0.9 megapixels). Many surveillance managers require much higher resolutions than that. There has been a tradition that camera resolutions have followed the consumer television market. Now that Ultra-HD TVs have been embraced by consumers, 4K Ultra-HD network cameras are now commonplace, providing a resolution of 3840 × 2160 or 8 megapixels.

4.7.1 The benefits of high resolution

A hi-res fixed camera can be used in two ways: It can enable viewers to see greater details (higher resolution) in an image, or it can be used to cover a larger part of a scene if the image scale (in pixels per area) is kept the same as a camera with lower resolution. The hi-res camera makes for easier identification of both people and objects and gives improved coverage of the most important areas of a scene. In addition, a single hi-res camera can take the place of multiple analog cameras and thereby reduce installation, operation, and maintenance costs. See Figure 4.55 and Figure 4.56.

4.7.2 High-resolution applications

All the camera types (except thermal cameras) discussed in Chapter 3 are available in high resolutions – and some are available only in high-resolution versions. Typically, fixed cameras have the highest resolutions, with some having 10–20 megapixels. PTZ cameras are typically available at up to 4K resolution (8.3 megapixels).

High resolutions are useful in most applications, including the following:

- *Retail*: Effective video surveillance can help drastically reduce theft and shrinkage. Hi-res cameras can provide an overview of a large part of a store (with no blind spots) or instead offer highly detailed images, for example, from the area around the checkout counter.
- *City surveillance:* High-resolution video streams are used to conclusively identify people and objects or to get a larger overview when viewing live or recorded video.
- *Government buildings:* Hi-res cameras provide the exceptional image detail necessary to facilitate the identification of people and to record evidence of any suspicious behavior.

4.7.3 The drawbacks of high resolution

Although higher resolution enables greater detail or more of a location to be seen, there are also drawbacks. Typically, high-resolution sensors are the same size or only slightly larger than lower-resolution sensors. This means that while there are more pixels on a high-resolution sensor, the size of each pixel is smaller than the size of the pixels on a lower-resolution sensor. The general rule is that the smaller the pixel size, the lower the light-gathering ability. This means that many high-resolution cameras are less light-sensitive than low-resolution cameras. But through technical advancements, many modern high-resolution cameras do have very good light sensitivity.

Not all multi-megapixel cameras (5 megapixel and above) can provide 30 fps at full resolution. Another factor to consider when using these cameras is that higher-resolution video streams increase the demands on network bandwidth and storage space for recorded video. This can be somewhat mitigated using video compression. (For more about compression, see Chapter 6.)

Another important consideration is the selection of lenses. A typical analog CCTV lens does not match the high-resolution sensor. But the lens should always be fitted to match the sensor, and this is critical for cameras with 5 megapixels or more. Cameras with 10 megapixels or more should use lenses made for professional photography.

4.7.4 High-definition television (HDTV) resolutions

HDTV provides up to five times higher resolution than standard analog TV. HDTV also has better color fidelity and a 16:9 format. Defined by SMPTE (Society of Motion Picture and Television Engineers), the two most important HDTV standards are SMPTE 296M that defines the HDTV 720p format and SMPTE 274M that defines the HDTV 1080p format.

HDTV 720p

- 1280 × 720 resolution
- 0.9 megapixels
- High color fidelity (color spaces defined by ITU-R BT.709)
- 16:9 aspect ratio
- Progressive scan
- Main compression standard is H.264, although H.265 can be used
- 25/30 Hz and at 50/60 Hz refresh rate, which corresponds to 25/30 fps and 50/60 fps (the frequency is country-dependent).

HDTV 1080p

- 1920 × 1080 resolution
- 2.1 megapixels
- High color fidelity (color spaces defined by ITU-R BT.709)
- 16:9 aspect ratio
- Interlaced or progressive scan

- Main compression standard is H.264, although H.265 can be used
- 25/30 Hz and 50/60 Hz refresh rate, which corresponds to 25/30 fps and 50/60 fps (the frequency is country-dependent).
- Also known as Full HD (FHD)

A camera that complies with the SMPTE standards indicates adherence to HDTV quality and should provide all the benefits of HDTV in resolution, color fidelity, and frame rate.

The HDTV standard is based on square pixels, as in computer screens, so HDTV video from network video products can be shown on HDTV screens or standard computer monitors. With progressive scan HDTV video, no conversion or deinterlacing needs to be applied when the video is processed by a computer or displayed on a computer screen. For more about deinterlacing, see section 4.4.2.

Table 4.9 and Table 4.10 list the basic HDTV image sizes in the European Broadcasting Union (EBU) and NTSC countries, respectively.

4.7.5 Ultra-HD resolutions

Developed in Japan and then standardized by the International Telecommunication Union (ITU) in 2012 and the Society of Motion Picture and Television Engineers (SMPTE) in 2013, Ultra-HD television (UHDTV) has two digital video formats.

Ultra-HD 2160p is mostly referred to as 4K, which refers to the horizontal resolution of approximately 4,000 pixels.

- 3840 × 2160 resolution
- 8.3 megapixels – four times more than HDTV 1080p (see Figure 4.57)
- Extra-high color fidelity (color spaces defined by ITU-R BT.2020)
- 16:9 aspect ratio
- Progressive scan
- Main compression standard is H.265, although H.264 is used temporarily
- Up to 100/120 fps
- Also known as UHDTV 2160p, UHD-1, 4K, or quadHD, although the term 4K was originally used for the cinema resolution of 4096 × 2160 and the aspect ratio 21:9

Table 4.8 Basic HDTV image sizes in the EBU

Size	Aspect ratio	Scan	Frame rate (fps/Hz)	Label
1280 × 720	16:9	Progressive	50	720p50
1920 × 1080	16:9	Interlaced	25a	1080i50
1920 × 1080	16:9	Progressive	25	1080p25
1920 × 1080	16:9	Progressive	50	1080p50

a 50 Hz field rate. Note that other frame rates can be used. The most common are 24, 25, 30, 50, and 60 fps.

Table 4.9 Basic HDTV image sizes in NTSC countries

Size	Aspect Ratio	Scan	Frame Rate (fps/Hz)	Label
1280 × 720	16:9	Progressive	60	720p60
1920 × 1080	16:9	Interlaced	30a	1080i30
1920 × 1080	16:9	Progressive	30	1080p30
1920 × 1080	16:9	Progressive	60	1080p60

a 60 Hz field rate. Note that other frame rates can be used. The most common are 24, 25, 30, 50, and 60 fps.

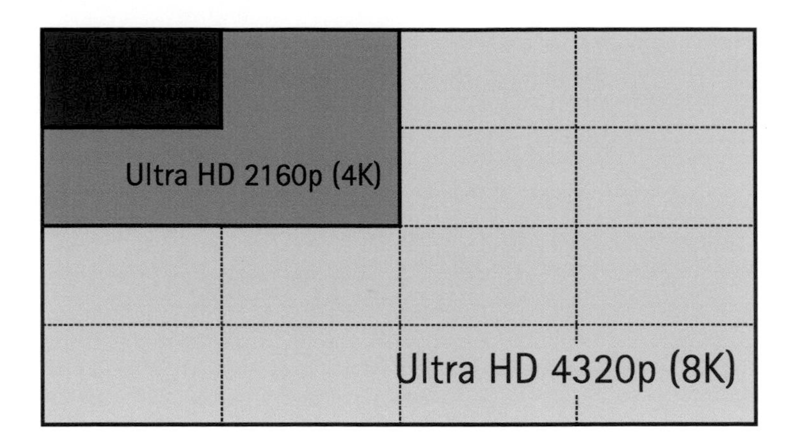

Figure 4.57 Ultra-HD 4320p (8K) has 16× the resolution of HDTV 1080p. Ultra-HD 2160p (4K) has 4× the resolution of HDTV 1080p.

Ultra-HD 4320p is mostly referred to as 8K, which refers to the horizontal resolution of approximately 8,000 pixels.

- 7680 × 4320 resolution
- 33.2 megapixels – 16 times more than HDTV 1080p (see Figure 4.50)
- Extra-high color fidelity (color spaces defined by ITU-R BT.2020)
- 16:9 aspect ratio
- Progressive scan
- Main compression standard is H.265, although H.264 is used temporarily
- Up to 100/120 fps
- Also known as UHDTV 4320p, UHD-2, 8K, Full Ultra-HD, or Super Hi-Vision (SHV)

Figure 4.57 shows a comparison between HDTV 1080p and the two Ultra-HD resolutions.

As in the HDTV standards, the "p" indicates progressive scan. In the home theater industry, the Ultra-HD promise of significantly better color fidelity has generated a lot of interest, along with the advantages of higher resolution. In the video surveillance industry, color also makes a difference, but the unmatched resolution means exceptional digital zoom capabilities and an extended field of view. These superpowers make Ultra-HD cameras ideal for capturing facial features and other fine details. Like HDTV, Ultra-HD is a standard that guarantees a certain video quality, whereas megapixel is merely a statement of the camera's resolution. Ultra-HD technology requires high-quality lenses, a lot of pixels, and large sensors that can capture enough light, as well as corresponding capacities in bandwidth and storage.

4.7.6 Aspect ratios

Aspect ratio is the ratio of the image's width to its height. Megapixel, multi-megapixel, HDTV, and Ultra-HD resolutions allow a greater flexibility in the choice of aspect ratio (Figure 4.58). An old-fashioned TV monitor, traditionally used to view analog surveillance video, displays an image with an aspect ratio of 4:3. Network video can offer the same but also other ratios such as 16:9.

The advantage of a 16:9 aspect ratio is that unimportant details in the upper and lower parts of the scene are not part of the image. Therefore, bandwidth and storage requirements can be reduced.

The aspect ratio 9:16 is also popular in video surveillance. This is simply a 16:9 image that has been rotated 90°. In 9:16, the full view of the camera can be utilized to monitor narrow scenes, such as store aisles, highways and freeways, school hallways, and train station platforms (see Figure 4.59). This aspect ratio is sometimes referred to as corridor format.

Figure 4.58 Different aspect ratios.

Figure 4.59 Corridor format is often used in retail environments.

4.8 BEST PRACTICES

- *Select the right camera for the lighting conditions:* A camera might deliver reasonable image quality in bright light conditions, but you still need to discover all the challenges of the surveillance area in order to meet them in the best possible way. Will the camera be used only in indoor or low-light environments, or will it need to cover both daylight and nighttime scenes? If possible, avoid backlight situations and extreme dynamic ranges altogether. If this is not possible, could curtains or blinds be used? Can the camera be moved to a better position? Can light be redirected, or can frontal lights be added? Can the sky be cut out of the image (to limit the dynamic range)? What types of WDR does the camera support?

- *Select the right camera for the surveillance needs:* Different camera types and resolutions have different strengths. If the main objective is to get an overview of a large area or to focus on several smaller areas within a larger area, a multi-megapixel camera is probably a good choice. Is the surveillance area narrow or wide? Multi-megapixel cameras and lenses with a wide field of view produce great overview images of large areas. Select a camera that offers a suitable aspect ratio so that as much as possible of the image fulfills the purpose.
- *Consider the lens:* A high-quality lens can deliver better images. Typically, only fixed cameras have interchangeable lenses, and some lenses feature auto-iris control that improves the dynamic range. With remote focus and zoom, focus and zoom can be adjusted from a computer, making installation and maintenance easier, faster, and cheaper. Is a fixed lens good enough, or could the required field of view change over time? If so, a varifocal lens is probably a smarter choice. Often, local authorities can make suggestions as to how best to optimize the view. If the possibility to track and zoom in on subjects is required, consider a high-resolution camera with digital PTZ, a varifocal lens, or even a camera with full mechanical PTZ capabilities. Make sure the lens fits the sensor, that the field of view is appropriate, and that the focal length and aperture give the right depth of field for the camera-to-object distance.
- *Use image-enhancement techniques based on requirements:* If used correctly and if the situation requires it, WDR and noise filters, as well as white balancing, sharpening, and contrast enhancement can significantly improve image quality. Different techniques serve different purposes. To find the right settings for the right scene, play around with the settings and learn how these enhancements affect the images.
- *Look for progressive scan:* Progressive scan makes it possible to identify details, such as license plates, of moving objects and subjects in freeze frames. Most surveillance cameras have progressive scan.
- *Understand file size and bandwidth requirements:* Network cameras use image compression. There is a trade-off between high-quality images and compressed images, which require much less bandwidth and storage. Learn how motion, changes in lighting, and other parameters affect the scene and the bandwidth consumption. Different compression settings have different types of allowances for controlling bitrate and frame rate. Learn how they work and what allowances the typical scenario for a specific camera can have to maintain the optimal rates.

Chapter 5

Thermal cameras

All objects and organisms generate some amount of heat. This heat is a form of light that is invisible to the eye and is also known as thermal infrared radiation. A thermal network camera creates images based on the principles of infrared radiation and detects temperature differences between the objects in the scene. Rather than a representation of light, the thermal image is a visual capture of heat. The more heat an object emits, the brighter it appears in a thermal image.

Thermal images give the most information when there are significant temperature differences in a scene. Images are generally produced in grayscale where dark areas indicate colder temperatures and light areas indicate warmer areas. Color palettes can be added to enhance different shades in the image.

Thermal cameras can detect people and objects in low-light environments, complete darkness, or other challenging conditions such as smoke-filled and dusty environments. Thermal imaging has proved to be a lifesaver in emergency situations. See Figure 5.1 for illustrations of thermal camera detection.

Because thermal imaging does not provide sufficient information to identify persons, thermal cameras are primarily used to detect irregularities and suspicious activities in their field of view. They are robust and cannot be blinded by strong lights nor are they affected by laser pointers. These qualities make them a great choice for a first line of protection. When detection occurs, they can immediately trigger further action, dramatically enhancing the effectiveness of a surveillance system.

Thermal cameras are perfect for perimeter or area protection. They are a powerful and cost-effective alternative to radio frequency intruder detection, fences, and floodlights. Since thermal cameras do not need light to produce images, they provide discreet surveillance in complete darkness. In situations where some light is still needed for identification purposes, they can reduce the need for excess illumination. Thermal cameras can also improve security in off-limits areas, for example, tunnels, railway tracks, and bridges.

Indoor uses include building security and emergency management. Thermal cameras can detect humans inside a building after business hours or in emergency situations, for example, in smoke-filled rooms. High-security buildings, nuclear power plants, prisons, airports, pipelines, and sensitive railroad sections can also benefit from thermal camera surveillance.

Bi-spectral cameras offer a combination of conventional and thermal camera technology. They can provide a very wide range of detection and surveillance and are ideal for mission-critical applications where 24-hour monitoring is required. See Figure 5.2 for an example.

Thermal imaging technology is not new, but until costs came down, it was rare to see practical applications outside the military, law enforcement, and high-security locations. The increased use

DOI: 10.4324/9781003412205-5

Figure 5.1 Complete darkness, haze, smoke, rain, snow, and even bright and blinding lights – a thermal network camera is still able to detect people and objects.

of thermal cameras today is a result of streamlined sensor production and lens material improvements, which are driving volumes and making prices more reasonable. Thermal imaging is now found in many areas, such as the aircraft industry, shipping, and critical infrastructure. The technology is also used in public services, such as firefighting and law enforcement.

This chapter gives an overview of the basic principles of thermal imaging, along with an overview of the components in a thermal camera. Calculation of detection range, integration with video analytics, and export regulations are other topics covered here.

5.1 HOW THERMAL IMAGING WORKS

An image from a conventional surveillance camera captures light reflected by the object being photographed. Light is electromagnetic radiation at wavelengths between approximately 400 and 700 nanometers (0.4–0.7 μm), which is interpreted as an image by light-sensitive silicon material in an image sensor.

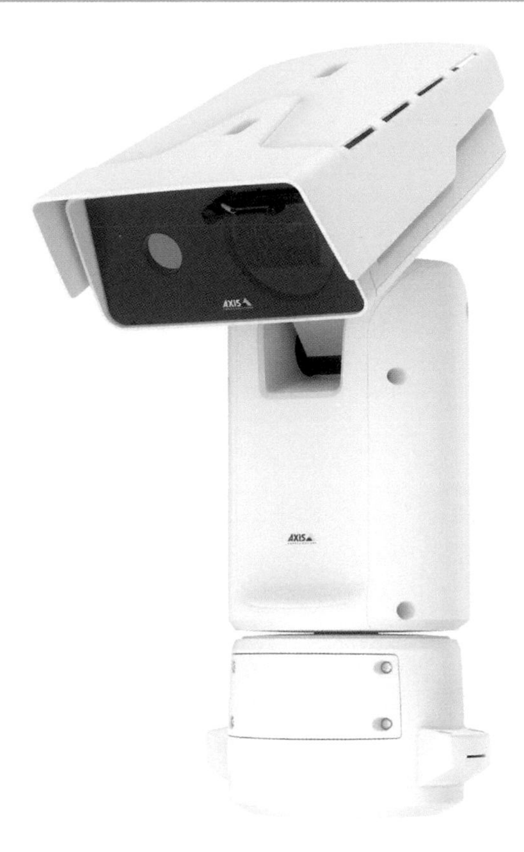

Figure 5.2 An example of a bi-spectral camera, which combines conventional and thermal imaging capabilities.

A thermal camera also collects electromagnetic radiation to create an image (see Figure 5.3), but it does so at higher wavelengths, up to around 14,000 nanometers (14 µm). Radiation in this part of the electromagnetic spectrum is referred to as infrared, or IR, which in turn can be divided into several subgroups.

5.1.1 The electromagnetic spectrum

The parts of the electromagnetic spectrum beyond visible light, the IR waveband, are often divided into the following subregions:

- *Short-wave infrared (SWIR), 1.4–3 µm*
- *Mid-wave infrared (MWIR), 3–8 µm*
- *Long-wave infrared (LWIR), 8–15 µm*
- *Far-wave infrared (FWIR), 15–1,000 µm (1 mm)*

Scientists define the LWIR waveband as thermal infrared, but in the thermal imaging industry, the MWIR waveband is also commonly referred to as thermal. However, the 5–8 µm part of the MWIR waveband is virtually unusable for thermal imaging purposes because of the high spectral absorption of the atmosphere in this range. The thermal imaging industry often divides the electromagnetic spectrum based on the response of various infrared detectors. Very long-wave infrared (VLIR) is added between LWIR and FWIR, and even the boundaries between the other ranges are slightly different. See Figure 5.4.

Microwaves have a wavelength greater than 1 mm. At the far end of the spectrum are radio waves, with a wavelength of 1 m or more. At the other end of the spectrum, wavelengths shorter than visible light are referred successively to as ultraviolet, x-rays, and gamma rays.

(a)

(b)

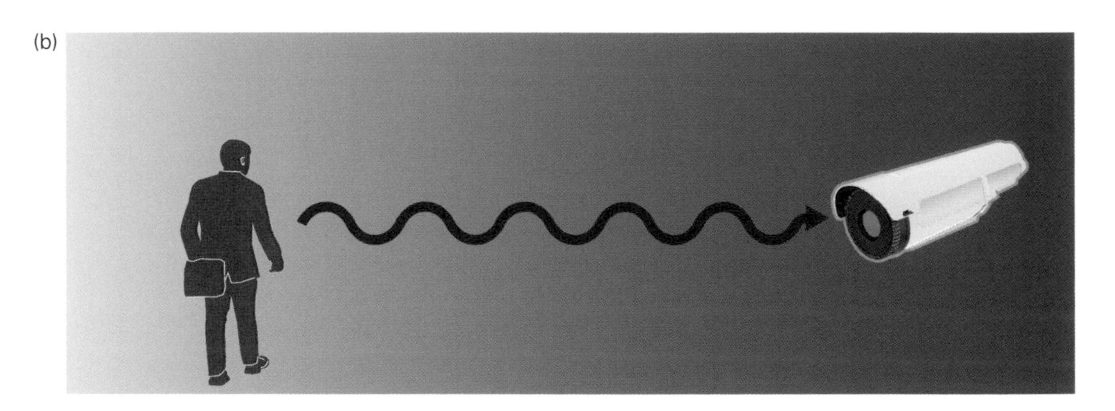

Figure 5.3 (a) A regular camera detects light reflected by an object. (b) A thermal camera detects thermal radiation, as emitted by all objects with a temperature above absolute zero, 0 K (−273 °C or −459 °F).

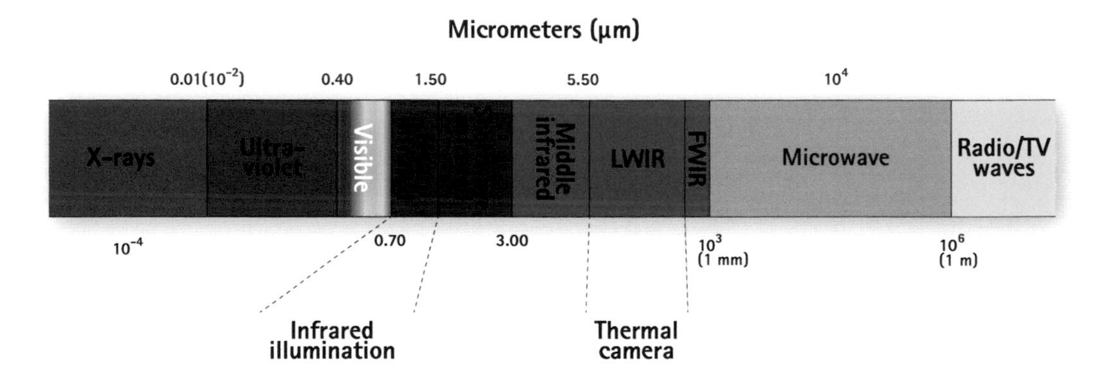

Figure 5.4 Different camera technologies use different ranges of the spectrum of light.

5.1.2 Using thermal radiation to create images

As explained earlier, what the human eye sees as images is in fact light reflected by different objects. No light means no reflection, and the eye is blind in these circumstances. Thermal images, on the other hand, are not dependent on visible light and can be created even in total darkness. Thermal imaging uses the thermal infrared radiation emitted by all objects – organic and inorganic – as a function of their temperature. This is true for all objects with a temperature above absolute zero, or 0 Kelvin (–273 °C or −459 °F). So even very cold objects, such as ice or an outdoor steel post in winter, emit thermal radiation.

The ability to emit absorbed energy is called emissivity (*e*) and is measured on a scale from 0 to 1. The maximum value, 1, only applies to a theoretical object called a black body. The duller and blacker a material, the more radiation it emits and the closer its value is to 1. Conversely, a more reflective material has a lower *e* value. For example, highly polished silver and brass have an emissivity of about 0.02 and 0.03 respectively. Iron has an emissivity of 0.14–0.035 if polished but 0.61 if it has rusted. Regular glass, which effectively blocks thermal radiation, has an emissivity of 0.92.

An object's thermal radiation also depends on its temperature. The hotter the object, the more thermal radiation it emits. Although we humans cannot always see it, we can still sense it. For example, we can feel the heat when inside a sauna or walking on hot tarmac. Objects with a sufficiently high temperature, such as burning wood or melting metal, also radiate visible light that indicates the temperature of its surface.

A camera's sensitivity can be defined as its capability to distinguish between temperature differentials. The greater the temperature difference in a scene, the clearer the thermal images will be. But the contrasts in a thermal image also depend on the emissivity of its objects.

5.2 THE COMPONENTS OF A THERMAL CAMERA

At first glance, a conventional network camera and a thermal network camera might seem identical. Many things are indeed similar, such as the compression and the networking features with Power over Ethernet. As with conventional cameras, different form factors are available for use in different environments and situations, for example, fixed indoor or outdoor cameras and cameras on pan-tilt heads or with zoom options. However, the two things that differ substantially are the lens and the sensor. This section explains these differences.

5.2.1 Sensors

The sensor in a thermal camera is an array of thousands of detectors that are sensitive to thermal infrared radiation. The sensor detects, records, and converts this thermal infrared information into electrical signals, which is what makes the video image. Detectors used for thermal imaging can be broadly divided into two types: cooled and uncooled infrared sensors.

Uncooled infrared image sensors are smaller and have fewer moving parts, making them less expensive than their cooled counterparts. Cameras with cooled sensors generally need to be serviced, although newer designs based on Sterling motors require less service. Also, the cooling medium must be refilled every 8,000–10,000 hours.

The individual elements in an uncooled sensor respond in different ways to the incoming infrared radiation, which results in a "drift" in individual pixel values. To remedy this, the sensor performs non-uniformity correction (NUC). A mechanical shutter blocks the sensor and gives it a standard temperature target, against which every pixel is corrected. This process occurs at regular intervals or when a specific temperature change takes place. A typical shutter for a thermal camera is shown in Figure 5.5.

5.2.1.1 Cooled sensors

Cooling is needed to reduce thermally induced noise. Otherwise, at higher temperatures, the sensors risk being flooded or "blinded" by their own thermal radiation. This equipment makes the detectors relatively bulky, expensive, and energy-consuming.

Cooled infrared sensors are usually contained in a vacuum-sealed case and are cooled to temperatures as low as 60 K to100 K (approx. −210 °C to −170 °C or −346 °F to −274 °F), depending on the type and level of performance desired. These extremely low temperatures are accomplished with so-called cryogenic coolers.

Figure 5.5 A mechanical shutter for a thermal camera. The shutter is used for temperature calibration of the sensor.

Although cooled sensor technology is both expensive and requires high maintenance, it does have its benefits. These detectors work in the mid-wave spectral band (MWIR), which provides better spatial resolution because the wavelengths are much shorter and deliver higher contrast than in the long-wave band. Hence, cooled detectors can distinguish smaller temperature differences and produce crisp, high-resolution images. See Figure 5.6 for an example of a cooled sensor.

Another advantage of cooled sensors is that the greater sensitivity also allows the use of lenses with high f-numbers (or f-stops). Consequently, cooled detectors are a better choice for long-range detection, that is, 5–16 kilometers (3–10 miles).

5.2.1.2 Uncooled sensors

The sensor in an uncooled thermal camera is stabilized at or close to the ambient temperature, using simpler temperature control elements or no control at all. Sensors of this kind operate in the long-wave infrared band (LWIR).

Uncooled sensors can be based on a variety of materials that all offer unique benefits. One common design is based on microbolometer technology. Typically, this is a tiny resistor (or thermistor) with highly temperature-dependent properties, mounted on a silicon element that is thermally insulated. The resistor is made of vanadium oxide (VOx) or amorphous silicon (α-Si). When the thermal infrared radiation hits the material, the electrical resistance changes.

Another kind of microbolometer is based on ferroelectric technology. Here, small changes in the material's temperature create large changes in electrical polarization. Ferroelectric microbolometers are made of barium strontium titanate (BST). Figure 5.7 shows an uncooled sensor.

Changes in scene temperature cause changes in the bolometer, which are then converted into electrical signals and processed into an image. The camera's sensitivity to thermal radiation, which determines its ability to distinguish different temperature differences in a scene, can be expressed as its NETD value (Noise Equivalent Temperature Difference). Most thermal network cameras have a NETD value of 50–100 mK (milliKelvin), though there are newer generations of bolometers that have a NETD as low as 20 mK.

5.2.2 Sensor resolutions

Resolutions in thermal cameras are generally much lower than in conventional network cameras. This is mostly due to the more expensive sensor technology in thermal imaging. The pixels are larger,

Figure 5.6 A cooled thermal sensor with cryogenic cooling unit.

Source: LYNRED. Copyright, all rights reserved.

Figure 5.7 An uncooled infrared sensor.

Source: LYNRED. Copyright, all rights reserved.

which affects the sensor size and the cost of materials and production. Typical resolutions for thermal cameras range from 160 × 128 to 640 × 480 (VGA), though higher resolutions are available.

In visual observation, larger image sensors deliver higher resolution and better image quality (see Figure 5.8).

Figure 5.8 The effect of sensor size on resolution and image quality.

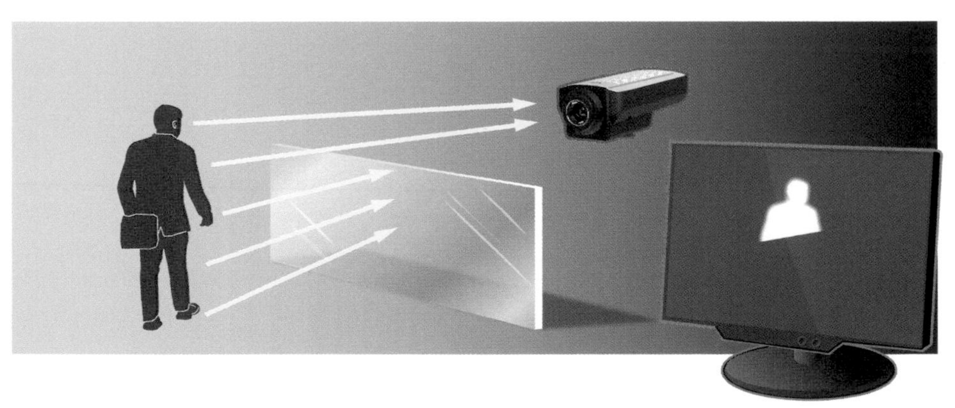

Figure 5.9 Glass prevents thermal imaging.

5.2.3 Lenses for thermal cameras

Because regular glass blocks thermal radiation, lenses in thermal cameras cannot use glass optics. Currently, germanium is the most used material for thermal camera optics. This very expensive metalloid, which is chemically similar to tin and silicon, blocks visible light while admitting infrared light.

Figure 5.10 Lenses for thermal cameras.

Table 5.1 Focal length and field of view.

Focal length	Horizontal field of view	Sensor size
7 mm	55°	384 × 288 pixels
10 mm	51°	384 × 288 pixels
	57°	640 × 480 pixels
13 mm	17°	160 × 128 pixels
	28°	384 × 288 pixels
19 mm	28°	384 × 288 pixels
	32°	640 × 480 pixels
35 mm	16°	384 × 288 pixels
	18°	640 × 480 pixels
	10.7°	384 × 288 pixels
60 mm	9°	384 × 288 pixels
	10°	640 × 480 pixels
	6.2°	384 × 288 pixels

Not all lenses are pure germanium. For example, some are made of a germanium-based material called chalcogenide glass, which admits a wider spectrum of infrared light. As with most materials, there are benefits and disadvantages. Chalcogenide glass contains cheaper materials and is moldable. However, the master mold requires a significant initial investment that can only be justified for large-scale production.

Thermal cameras use different lens mounts than conventional network cameras. The mount needs to be wider to fit the sensor, which is typically larger than a conventional sensor. A TA-lens has an M34×0.5 screw mount, allowing for sensors as large as 13 mm in diameter. Figure 5.10 shows examples of lenses for thermal cameras.

5.2.3.1 Calculation of focal length

The focal length of a lens is defined as the distance between the entrance lens (or a specific point in a complex lens assembly) and the point where all the light rays converge at a point, which is normally the camera's image sensor.

Like conventional lenses, thermal lenses come in different focal lengths, usually specified in millimeters (for example, 35 mm). A longer focal length results in a narrower field of view. The field of view depends on the focal length and the diameter of the sensor. Because varifocal and zoom lenses need more lens material (more germanium), they often become too expensive to justify production and purchase, which is why fixed lenses are more common. Table 5.1 shows the relationship between focal length and field of view (angle of view).

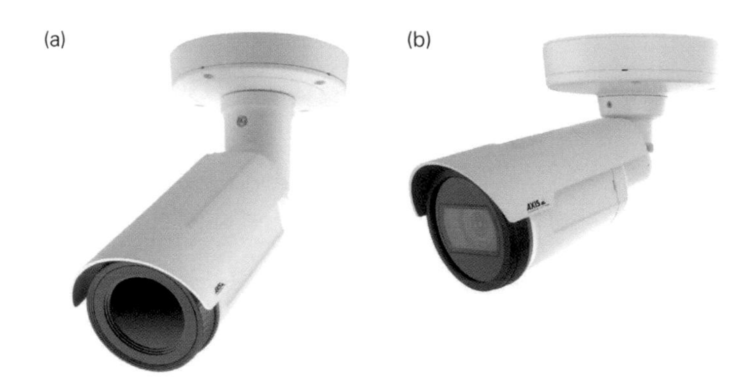

Figure 5.11 (a) Typical enclosure of a thermal camera. (b) A conventional camera enclosure. These enclosures look very similar except for the material used in the front window.

While focal length is specified for lenses, we sometimes need to know which lens to use for a specific application. Nomographs are used to determine the relationship of the focal length of the lens, the number of pixels across the object, and the range. See section 5.4.1 for more information.

5.2.4 Thermal enclosures

In some situations, a thermal camera needs a protective enclosure to protect it against outdoor conditions. Naturally, the same transparency to infrared light needs to apply to the enclosure's window, so germanium is used here too.

For more information about environmental conditions, enclosures, and operating ranges, see Chapter 18.

Figure 5.11 shows an outdoor-ready thermal camera and an outdoor-ready conventional camera.

5.3 PRESENTING THERMAL IMAGES

The most common presentation of thermal images is in black, white, and gray. The different temperature values are then translated into 256 grayscale values. The most common presentation (or palette) is *white-hot*, in which heat sources appear white against lower-temperature gray and black backgrounds. In some cases, *black-hot*, where sources of heat appear black, may be easier to use. Most cameras can switch between palettes.

Thermal images are sometimes associated with bright, intense colors. This is solely because the human eye is better at distinguishing different shades of color than different shades of gray, so adding color can make it easier to see differences in thermal images. These so-called pseudo-colors are created digitally. Each color or nuance represents a different temperature, usually ranging from white and red for higher temperatures to green, blue, and violet for lower temperatures. Figure 5.12 compares different kinds of thermal image presentation.

5.3.1 Thermometric cameras

Thermometric cameras are based on thermal imaging and use the same sensor technology as other thermal cameras. They can be used for remote temperature monitoring and include the possibility to set temperature alarms, of which there are two basic types. One is triggered when the temperature rises above or falls below the set temperature limit but also if the temperature changes too quickly. The other type is a spot temperature alarm, where the camera measures the temperature of a specific area in the image. Isothermal palettes highlight the critical temperatures so that they stand out from the rest of the scene. Figure 5.13 shows an example of an isothermal palette applied at an industrial installation.

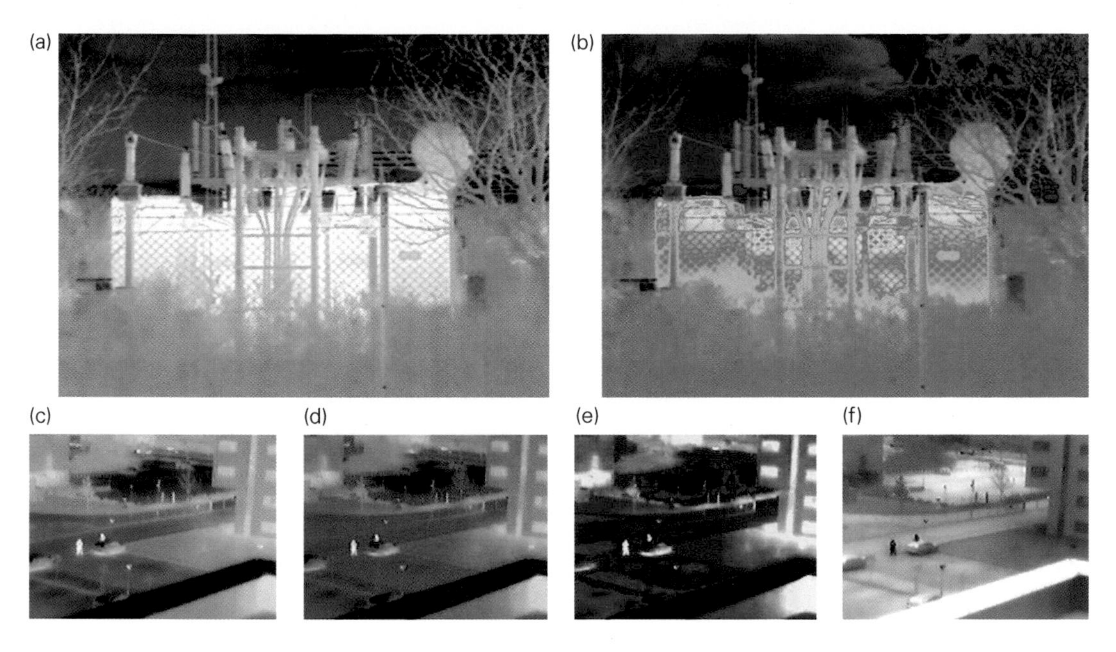

Figure 5.12 Thermal images with five different palettes: (a, c) white-hot, (f) black-hot, and (b, d, and e) different types of pseudo-color.

Figure 5.13 The isothermal palette makes it possible to highlight the temperature span and easily see if a surface reaches a hazardous temperature.

5.4 DETERMINING DETECTION RANGE

The resolution required to detect an object is given in pixels and can be determined using Johnson's criteria. John Johnson, a US military scientist, developed this method for predicting the performance of sensor systems during the 1950s. Johnson measured the ability of observers to identify scale model targets under various conditions and determined the criteria for the minimum required resolution. These criteria give a 50% probability of an observer detecting, recognizing, or identifying an object at the specified level. This object could be a person, typically defined with a critical dimension (width) of 0.75 m (2.46 ft) across, or a vehicle, typically defined with a width of 2.3 m (7.55 ft) across.

Johnson's criteria assume ideal conditions. In reality, the weather conditions at the site affect the thermal radiation emitted from the object and lower the effective detection range. According to Johnson's definition, the detection range in Table 5.2 ideally requires a temperature difference of 2 °C (3.6 °F) between the target object and the background. However, certain weather conditions, such as rain, snow, and fog, attenuate the radiation emitted from the object because the heat that radiates from the object scatters when it hits particles in the air. Johnson's criteria can be used as a guide, but the camera should always be evaluated in the intended location and environment to avoid performance and reliability issues. The levels of Johnson's criteria used for thermal network cameras are as follows:

- *Detection:* At least 1.5 pixels across the object are needed for the observer to detect it.
- *Recognition*: At least 6 pixels across the object are needed for the observer to distinguish the object, for example, a person in front of a fence.
- *Identification*: At least 12 pixels across the object are needed for the observer to distinguish both the object and its characteristics, such as a person holding a crowbar.

Johnson's criteria were developed under the assumption that the visible information is processed by a human observer. If the information is instead processed by an algorithm, there will be specific requirements on the number of pixels needed across the target for reliable operation. All video analytics algorithms need a certain minimum number of pixels to work. The exact number may vary, but as a rule of thumb, at least six pixels across the object are required, which is the same as Johnson's criteria for recognition. Even if a human observer can detect the object, an application algorithm often needs more pixels to successfully complete the same task.

Table 5.2 Example of a thermal camera range chart. It specifies at which distances humans and vehicles can be detected, recognized, and identified.

	Focal length	Horizontal field of view	Person: 1.8 × 0.5 m (5′ 11″ × 1′ 8″) Critical dimension: 0.75 m (2′ 6″)		Vehicle: 1.4 × 4.0 m (4′ 7″ × 13′ 2″) Critical dimension: 2.3 m (7′ 7″)	
	mm	degrees	meters	yards	meters	yards
Detection (1.5 pixels across target) An observer can see an object	10	51	220	241	660	722
	19	28	390	427	1200	1,312
	35	16	700	766	2,200	2,405
	60	9	1,200	1,312	3,700	4,046
Recognition (6 pixels across target) An observer can distinguish an object	10	51	55	60	170	186
	19	28	100	109	300	328
	35	16	175	191	550	601
	60	9	300	330	920	1,006
Identification (12 pixels across target) An observer can distinguish a specific object	10	51	25	37	85	93
	19	28	50	55	150	164
	35	16	90	98	270	298
	60	9	150	165	460	503

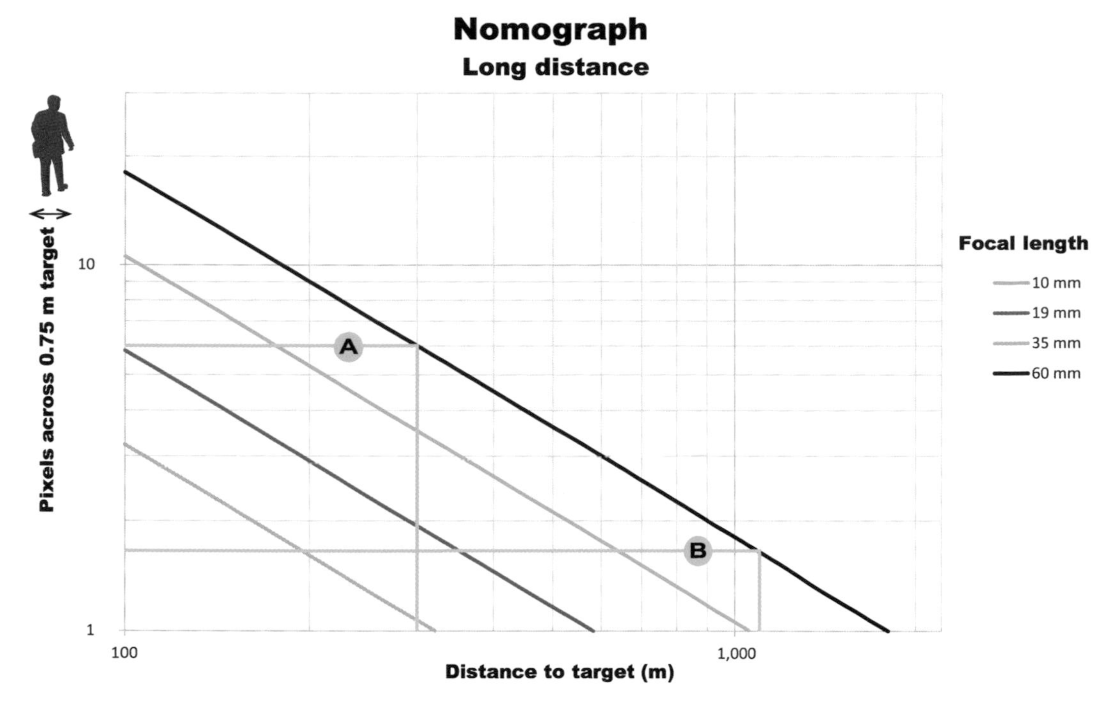

Figure 5.14 An example of a nomograph.

5.4.1 The nomograph

A nomograph is used to find the number of pixels available at a given range. This is a two-dimensional diagram that shows the relationship between the focal length of the lens, the number of pixels across the object, and the range. For example, if the number of pixels required and the distance to the object are known, it is then possible to calculate which lens or camera to use. In the same way, if the camera and the number of pixels required are known, the distance at which the camera can be used to detect an object is indicated by the nomograph. See Figure 5.14.

An example: A thermal network camera with a 60 mm lens is aimed at a person, with a critical dimension of 0.75 m (2 ft 5½ in) across the object. The nomograph in Figure 5.14 shows that the object is recognizable at 300 m (328 yd) and 6 pixels across the object (A). If only detection is required, the range is 1,200 m (1,312 yd) and 1.5 pixels across the object (B).

5.4.2 Environmental considerations

Remember that Johnson's criteria are only valid in ideal conditions. The weather conditions on-site affect the thermal radiation emitted from the object and decrease the effective detection range. The detection range used in the nomograph earlier ideally requires a temperature difference of 2 °C between the target object and the background.

Environmental factors that affect the thermal camera's ability to produce an image include weather conditions and the temperature difference between the object and the background. An object with almost the same temperature as the background, such as a human body on a hot summer day, is harder to distinguish from its background than an object that is considerably hotter or colder, such as a car with its engine running on a cold winter day.

The two most important environmental factors that affect the camera's image of an object are absorption and scattering. These factors reduce the thermal radiation that reaches the camera and so lessen the distance at which the camera can detect an object. Scattering has a greater effect on the loss of thermal radiation than absorption.

Table 5.3 Different environmental conditions cause different levels of attenuation

Weather conditions and attenuation				
Heavy rain	Light rain	Urban pollution	Dense fog	Fog
11 dB/km 17.6 dB/mile	4 dB/km 6.4 dB/mile	0.5 dB/km 0.8 dB/mile	80 dB/km 128 dB/mile	10 dB/km 16 dB/mile

Table 5.3 describes examples of attenuation in various environmental conditions. For example, on a foggy day the attenuation is 10 dB/km or 1 dB/100 m. If the thermal camera is placed 300 m (980 ft) away, the attenuation for that distance totals 3 dB, which means that the thermal camera receives 50% less thermal energy than in optimal conditions. The thermal energy reduction can be calculated with the following formula, where *a* is attenuation as a negative decibel value:

$$Thermal\ energy\ reduction = 100\% \times 10^{\frac{a}{10}}$$

$$Thermal\ energy\ reduction = 100\% \times 10^{\frac{-3}{10}} = 50.1\%$$

5.4.2.1 Absorption

Water vapor and carbon dioxide in the air are the primary causes of absorption. Heat radiated from the object is absorbed by water vapor and carbon dioxide and loses some of its intensity before reaching the camera. Absorption usually affects the background more than the objects in the foreground.

The water vapor content of the air affects image quality even in sunny and clear weather. In winter, if all other weather conditions are the same, the water vapor content of the air is lower than in summer. Because water vapor content is lower in winter, less thermal radiation is absorbed by the water molecules. More thermal radiation reaches the thermal network camera, and the result is an image with better quality than it would have on a summer day.

5.4.2.2 Scattering

The thermal radiation the object emits is dispersed when it hits particles in the air, which is called scattering. The loss of radiation is directly related to the size and concentration of particles, droplets, or crystals in the air, such as smog, fog, rain, or snow.

Fog appears when water vapor in the air condenses into water droplets. Droplet sizes vary with different kinds of fog. In dense fog, the water droplets are larger due to piling (or accretion), thus scattering thermal radiation more than light fog. Also, fog scatters thermal radiation to a greater extent than both smog and haze because of the greater size and concentration of its water droplets.

The effects of scattering are often mentioned in relation to thermal images, but this does not necessarily mean that a conventional camera performs better in these conditions.

5.5 INTEGRATING THERMAL CAMERAS WITH VIDEO ANALYTICS

A conventional camera reacts to changes in the captured image and can, for example, be affected by shade and backlighting. A thermal camera detects the thermal radiation from the object, which is a more static parameter compared to visual changes in an image. Therefore, a thermal camera is an especially reliable platform for integrating intelligent video applications (analytics) such as the following:

(a)

(b)

Figure 5.15 (a) An image from a conventional camera in resolution 1280 × 800 pixels. (b) An image from a thermal camera with resolution 384 × 288 pixels. These images were taken on the same day, and despite the difference in resolution, the thermal camera provides better detection capabilities.

(a)

(b)

Figure 5.16 (a) An image taken on a sunny day by a conventional network camera. (b) The same scene captured by a thermal network camera. Even in broad daylight, the thermal image delivers contrast in areas where a conventional color image does not.

- *Motion detection:* When a thermal camera detects a moving object, it can automatically perform or relay actions, such as sending an alarm to an operator, turning on floodlights, or triggering a conventional camera to capture visual information about the incident. The camera only records video during the actual incident, which means storage space is minimized. This facilitates video analysis and saves valuable operator time.
- *Crossline detection or perimeter detection:* A virtual line can be placed in the image of the thermal camera (see Figure 5.17). The virtual line acts as a tripwire. As in the motion detection example, the thermal camera can trigger another camera if an object crosses the virtual tripwire.

When motion is detected or a virtual line is crossed, the camera can automatically trigger an alarm to the operator and at the same time trigger a PTZ camera to supply video to the operator. Based on the information, the operator can decide the action to take.

Integrating thermal cameras with analytics has many advantages. However, to get the optimum use of thermal cameras, other things need to be considered than when using conventional cameras. The definition of detection range, the number of pixels across the object, and the surrounding environment are all such parameters, and they are of particular importance when integrating with analytics. For more information, see section 5.7 and Chapters 16 and 17.

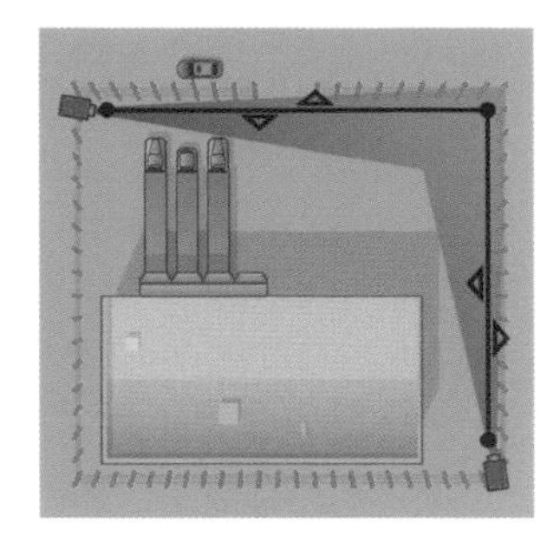

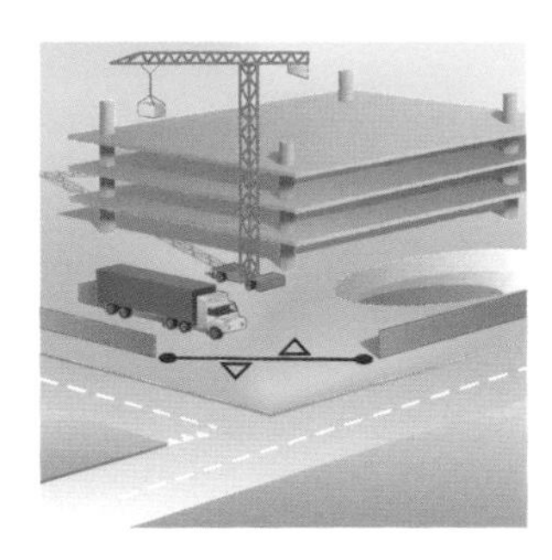

Figure 5.17 Crossline detection sets up a virtual line. When the line is crossed, the camera can be triggered to perform actions, such as turning on floodlights and sirens or triggering another camera to zoom in on the scene.

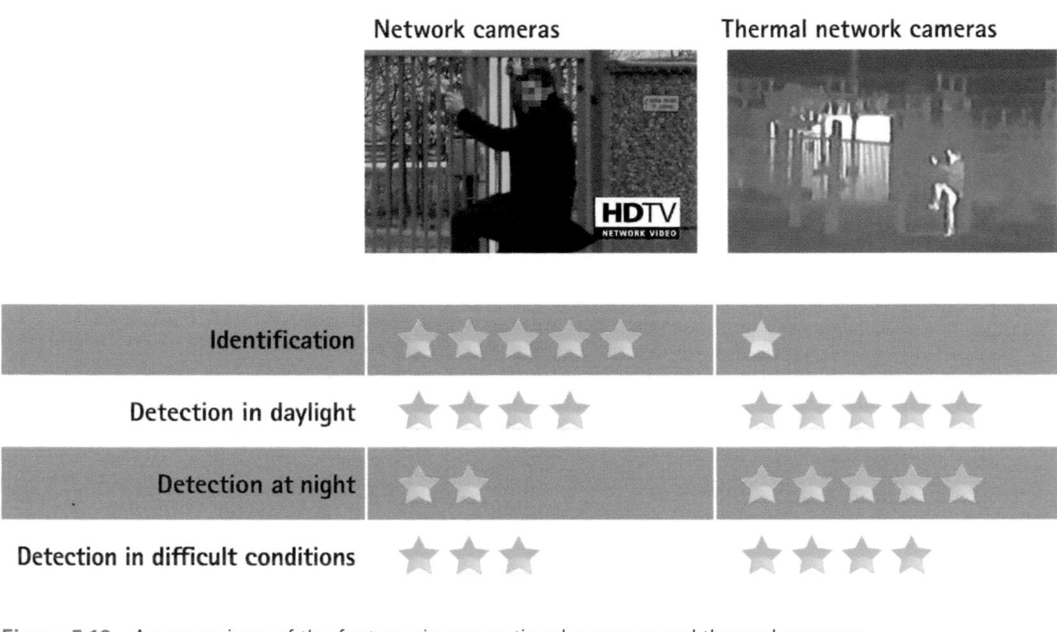

Figure 5.18 A comparison of the features in conventional cameras and thermal cameras.

5.6 EXPORT REGULATIONS FOR THERMAL TECHNOLOGIES

Technical aspects, legal considerations, and other issues present challenges when integrating thermal cameras into the conventional video surveillance market.

Products and technologies that can be used both for military and commercial purposes are called dual-use goods. Exports of such items are regulated in the international Wassenaar Arrangement from 1996, which, among other things, aims to promote transparency and greater responsibility in transfers of conventional arms, as well as dual-use goods and technologies.

For a thermal camera to be freely exported, its maximum frame rate cannot exceed 9 frames per second (fps). Thermal cameras with a frame rate of up to 60 fps can be sold in most 42 Wassenaar Arrangement member countries on the condition that the buyer is registered and can be traced and that the seller fulfills all country-specific export requirements, including license or license exception. Thermal cameras may never be sold to any embargoed or sanctioned destination.

5.7 BEST PRACTICES

Thermal imaging is becoming an integral part of video surveillance systems and is getting cheaper, making thermal imaging affordable for an ever-growing market. Thermal cameras can be an excellent complement in many situations where conventional cameras are lacking. See Figure 5.18 for a comparison of features in conventional and thermal cameras.

Best practices for thermal cameras include the following:

- *Situations in total darkness:* Thermal cameras are unparalleled when light is completely absent. They can also be an option in areas that are very difficult to illuminate effectively, such as seafronts, ports, or other large expanses of open water. Artificial light not only runs the risk of revealing where the cameras are placed, making them easier to avoid or vandalize, but it can also project shadows that help intruders avoid detection.
- *When only detection is required:* For privacy reasons, it is sometimes desirable to detect persons but not to identify them. Examples are train platforms or fenced-off areas where you simply want to detect when an object enters the area, a perfect use-case for thermal cameras.
- *When lighting is an issue:* Spotlights can blind as well as illuminate. Cameras that do not rely on light can be the preferred solution in railway tunnels, airstrips, or on streets. Thermal cameras cannot be blinded by bright lights or laser beams.

A thermal camera can meet demanding surveillance requirements, but it is essential to determine the correct lens and resolution for the application by conducting your own field tests at the intended site.

The advantages of a thermal camera can be maximized when combined with video analytics solutions that can help analyze and use the thermal images. For example, thermal cameras can be used to detect the presence of persons in dark areas outside a building, and an intelligent video system can warn security staff when this occurs. This means that personnel do not spend unnecessary time monitoring video streams with no activity.

Chapter 6

Video compression technologies

The amount of data delivered by a modern image sensor is huge, and when digitized and transmitted, this data can consume as much as 0.5 Gbit/s (gigabits per second) of bandwidth for a resolution of 1080p at 30 frames per second.

It is obviously not practical to transmit this amount of data over an IP-based network, and it is also a challenge to store it cost-effectively. Therefore, image and video compression techniques are used to reduce the bitrate. The goal is to drastically decrease the amount of data that must be sent but with as little negative effect as possible on the image and video quality. Depending on the purpose, different compression formats can be used. These formats are usually referred to as codecs. Codec is the abbreviated form of *CO*mpressor-*DEC*ompressor or *CO*ding-*DEC*oding.

The development of video compression has been instrumental in the adoption of network video and the capabilities and scalability of today's video surveillance systems. The first network cameras used JPEG compressed still images, typically, at 1 megapixel resolution, which in today's systems would have generated around 1 TB of data per camera every day. Using the latest H.265 compression combined with intelligent video compression generates only around 5 GB per day, a reduction of 99.5%. Combined with video analytics that generate metadata, this reduction becomes even greater.

This chapter introduces basic compression techniques, the most common compression formats (JPEG and H.264/H.265), and the standardization organizations behind them, as well as new trends and best practices in the field.

6.1 THE BASICS OF COMPRESSION

The primary purpose of compression techniques is to reduce the amount of image data by discarding unnecessary information that is not needed by a human viewer. For example, there is little surveillance benefit in a video where a large portion of the image shows a wall or the sky.

All compression techniques are based on an understanding of how the human brain and eyes work together to form a complex visual system. An image can be optimally compressed, but if compressed too much, artifacts (blur, noise, trails, blocklike effects, or low color fidelity) may become visible in the video. To effectively be able to pause or zoom in on a recorded image in a video sequence, recorded video should have a slightly higher bitrate (size) than that required for live viewing.

Some of the techniques commonly employed to reduce the size of images and video sequences include the following:

- *Quantization* reduces color nuances within an image.
- *Subsampling* reduces color resolution.

DOI: 10.4324/9781003412205-6

- *Transform coding followed by quantization* removes small invisible parts from the image.
- *Intra-prediction* predicts how parts of an image will look based on adjacent parts in the same image.
- *Run-length* coding or prediction removes repeated pixel values.
- *Entropy coding* is a lossless data compression scheme (such as Huffman or arithmetic coding) that efficiently encodes pixels.
- *Inter-prediction* compares adjacent images and removes unchanged details.

Some compression techniques use so-called intra-frame compression, where frames are evaluated and compressed one by one. Other techniques use inter-frame compression, where several adjacent frames are compared to reduce the amount of image data.

6.1.1 Image and video compression

The video sequence in Figure 6.1 shows a person running from left to right and a house and trees in the background. Image compression techniques code each image in the sequence as a separate, unique image.

In Figure 6.2, the same sequence is encoded using *video* compression techniques, where the static parts of the image only need to be included in the first frame (a). In the following two frames (b, c), the area with motion (the running person) is the only part that needs to be included. Because an encoded sequence that uses a video compression technique contains less information, less bandwidth and storage are required. When the encoded sequence is displayed, the images appear to the human eye just as in the original video sequence because the background is copied into each frame by the decoder.

For more information about frame types, see section 6.5.1.

Figure 6.1 Motion JPEG encodes and sends the three images (a-c) in this sequence as separate, unique images (I-frames) with no dependencies on each other.

Figure 6.2 (a) The video encoder only sends the first image (I-frame) as a complete image. (b, c) In the following P-frames, references are made back to the I-frame for the static elements (the house and trees). Only the runner is coded using motion vectors, so the amount of information sent and stored is less.

6.1.2 Lossless and lossy compression

The two basic categories of compression are lossless and lossy compression. Lossless compression reconstructs the exact original data from the compressed data. This might be advantageous in some cases, but it does not provide significant data reduction. Graphics Interchange Format (GIF) is an example of a lossless image compression technique. Because of its limited compression abilities, GIF is not suitable for video surveillance.

Lossy compression, on the other hand, means that so much data will be discarded that the original information cannot be fully reconstructed when the video is decompressed. It is the absence of the discarded data that creates the artifacts mentioned earlier.

6.1.3 Block transform

The compression algorithm can be pixel-based, line-based, or block-based with a selectable block size. The size of a block is measured in number of pixels. For example, the JPEG format (named for the Joint Photographic Experts Group) uses 8 × 8 blocks, so each block contains 64 pixels. The block is the smallest image part that algorithms can use to make calculations. Many popular algorithms use block transform to pre-process the data before compression to sort important information first and less important information last.

6.1.4 Prediction

Prediction is another image and video compression technique. It means forecasting the next pixels based on other nearby pixels. If prediction is performed well, an image with a blue sky can be encoded by transferring one blue pixel along with instructions to use the same pixel to create a complete sky section in the image.

Prediction can also be extended to predict the next image once the first image is sent. Image-to-image prediction is used in video compression algorithms but not in still image compression algorithms. This is the basic difference between the two types of algorithms. For more information about image-to-image prediction, see section 6.5.1.

6.1.5 Latency

In compression, one or more mathematical algorithms are used to remove image data. Similarly, to view a video file, algorithms are applied to interpret and display the data on a monitor. This process requires a certain amount of time, and the resulting delay is known as compression latency. Given the same processing power, the more advanced the compression algorithm, the higher the latency. But in modern encoders, the higher frame rate reduces the effect of increased latency, so the computing power available in today's network cameras and servers make latency less of a problem. However, the latency of a network and the processing power needed to avoid it must still be considered when designing a network video system.

In some contexts, such as the compression of studio movies, compression latency is irrelevant because the video is not viewed live. Low latency is essential in video surveillance applications where live video is monitored, especially when PTZ (pan, tilt, zoom) cameras are used.

Normally, video compression techniques have higher latency than image compression techniques. However, if a video compression technique such as H.264 or H.265 is used, latency can be minimized by using a single-pass, network video–optimized encoder that avoids using double image references (B-frames) for prediction. For each B-frame that is added between P-frames in a video at 30 fps, 33 milliseconds must be added to compensate for out-of-order encoding. For more on frame types, see section 6.5.1.

6.1.6 Jitter

Jitter is an artifact that causes parts of images to appear on the client monitor at the incorrect time. This can occur when video is delayed on the network or if the frame rate differs between the camera and the monitor. A video stream from the camera contains timestamps for when the image was captured. This information is transmitted to the client with the video so that the client can display the video correctly. The timestamps are also used to synchronize audio with the corresponding video playback.

The screen frequency must also be the same as the camera capture frequency. In Europe, a 50 Hz or 100 Hz video card and monitor must be used. In the USA, a 60 Hz or 120 Hz video card and monitor must be used.

6.1.7 Compression ratio

Compression ratio is defined as the ratio of the compressed bitrate to the uncompressed original bitrate. A 50% compression ratio means that 50% of the original data has been removed, and the resulting video stream has only half the bitrate of the original.

With efficient compression techniques, significant reductions in bitrate can be achieved with little or no adverse effect on visual quality. The extent to which image modifications are perceptible depends on the amount of data that has been discarded. Often, 50–95% compression is achievable with no visible difference, and in some scenarios, a compression ratio of more than 98% is possible.

6.2 COMPRESSION STANDARDS

Standards are important for ensuring compatibility and interoperability. They are particularly relevant to video compression because video can be used for different purposes and must sometimes be viewable many years after the recording date.

When storage was still relatively expensive and standards for digital video compression were new, many vendors developed proprietary video compression techniques. Today, most vendors use standardized compression techniques because they are just as good or better than the proprietary techniques. The increased use of standards means that users can pick and choose from different vendors rather than being limited to a single supplier.

6.2.1 ITU and ISO

Two organizations are significant in the development of image and video compression standards: the International Telecommunication Union (ITU) and the International Organization for Standardization (ISO).

ITU is not a formal standardization organization. It stems from the telecommunications world and releases its documents as recommendations, for example, the ITU-R Recommendation BT.601 for digital video.

ISO is a formal standardization organization that cooperates with the International Electrotechnical Commission (IEC) to develop standards within areas such as IT. ISO is a general standardization organization, and IEC is a standardization organization dealing with electronic and electrical standards. Often, these two organizations are referred to as a single body: ISO/IEC.

6.2.2 History of compression formats

The two basic compression standards are JPEG (named for the Joint Photographic Experts Group) and MPEG (named for the Moving Picture Experts Group). Both are international standards set by ISO, IEC, and contributors from the United States, Europe, and Japan, among others. JPEG and MPEG are also recommended by ITU, which has helped to establish them further as global standards for digital image and video encoding.

The development of JPEG and MPEG standards began in the mid-1980s when the Joint Photographic Experts Group was formed. Seeking to create a standard for color image compression, the group's first public contribution was the release of the first part of the JPEG standard in 1991.

In the late 1980s, the Moving Picture Experts Group was formed. The purpose was to create a coding standard for moving pictures and audio. Since then, the group has developed the MPEG-1, MPEG-2, and MPEG-4 standards.

At the end of the 1990s, a new group called the Joint Video Team (JVT) was formed. It included both the VCEG and MPEG. The purpose was to define a standard for the next generation of video coding. The work was completed in May 2003 and resulted in the MPEG-4 AVC/H.264 standard.

JVT added the new standard to MPEG-4 as a separate part (Part 10) called *Advanced Video Coding*, from which the commonly used abbreviation AVC is derived. The codec was simultaneously launched as a recommendation by the ITU (*ITU-T Recommendation H.264, Advanced video coding for generic audiovisual services)* and as a standard by the ISO/IEC (*ISO/IEC 14496–10 Advanced Video Coding*).

The same organizations that worked on H.264 also worked on the updated compression standard, with a 50% reduction in the bitrate for the same subjective image quality as H.264. This standard is called H.265 or High Efficiency Video Coding (HVEC). In 2013, it was published as an ITU standard and later became an MPEG standard.

6.3 COMPRESSION FORMATS

This section describes some of the compression formats that are or have been relevant to video surveillance. For more information about the technical aspects of the format groups, see section 6.4 and 6.5.

6.3.1 JPEG

The JPEG standard (ISO/IEC 10918) is the most widely used image compression format in smartphones and digital cameras today. It is also supported by all web browsers, which makes it widely accepted and easy to use. Users can choose between high image quality at a fairly low compression ratio or very a high compression ratio with lower image quality. The low complexity of the JPEG technology allows cameras and viewers to be produced cheaply.

Normally, there is no visible difference between a JPEG compressed image and the original uncompressed image. However, if the compression ratio is pushed too high, artifacts in the form of blockiness appear (Figure 6.3).

The JPEG compression standard contains a series of efficient compression techniques. The main technique that compresses the image is the discrete cosine transform (DCT) followed by quantization that removes the unnecessary information (the "invisible" parts). The compression level can

Figure 6.3 Original image (left) and JPEG compressed image (right) using a high compression ratio that results in blockiness.

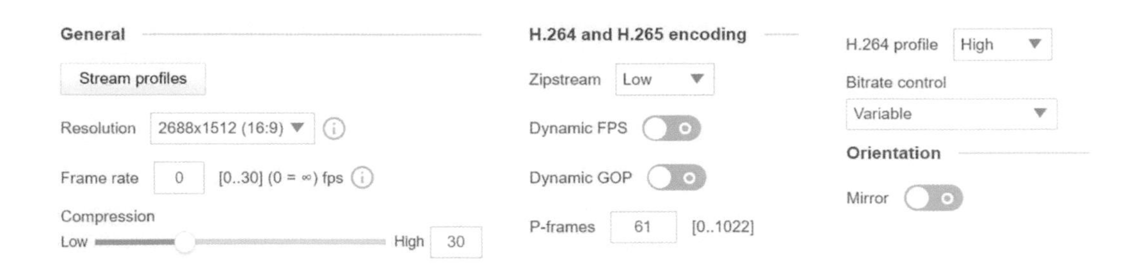

Figure 6.4 A typical interface for a network camera, in which the amount of compression is set by a (nonlinear) number between 0 and 99, where 30 is the recommended default. Depending on the level of detail and complexity in a scene, different compression ratios will be achieved based on this number.

usually be set as a percentage, 1–99%, where 99% gives the highest compression level and the smallest bitrate but also the most artifacts (see Figure 6.5).

For a description of the JPEG compression process, see section 6.4.

6.3.2 Motion JPEG

Motion JPEG (MJPEG) is a digital video sequence represented as a series of JPEG images. The advantages are as for single JPEG still images: flexibility in terms of both quality and compression ratio. Additionally, because there is no dependency between the frames, MJPEG is robust. This means that if a frame is dropped during transmission, the rest of the video will be unaffected.

The main disadvantage of Motion JPEG is that it makes no use of video compression techniques, as it simply uses a series of still images. This means a lower compression ratio for Motion JPEG video sequences than for other compression techniques such as MPEG. MJPEG is popular in situations where individual frames in a video sequence are required (for example, for analysis) and where lower frame rates (typically 2 fps or lower) are used.

6.3.3 MPEG-4

MPEG-4 (ISO/IEC 14496) is a later generation of MPEG and is based on the earlier MPEG-1 and MPEG-2.

The most important features of MPEG-4 include the support of low bandwidth–consuming systems and use cases that require high-quality images and virtually unlimited bandwidth. The MPEG-4 standard allows for any frame rate, whereas the preceding MPEG-2 was locked at 25/30 fps (PAL/NTSC).

In video surveillance contexts, MPEG-4 usually means MPEG-4 Part 2. This is the classic MPEG-4 video streaming standard, also called MPEG-4 Visual.

6.3.4 H.264

MPEG-4 Part 10 AVC/H.264 (hereinafter H.264) first became available in network video products in 2008. H.264 was jointly defined by standardization organizations in the telecommunications and IT industries and is currently the most widely adopted standard.

The standard addresses several weaknesses in earlier MPEG standards, such as providing good video quality at substantially lower bitrates, better error robustness, and better video quality at an unchanged bitrate. The standard is designed to give lower latency and better quality for higher latency.

An additional goal of H.264 was to give enough flexibility so that the standard could be used in a wide range of situations and applications with very different demands on bitrate and latency.

Figure 6.5 (a) JPEG image with no compression (frame size 71 KB). (b) JPEG image with 50% compression (frame size 24 KB). (c) JPEG image with 90% compression (frame size 14 KB). (d) JPEG image with 95% compression (frame size 11 KB).

Table 6.1 Examples of video services and their bitrate and latency targets

Service/Application/Industry	Bitrate	Latency
Entertainment video, including broadcast, DVDs, satellite, and cable TV	1–10 Mbit/s	High
Telecom services	<1 Mbit/s	Low
Streaming services	Low	High

H.264 helped accelerate the adoption of megapixel/HDTV cameras because the highly efficient compression technology can reduce large file sizes and bitrates without compromising image quality.

6.3.5 H.265

In 2013, ITU approved the High Efficiency Video Coding (HEVC) standard, also referred to as H.265, and a second version was published in early 2015.

The main advantage of H.265 compared to H.264 is the great advance in compression ratio. It improves compression by a factor of two for the same video quality. The standard also supports a broader range of resolutions, from QVGA (320 × 240) to 4320p. However, it only supports progressive or noninterlaced scanning, which codes the entire frame at once. For more information about image scanning techniques, see section 4.4.

While the standard has yet to become fully established in the surveillance industry, H.265 offers some significant advantages and is being adopted more and more. Higher resolution can be achieved at lower data rates, lowering minimum bandwidth requirements and reducing storage needs. However, there are trade-offs. While H.265 provides savings in network bandwidth and storage costs, it requires higher-performance network cameras and monitoring stations.

6.4 MORE ON JPEG COMPRESSION

The goal of JPEG compression is to deliver the highest possible quality for a given compression ratio. The more important data must be distinguished from the less important data in the image. This process is described in the following steps.

1. *Divide the image into macro blocks.* The image is divided into smaller images, in which each macro block is 16 × 16 pixels. These blocks can be manipulated in real time by fast processors.
2. *Divide image into components.* The image is then divided into components of 8 × 8 pixels, and the color information is separated from the texture. It is possible to compress color blocks more without creating visible artifacts. The color spaces Y (brightness) and CB and CR (color) are normally used for this division.
3. *Discrete cosine transform (DCT).* DCT is a form of lossy compression that converts each frame of the image into the transform domain, thus concentrating (sorting) the image information. Simple structures in the image are given low values, and complex structures receive high values.
4. *Quantization.* The resulting set of values from DCT describes the image content. The process of quantization compresses and eliminates visual data that is almost imperceptible to the human eye (thanks to the sorting). Reducing the number of symbols in the data stream for the image means the stream can be compressed more, and this in turn means the image file size can be reduced. Quantization cannot be reversed, and therefore, image quality is affected.
5. *Differential pulse-code modulation (DPCM) coding of DC component.* DPCM encodes the difference between each current and previous 8 × 8 block. The differences are likely to be small, so encoding the signal requires fewer bits, thus increasing the rate of throughput.
6. *RLE, Huffman, VLI encoding.* A combination of techniques such as Huffman encoding, run-length encoding (RLE), and variable length integer (VLI) encoding is used to remove redundancies and reduce the amount of data needed to store and reproduce the image.
7. *Bitstream generation, byte stuffing.* The bitstream is created and a special step called byte stuffing adds a "00" to each byte with a value of "FF". JPEG uses "FF" as a marker, and a decoding program can misunderstand the "FF" unless it has the "00".

6.5 MORE ON MPEG COMPRESSION

The MPEG standard H.264 is the compression standard of choice for most video surveillance applications today, along with its successor H.265, which has some advantages but also some disadvantages. These fairly complex and comprehensive standards have some characteristics that are useful to understand. The following subsections outline these characteristics.

6.5.1 Frame types

The basic principle for video compression is image-to-image prediction. The first image in the stream is called an I-frame and is self-contained, having no dependency outside that image. The following frames can use part of the first image as a reference (see Figure 6.6). An image predicted from one reference image is called a P-frame, and an image that is bi-directionally predicted from two reference images is called a B-frame.

- *I-frames:* Intra-predicted, self-contained.
- *P-frames:* Predicted from last I- or P-reference frame.
- *B-frames:* Bi-directional, predicted from two references – one in the past and one in the future – and thus out-of-order decoding is needed.

The video decoder (the "playback" algorithm) restores the video by decoding the bitstream frame by frame. Decoding must always start with an I-frame, which can be decoded independently, whereas P-frames and B-frames must be decoded together with the current reference image or images.

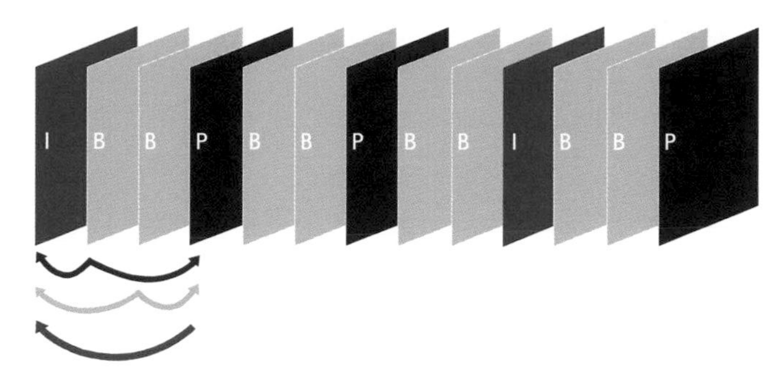

Figure 6.6 A typical sequence of I-, B-, and P-frames. A P-frame may only reference preceding I- or P-frames, while a B-frame may reference both preceding and subsequent I- or P-frames.

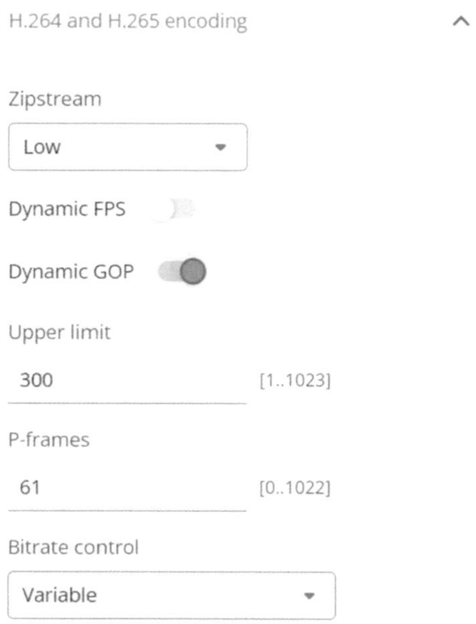

H.264 and H.265 encoding

Zipstream

Low

Dynamic FPS

Dynamic GOP

Upper limit

300 [1..1023]

P-frames

61 [0..1022]

Bitrate control

Variable

Figure 6.7 The interface in a network camera allows the GOP settings to be adjusted to fit the surveillance context. This example shows the upper frame limit for the dynamic GOP, as well as the number of P-frames (between two I-frames) for the regular, fixed GOP.

6.5.2 Group of pictures

One parameter that can be adjusted in H.264/H.265 is the group of pictures (GOP) length and structure (Figure 6.7). The GOP is normally repeated in a fixed pattern, for example:

GOP = 4 (IPPP IPPP . . .)
GOP = 15 (IPPPPPPPPPPPPPP IPPPPPPPPPPPPPP . . .)
GOP = 8 (IBPBPBPB IBPBPBPB . . .)

The appropriate GOP depends on the surveillance context. Decreasing the frequency of I-frames decreases the bitrate. Removing the B-frames reduces the latency.

There is a trend towards using longer GOPs of up to 64, 128, and higher because H.264/H.265 offers a technique for avoiding error accumulation. This reduces the required number of I-frames (which consume a higher bitrate) and makes encoding more efficient. The video management software can set limits for the GOP length and requires one frame per second to maintain access points in the system. Keeping these points available comes at the expense of bitrate.

6.5.3 Constant, maximum, and variable bitrates

Another central aspect of MPEG is the ability to select a bitrate control mode. Most MPEG and H.264 systems offer constant bitrate (CBR), maximum bitrate (MBR), or variable bitrate (VBR). The optimal selection depends on the surveillance situation and the network infrastructure. Different camera manufacturers use different technologies and names, but the basic principles are similar.

When the available bandwidth is limited and stable continuous streaming is essential, the preferred mode is normally CBR or MBR, the latter being sometimes called VBR with a cap.

The goal of CBR is to stay at the target bitrate no matter what happens in the scene. The quality is consistent when the scene complexity is low. But when activity increases, the camera either drops frames or increases compression. Some CBR techniques also allow bit-padding to fill the bitrate gaps in videos of low-complexity scenes. Bit-padding is a waste of bandwidth and storage for no gain in video quality.

The goal of MBR is to stay below the target bitrate and to continue to stream video no matter what happens in the scene. Like CBR, MBR uses compression and frame rate to control the stream. It works mainly with compression but also drops frames if it needs to. MBR reacts quickly to scene changes, so when the scene complexity is high, it actively limits bitrate overshoots. When the scene complexity is low, the image quality is high, and the bitrate stays below the target. There is no bit-padding to reach the target bitrate. When bandwidth is limited, MBR is typically the best choice for video surveillance.

Some cameras with MBR or CBR can prioritize the reduction of frame rate or quality. Not setting a priority normally means the frame rate and image quality are affected more-or-less equally.

With VBR, the image quality is consistent regardless of the scene complexity. VBR is suitable in video surveillance situations where images must have a high level of detail, especially with motion in the scene. Because the bitrate can vary a lot with VBR, the network infrastructure for such a system needs to have greater bandwidth and storage capacity. In other words, VBR gives crisp details at the price of bandwidth and storage.

The mode and settings that work best depend on the scenario, quality requirements, and network infrastructure:

- *Is the scene complexity high or low?* High-complexity scenes require more bandwidth. Medium or high motion means high complexity. Little or no motion means low complexity. Because nighttime video tends to have more noise, it usually needs more bandwidth than daytime video.
- *Is frame rate or image quality more important?* A very compressed image has fewer details, making identification more difficult. A very low frame rate means that some activity may go undetected because too many critical frames are dropped.
- *What is the available bandwidth and storage?* If ample bandwidth and storage is available and details in the video are essential, use VBR or set the target bitrate very high. Limiting the bandwidth too much makes image quality suffer. Limiting the bandwidth too little wastes money on storage.

6.5.4 Average bitrate

Average bitrate (ABR) is a sophisticated bitrate control scheme designed to improve video quality while also keeping the storage within the system's limits. While its primary goal is not to save bandwidth, ABR maintains a bitrate budget over time. A video stream is assigned a certain amount of storage, and the ABR controller adapts the video quality to make the whole stream fit.

Because ABR is constantly monitoring the current bitrate, the camera keeps track of the amount of consumed storage and continuously projects the bitrate target to ensure optimal video quality over the period. Unused storage from earlier idler periods can be used for maintaining high video

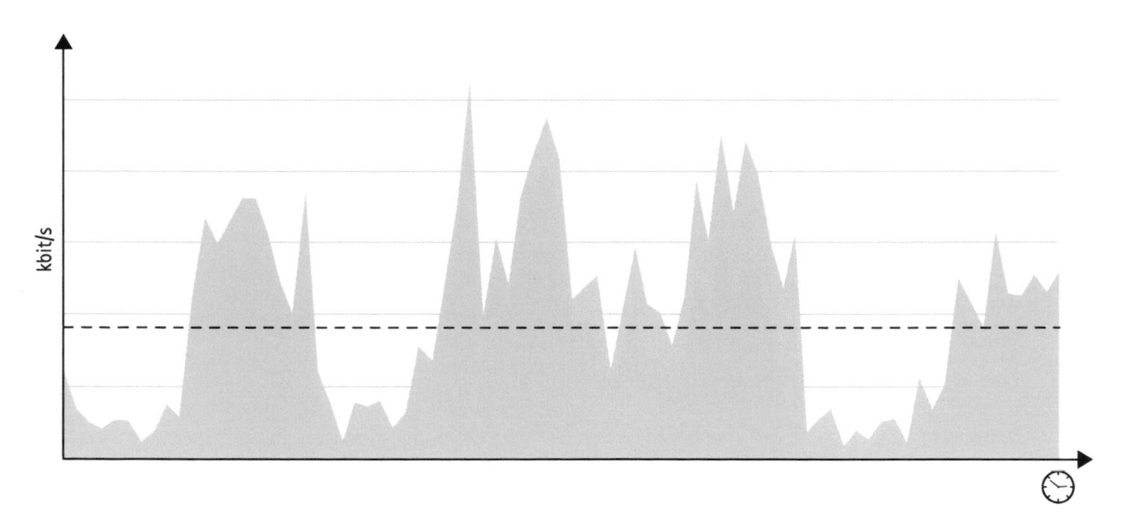

Figure 6.8 With ABR, high image quality can be maintained during peak events thanks to savings from earlier idle periods. Averaged over a set period of time, the bitrate goal must be fulfilled.

quality in later busier periods, and the bitrate budget is adhered to. ABR works with continuously recorded streams without scheduled pause periods.

6.5.5 Profile@Level

Because both MPEG-2 and MPEG-4 cover a wide range of image sizes, frame rates, and variable bandwidth usage, MPEG-2 introduced a concept called Profile@Level. This was created to make it possible to only support subsets of the standard and to communicate compatibilities between systems. Examples of common profiles are MPEG-2 Main profile at Main Level (MP@ML), MPEG-4 Main profile at L3 Level, and H.264 Main profile at Level 5.

The Internet Streaming Media Alliance (ISMA) uses the Profile@Level definitions to ensure that devices streaming video over the internet are compatible.

6.5.6 The Baseline and Main profiles

The two most popular profiles in H.264 encoding for video surveillance applications are the Baseline and Main profiles. The Baseline profile for H.264 uses only I- and P-frames, while the Main profile may also use B-frames in addition to I- and P-frames. The Baseline profile allows network video products to have low latency. In video products with more powerful processors, the Main profile is sometimes used without B-frames to enable higher compression and at the same time low latency and maintained video quality. Using Main profile H.264 compression, VGA-sized video streams can be reduced by 10–15% and HDTV-sized video streams can be reduced by 15–20%, compared to using Baseline profile H.264 compression. See Figure 6.9.

The most significant part of the Main profile is the entropy coder called context-adaptive binary arithmetic coding (CABAC), which replaces the simpler context-adaptive variable-length coding (CAVLC) used in the Baseline profile. CAVLC and CABAC produce the same video quality, but CABC can do it at a reduced frame rate without introducing new artifacts.

The standardization organization continues to add new profiles to H.264. However, not all profiles are suitable for video surveillance.

6.5.7 Improving H.264 for surveillance

The H.264 standard defines various profiles, that is, sets of capabilities for use in specific types of applications. A profile uses a subset of the coding and decoding tools defined by the standard. When video is compressed according to a specific profile, this defines which tools the decoder

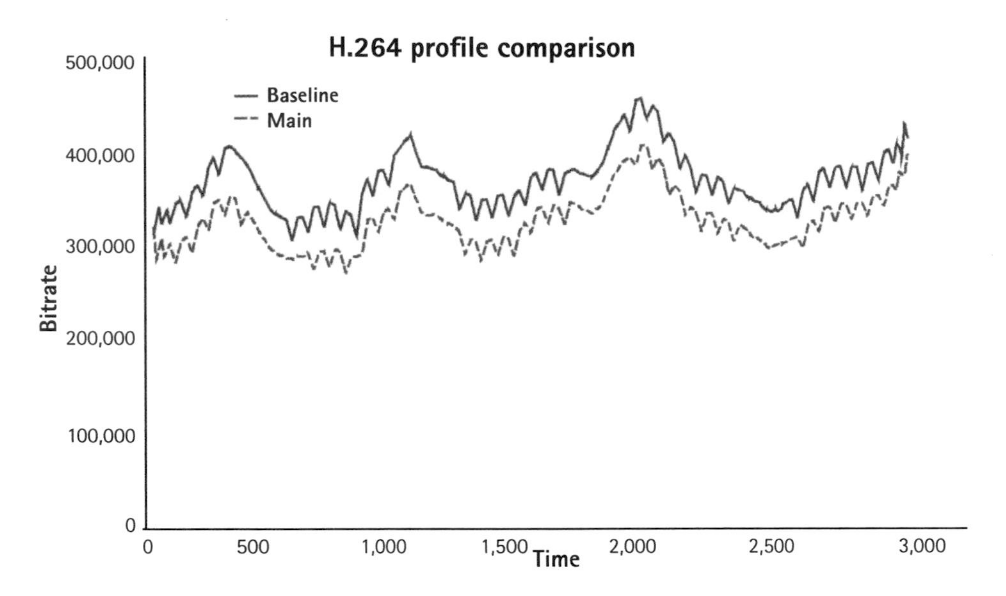

Figure 6.9 A comparison of bitrates generated by Baseline profile and Main profile H.264 compression. Main profile can offer higher compression with low latency while maintaining image quality.

must use to decompress the video. A decoder may support some profiles while not supporting others. The most common profile in use is the High profile. Because the H.264 standard dictates the method of decoding of video rather than the encoding, developers have the freedom to improve the encoding solutions, as long as they maintain playback compatibility (see also section 6.6).

A common goal when seeking to improve video compression is to use higher resolution and increase forensic usability while also reducing storage costs. An intelligent video compression method (or smart codec) can ensure that important details in the video stream get enough attention, while unnecessary data is removed.

One such intelligent video compression technology is a standard-compatible encoder implementation that analyzes the video stream to reduce bandwidth and storage requirements by an average of 50% or more when compared to standard methods. Important details and motion are preserved at high quality, while the compression enhancement filters the rest of the image information more aggressively to make optimal use of the available bandwidth.

This intelligent video compression technology consists of a collection of algorithms that analyze the video stream in real time:

- *Dynamic ROI (region of interest):* Identifies regions of interest based on objects, people, or motion in the scene and applies the correct level of compression from a forensic perspective. Depending on what happens in the scene, an ROI automatically changes shape and size, appears and disappears, splits and merges. Therefore, dynamic ROI is better prepared for unexpected events than a traditional ROI implementation, where the region is set manually.
- *Dynamic GOP (group of pictures):* By switching between a maximum GOP value for busy scenes and a default value for low-activity scenes, the bitrate can be drastically reduced.
- *Dynamic FPS (frames per second):* Reduces the frame rate when there is little or no motion in the scene, which in turn reduces the bitrate. The camera captures and analyzes video at full frame rate, but if there is no change in the scene, it will only encode at a very low frame rate.

Again, these improvements fully comply with the standard because they only change how the video is encoded, not how it is played back. Some improved implementations of H.264 could yield bitrate savings of 50% or more and, therefore, become a more viable option than early implementations of H.265.

Figure 6.10 The same scene captured with (left) H.264 and (right) improved H.264. In this case, improved H.264 results in bitrate savings of nearly 60% with similar quality video from a video surveillance perspective.

Another benefit of an improved implementation of H.264 or H.265 is that because it is just a firmware upgrade away, existing network cameras can be used, and the same video management system and monitoring station can be used with no changes. Figure 6.10 shows an example of an implemented improvement of H.264.

6.5.8 Licensing

MPEG-4 H.264 and H.265 are subject to licensing fees, which any company manufacturing products using these compression standards must pay. MPEG LA, an independent license administrator, manages the licensing fees. For most network video products, one or more licenses have been paid for by the manufacturer, which means that the video can be viewed on one, or possibly multiple, monitoring stations.

A user planning to view the video at more monitoring stations than the product is licensed for must purchase additional licenses to match the number of stations. If the manufacturer has not paid the license fees, it most likely means that the manufacturer does not fully follow the compression standard, which in turn limits the compatibility with other systems.

6.5.9 Backward compatibility

MPEG-2 and later standards are not backward compatible. This means that H.264 encoders and decoders do not work with MPEG-2 or previous versions of MPEG-4 unless they are specifically designed to handle multiple formats. The same applies to devices designed for H.265. These could be designed to support older standards, although the more costly hardware required makes it more likely that the manufacturer will save on these functions. However, various solutions are available where streams encoded with newer standards can sometimes be packetized inside older standardization formats to work with older systems.

6.6 COMPARING STANDARDS

When comparing the performance of MPEG standards such as H.264 and H.265, it is important to note that results may vary between encoders that use the same standard. This is because encoder designers may implement different subsets of the standard. As long as the output of an encoder conforms to a standard's format and decoder, it is possible to make different implementations. This helps to optimize the technology and reduce the complexity in implementations. However, it

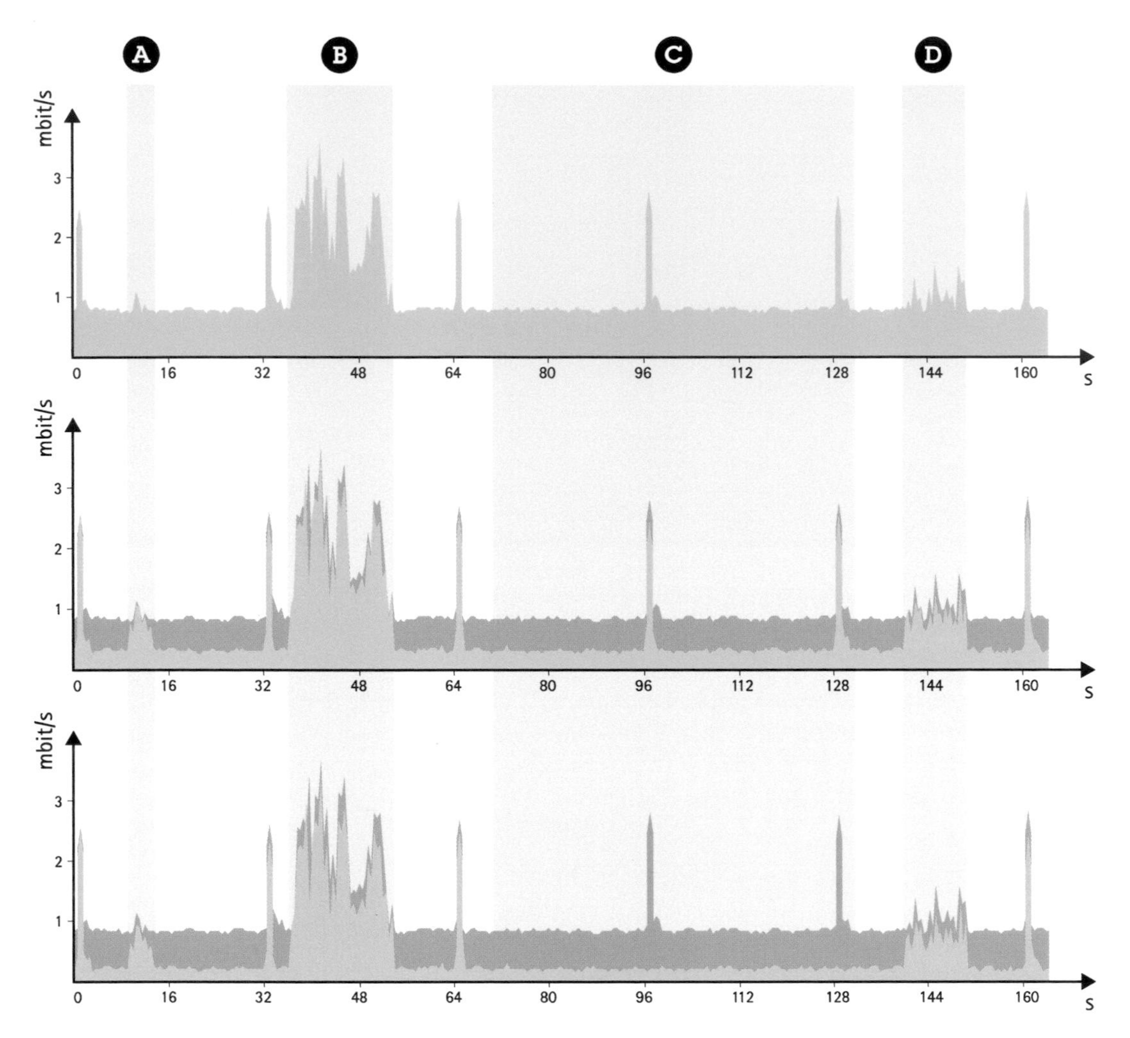

Figure 6.11 Plotting the instantaneous bitrate from a video with four different motion events (A, B, C, and D.) The top graph uses regular H.264. The middle graph shows the bitrate when using a smart codec at a low level and a dynamic GOP. The bottom graph shows the bitrate using a smart codec at a high level and a dynamic GOP.

also means that an MPEG standard cannot guarantee a given bitrate or quality, and comparisons cannot be properly made without first defining how the standards are implemented in an encoder.

Unlike an encoder, a decoder must implement all the required parts of a standard to decode a compliant bitstream. This means that only the decoder is truly standardized. The standard specifies exactly how a decompression algorithm should restore every bit of a compressed video. If video quality is a concern, the user should test several products to make sure the quality matches the purpose.

Figure 6.11 shows bitrate comparisons when using improved H.264 (a smart codec).

6.7 BEST PRACTICES

One compression standard and configuration does not fit all situations. When designing a network video application, consider the following questions:

- *What is the required frame rate?* Is the same frame rate required all the time? At less than 2 fps, consider using MJPEG and controlling the frame rate by video motion detection. For higher frame rates, H.264 or H.265 is normally best because it saves bandwidth and storage.

- *What is the available network bandwidth?* In scenarios with very low bandwidth, H.264 and H.265 compression using CBR or MBR may be the only option, but image quality will be degraded when motion occurs in the scene.
- *What is the allowed level of image degradation (artifacts) due to compression?* Compression ratios well above 90% can be used if the scene is not too complex.
- *What is the acceptable level of latency?* If video is not monitored live but only recorded, latency might not be an issue. When controlling PTZ cameras, it is important to have low latency.
- *Bitrate control?* Which type of control best suits your parameters? Can ABR or VBR be used, or is CBR or MBR the best choice? If CBR or MBR is used, is it possible to set the priority for whether frame rate or image quality should be reduced?
- *How robust?* How reliable or secure does the system need to be? Is it acceptable that video might be lost for 0.5 seconds if a frame is dropped on the network?
- *Is system openness and interoperability essential?* If so, make sure the chosen products follow the standard 100%.

Chapter 7

Audio technologies

Much of our learning and our relations with others are conducted through audiovisual cues – that is, what we see and hear. We are often first alerted to events by what we hear, after which we verify the events visually.

7.1 USE CASES FOR AUDIO

Integrating audio with a video surveillance system can be an invaluable addition to the system's capabilities to detect, interpret, and manage events and emergencies, as well as to analyze, process, and store the audio data.

7.1.1 Listen and interact

Perhaps the most basic and intuitive use case is audio surveillance with direct operator interaction. One example is hearing a suspicious conversation or noise and sending a security guard to investigate. Another is in a hospital, where a nurse could hear if a patient is in distress.

These use cases involve operators having access to the audio environment from a control room or via an app on a mobile device. The human ear captures sound, and the brain extracts the information that is relevant. If used in conjunction with video surveillance, audio adds another dimension of information for decision-making. In some cases, audio will be the only dimension, for example, if the audio source is outside the camera's field of view or if the lighting conditions are challenging.

Interacting with the help of audio can also allow users to listen in on an area and communicate with visitors or to give orders and warnings to intruders. For example, if a person in a camera's field of view demonstrates suspicious behavior, such as loitering near an ATM or is seen entering a restricted area, a security guard can give a verbal warning to that person. In a situation where a person has been injured, the person can be greatly comforted by a voice, keeping them calm and assuring them that help is on the way. Audio can also be used in access control applications – that is, a remote "doorman" can communicate with visitors at an entrance.

7.1.2 Listen and witness

Audio surveillance can also be used for the purpose of direct testimony based on witnessed (heard) events. For example, upon hearing an escalating argument with incriminating speech, an operator can send guards to the scene but can also later bear witness to what was said and heard. The audio can, of course, also be recorded, avoiding the need to rely on a witness's memory.

DOI: 10.4324/9781003412205-7

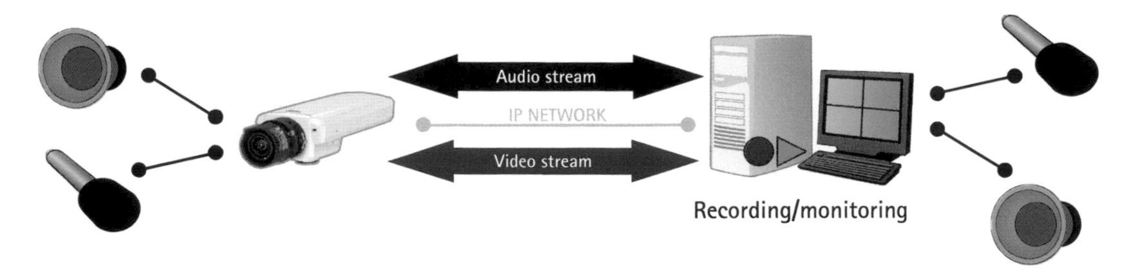

Figure 7.1 A network video system with integrated audio support. Audio and video streams are sent over the same network cable.

7.1.3 Record and store

If appropriate, capturing and recording audio data can provide additional evidence, such as incriminating speech or gunshots. Recorded audio can provide proof of who said what, how many shots were fired, or similar events of forensic interest.

When audio is recorded in a forensic context, care should be taken to conserve the original data and avoid the processing that may occur in other contexts. In legal proceedings, any type of processing could be considered as tampering with evidence. Voice-enhancing algorithms can be used to improve the audibility of recorded speech and thus the forensic value, but any such processing should only be performed on copies of the original material.

7.1.4 Automatic intervention

Consider a surveillance system that has no audio. Outside the camera's field of view, a cry for help, the sound of breaking glass, a gunshot, or an explosion in the vicinity would not be noticed by this video-only system. The ability of audio to cover a 360° area enables a video surveillance system to extend its coverage beyond the camera's field of view. Audio triggers can instruct a PTZ camera, or its operator, to verify an audio alarm visually. It can also automatically trigger a speaker to play a pre-recorded message, for example, to warn an intruder that the area is under surveillance. Because it gives more information and speaks to two sensory systems, an audiovisual surveillance system increases the effectiveness of the security solution.

7.2 AUDIO IN SURVEILLANCE SYSTEMS

The implementation of audio in video surveillance systems is increasing with the proliferation of network video. Network video enables easier implementation of audio than analog CCTV, where dedicated audio and video cables must be run from the camera and microphone to the viewing station or recording location. In a network video system (Figure 7.1), a camera with audio support sends both the audio and video over the same network cable. This eliminates the need for extra cabling and makes audio and video synchronization much easier.

Support for audio can be found in many types of network cameras, including fixed box cameras, PTZ cameras, and fixed dome cameras. Many manufacturers recognize the importance of audio, and it is becoming a common feature in surveillance installations.

Some video encoders also have built-in support for audio, which means they can provide audio functionality in analog installations. This may be useful in an application where a specialty camera is used or if analog cameras are already installed (Figure 7.2). Another audio device for network video applications is the audio module, which only has audio support. The audio module also has I/O ports that can be located far from the network camera. For example, in a city surveillance application, the audio module and microphone are located close to street level, while the PTZ camera is located high up on a pole.

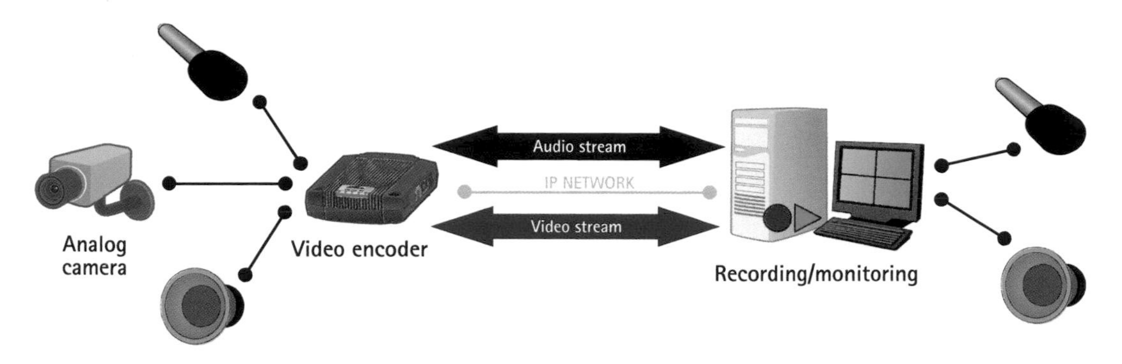

Figure 7.2 Some video encoders have built-in audio, which makes it possible to add audio when analog cameras are used in an installation.

When planning to implement audio, its application should be clear because it affects which products should be selected. The following sections of this chapter discuss audio transmission modes, audio equipment, acoustical adjustments, audio detection alarms, codecs and bitrates, audio and video synchronization, and a summary of the factors that affect audio quality. To read about how the use of audio can sometimes be restricted or regulated by local legislation or codes of practice (see Chapter 18).

7.3 AUDIO MODES

Depending on the application, audio might need to be transmitted in one or both directions. This can be done either simultaneously or in one direction at a time. There are three basic modes of audio communication: *simplex*, *half-duplex*, and *full-duplex*.

7.3.1 Simplex

Simplex means that audio can be sent in one direction only. Audio is sent either from the camera – which is most often the case – or from the user. Situations where audio is sent only from the camera include remote monitoring and video surveillance applications (Figure 7.3). Applications where audio is sent only from a user or operator include situations where there is a need to give instructions to a person seen by the camera, for example, in a parking lot where the operator can use audio to scare off a potential car thief (Figure 7.4).

Figure 7.3 In simplex mode, audio is configured to be sent in one direction only. In this case, from the camera to the operator.

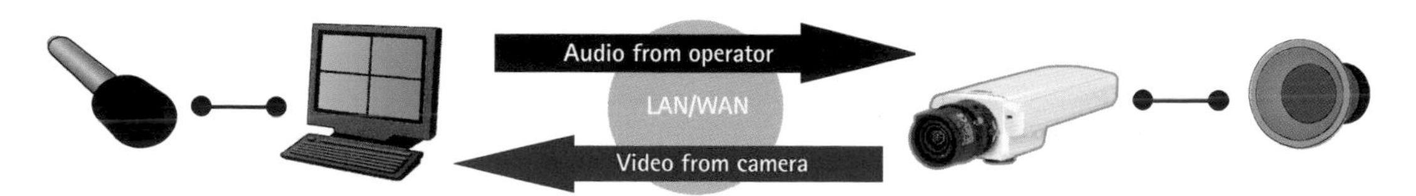

Figure 7.4 Also in simplex mode, this time, the operator sends audio to the camera.

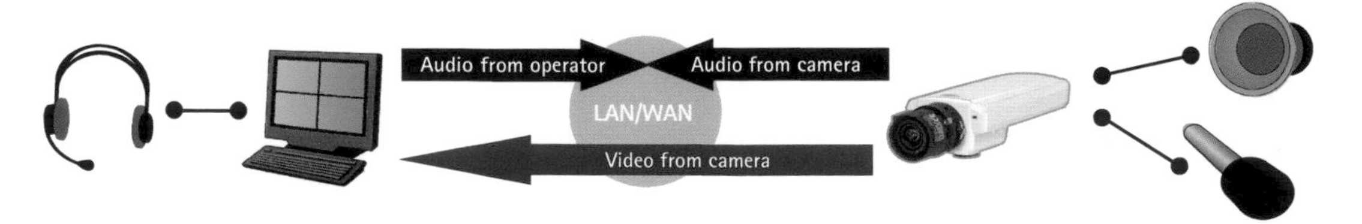

Figure 7.5 In half-duplex mode, audio can be sent in either direction but only in one direction at a time.

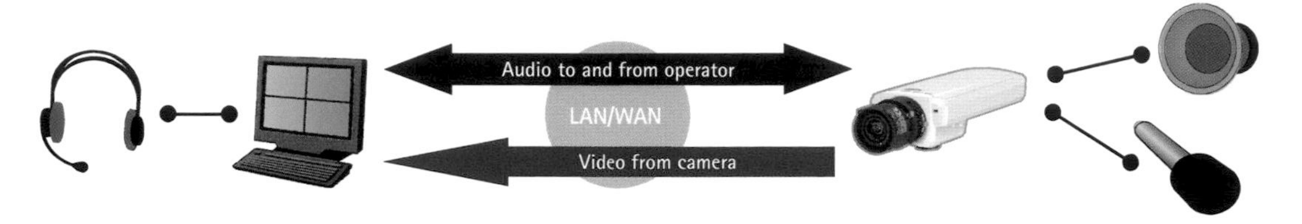

Figure 7.6 In full-duplex mode, audio is sent to and from the operator simultaneously.

7.3.2 Half-duplex

Half-duplex means that audio can be sent and received in both directions – from the camera and the operator – but only in one direction at a time (Figure 7.5). Just like using a walkie-talkie, an operator can press and hold down a push-to-talk button to speak. Releasing the button enables the operator to receive audio from the camera. Other methods can also be used to transmit in each direction, for example, voice-operated techniques. With half-duplex, there is no risk of echo (see section 7.5.3).

7.3.3 Full-duplex

Full-duplex means that users can send and receive audio (talk and listen) at the same time (Figure 7.6). This is like a conversation over the phone. While full-duplex has the advantage of simultaneous audio in both directions, it also requires more bandwidth.

7.4 AUDIO EQUIPMENT

When a network camera or a video encoder has support for audio, it often includes a built-in microphone but rarely a built-in speaker. While a built-in microphone may function well in some surveillance contexts, in many cases, an external microphone is a better solution. This section gives some guidance when selecting external audio equipment.

7.4.1 Sound input (microphones)

A camera or encoder with mic-in or line-in support gives users the option of using a microphone other than the one built-in, for example, another type or quality of microphone. In addition, an external microphone can be located some distance from the camera. Sound and video are often best captured from different (and often incompatible) positions. A corner close to the ceiling might be suitable for video but is normally not optimal for sound. In most cases, you get the best sound performance when placing a microphone close to the source of the sound, such as over a checkout counter.

Audio sources can have different output signal levels. For example, there are microphones with and without built-in amplification. In many cases, a network camera supports the use of both, allowing connection of a microphone without an amplifier to mic-in and a microphone with an amplifier to line-in. Often, the same jack is used, and the choice between mic-in and line-in is made in the software. Line-in means the camera can be connected to devices that deliver an amplified audio signal (known as a line signal). Examples of devices that provide line signals include mixers, which

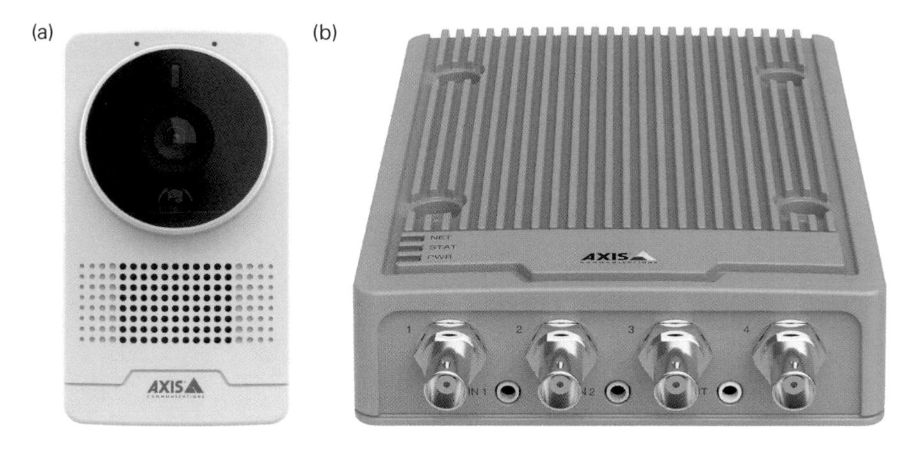

Figure 7.7 (a) A camera with built-in microphone and speaker. (b) A video encoder with inputs for external microphones and speakers.

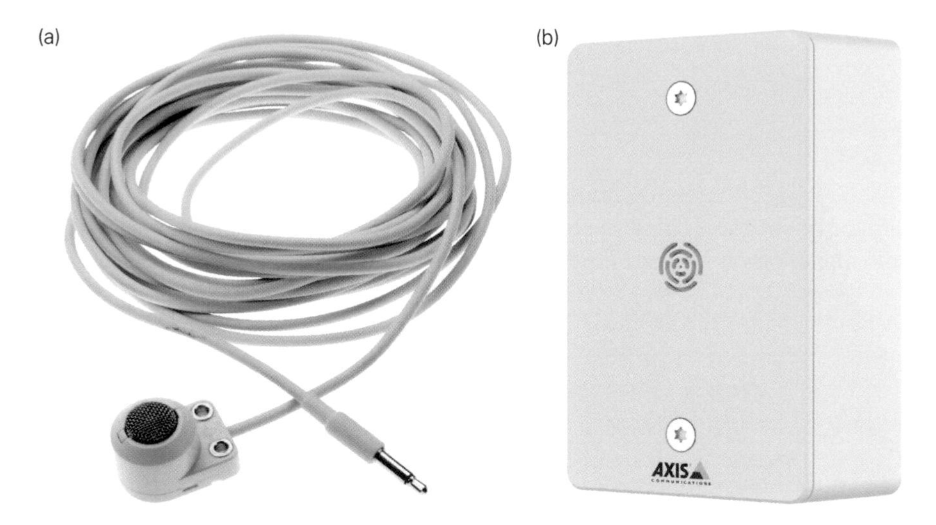

Figure 7.8 (a) Condenser microphone with a 3.5 mm connector. (b) A vandal-proof microphone for outdoor use.

allow a camera to connect to several microphones, and microphone amplifiers, which connect to microphones that lack built-in amplification.

There are three main types of microphones: *condenser*, *electret condenser*, and *dynamic*. These differ in the way they convert sound into electrical signals. See the next sections for more on these.

7.4.1.1 Condenser microphones

Of the three main types of microphones, the condenser has traditionally been considered to offer the highest audio sensitivity and quality, though some back-electret microphones can deliver similar quality. The condenser microphone is often used in professional recording studios and can be used in video surveillance applications that require high audio quality. A condenser microphone uses a balanced interface together with a so-called phantom power supply, which supplies power to the microphone. Sometimes a condenser microphone uses an XLR connector, a circular connector with three pins often found on professional audio equipment. If an XLR connector is not present, a condenser microphone can still be connected through an adapter.

7.4.1.2 Electret condenser microphones

This type of microphone is common in headsets and computer microphones (Figure 7.9) and often uses a 3.5 mm audio connector. It offers a high level of sensitivity and is less expensive than a

Figure 7.9 A computer headset with a microphone based on electret condenser technology.

condenser microphone. The electret condenser microphone normally needs a voltage of 1–10 V, which is supplied by so-called "plug-in power". If an external electret microphone is used, the camera can supply the microphone with the necessary power.

7.4.1.3 Dynamic microphones

The dynamic microphone is rarely used in the video surveillance industry because its sensitivity is not high enough. It also has a poor ability to reproduce low frequencies when the source of the sound is not close to the microphone.

7.4.1.4 Directional microphones

Microphones are made with different polar patterns, also called pickup or directional patterns. The pattern types include omnidirectional, which picks up audio equally from all directions, and unidirectional, such as cardioid (meaning heart-shaped) and supercardioid, which have high audio sensitivity in one specific direction. To pick up sounds from a location far from a camera, a specialized unidirectional microphone called a shotgun microphone can be used. This will amplify sounds from the location of interest and attenuate sounds from all other directions, reducing the impact of noise from other sources, such as ventilation and machinery.

7.4.2 Audio output (speakers)

There is a wide array of speakers available that can be used with network cameras, for example, PC speakers are often used. The power of passive speakers is measured in watts, which indicates how much power they can handle and, in combination with the sensitivity specification, often relates to how loud the speakers can be.

An active speaker, which has a built-in amplifier, can be connected directly to a network camera. If the speaker has no built-in amplifier, it must first connect to an external amplifier connected to the camera. Network devices with a built-in amplifier can be directly connected to passive speakers, but power specifications are usually low, so the result is often not very loud.

7.5 ACOUSTICAL ADJUSTMENTS

There are several adjustments (Figure 7.10) that can be made to get the best audio performance from an installation. For example, there are methods for reducing the volume of unwanted sounds

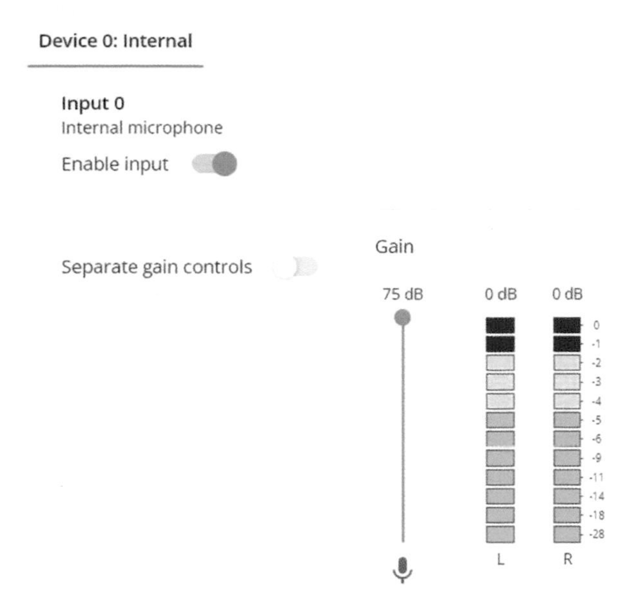

Figure 7.10 Examples of the gain setting for a microphone on a network camera.

and for amplifying sounds that are too quiet, for example, soft voices. This section discusses several of the most common adjustments.

7.5.1 Volume and gain

Since the audio level varies with the connected source, for example, line-level from a mixer or mic-level from a microphone, we need a way of amplifying low levels while still handling high levels. To optimize the audio quality, the audio gain setting on the network camera should be adjusted so that the signal is not clipped or distorted. It is crucial that the gain level is chosen wisely and that the necessary gain is applied in the signal path as early as possible, preferably at the microphone if available.

7.5.2 Audio processing

The term *audio processing* covers several types of automatic processes that work together to correct and control audio signals. For example, it can include equalization and automatic gain control. Equalization adjusts the balance between different frequencies, and automatic gain control continuously adjusts amplification (more for low and less for high signals) to get similar results regardless of the loudness or distance to a sound source. After the audio processing is complete, echoes have been cancelled and noise reduced, and the audio signal is free from reverberation and other distortion.

7.5.3 Voice enhancement

Voice enhancement (see Figure 7.11) is a specific type of noise reduction algorithm that aims to increase speech intelligibility – it amplifies voices and reduces everything else. This is used in scenarios where it is important to clearly hear what is being said, for example, in a hospital room/or a courtroom.

7.5.4 Noise cancellation

Noise cancellation (see Figure 7.11) can be used to reduce background noise. This feature is configured using two parameters: *threshold* and *attenuation*. Threshold is used to define the level under

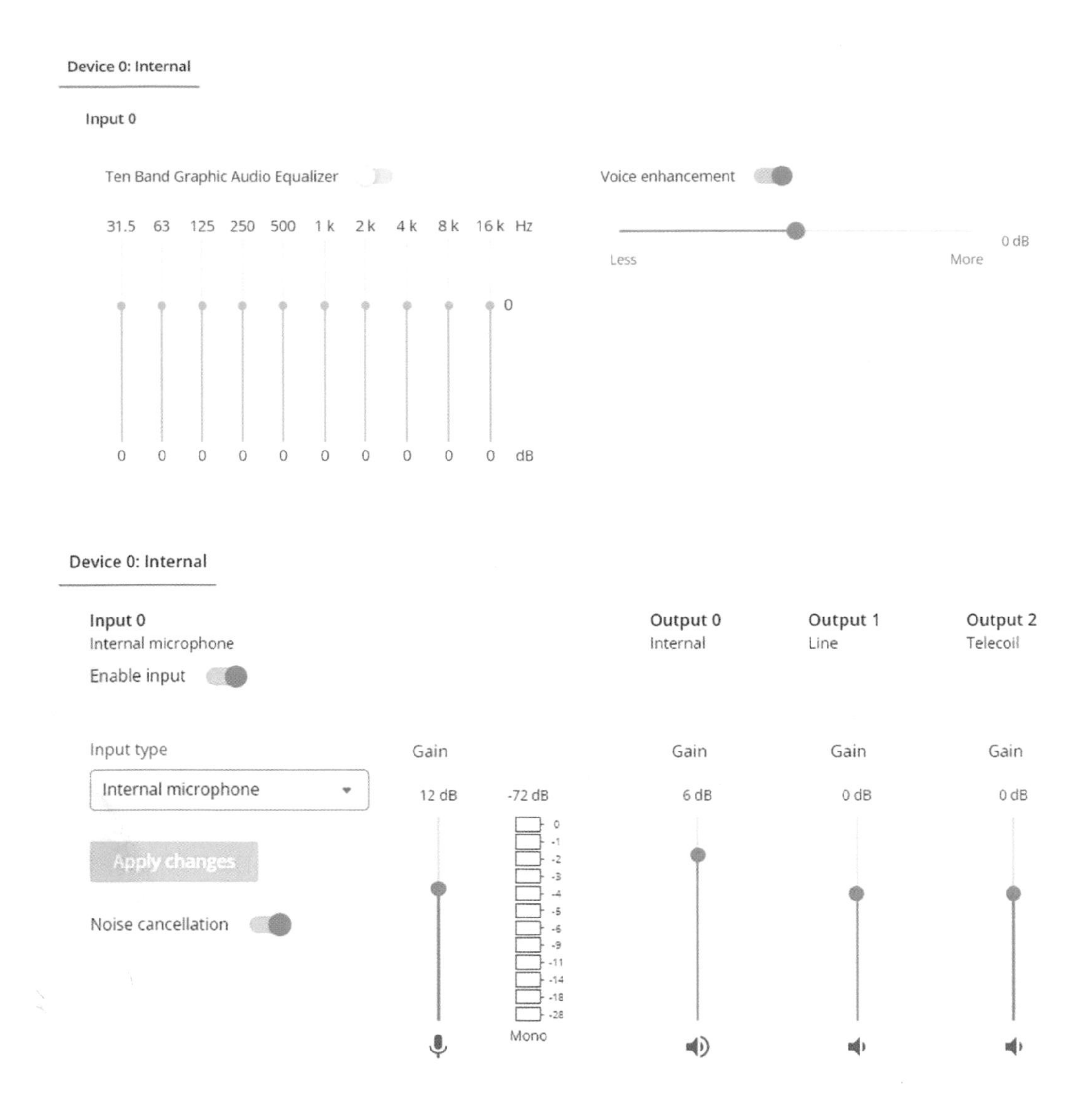

Figure 7.11 Examples of a user interface showing settings for audio processing and noise cancellation.

which noise will be reduced. Then attenuation can be used to choose the degree of noise cancellation. Full background noise cancellation may not be desirable because a listener might interpret it as a break in the connection. Because it can decrease the quality of the audio, noise cancellation should be used carefully. There are also other methods to reduce overall noise in audio during recording of the desired signals.

7.5.5 Echo cancellation

In full-duplex mode, the microphone will not only capture the desired incoming sound but also the sound generated from the speaker, creating what is known as feedback. This can be reduced by directing the microphone away from the speaker and using echo cancellation. A camera with echo cancellation (see Figure 7.3) has a signal processor with a short-term memory that remembers the audio signals that have just been sent from a loudspeaker or that leak between neighboring wires. If the microphone picks up these audio signals, the device recognizes them as an echo and removes them.

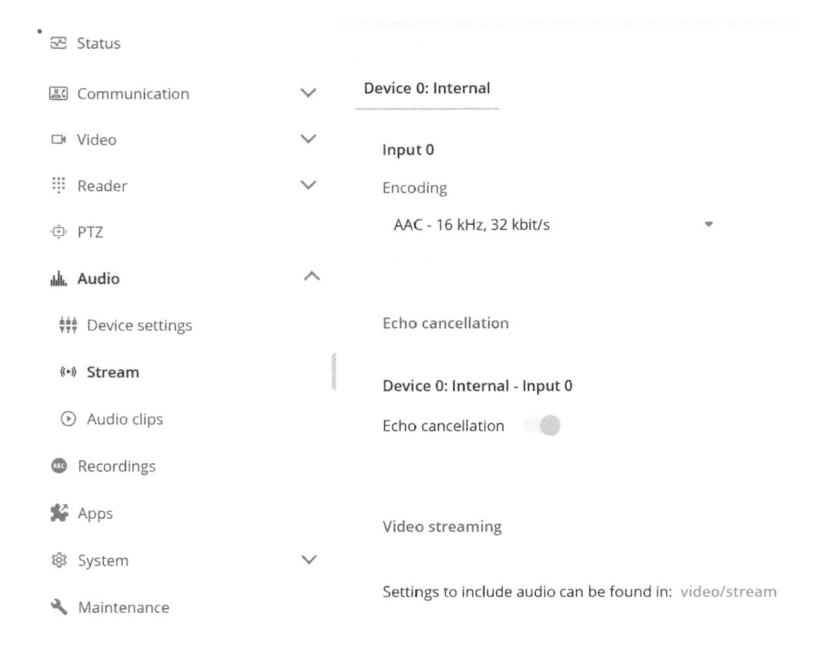

Figure 7.12 A user interface showing the setting for echo cancellation.

7.6 AUDIO ANALYTICS

Just as you can use several types of video analytics for automatic event detection and alarms based on visual detection, audio analytics can monitor the audio streams and react when something stands out. Audio analytics software can trigger alarms when certain sounds are detected, and they can also determine where a sound came from to automatically redirect a PTZ camera towards the source of the sound. In a hospital, audio analytics can be used to detect high sound levels that imply that a patient is in distress. This use case can also benefit from sound visualization analytics that makes it easier to simultaneously monitor sound from multiple locations.

Audio analytics applications do not generally record sound continuously. They typically only process the audio to search for specific patterns, levels, or frequencies. When analytics run on the edge (in the camera), no digital audio data needs to leave the camera, only the metadata or triggers resulting from the analytics. In many cases, there is no need to record audio to achieve the goal, which may help manage privacy concerns and comply with regulations regarding personal data.

7.6.1 Audio detection alarm

Just as video can be analyzed in a network camera, if the camera also has audio capabilities, then audio can be processed too. In areas too dark for video motion detection to work properly, an audio detection alarm can be used as a complement to motion detection. It can also be used to detect activity in areas outside the camera's view, as well as to reinforce the overall situational awareness where you see activity but cannot determine if it is unwanted without also hearing what is happening.

When sounds are detected, such as breaking glass or voices in a room, these can trigger a network camera to send and record video and audio, to send email or other alerts, and to activate external devices such as alarms. Similarly, alarm inputs such as motion detection and door contacts can be used to trigger video and audio recordings.

In a PTZ camera, audio alarm detection can trigger the camera to turn automatically to a preset location, such as a specific door or window (see Figure 7.13).

Depending on requirements, audio detection can be used all the time, at specific times, or turned off altogether. It can be set to trigger an event if the incoming sound rises above, falls below, or passes a certain level of intensity.

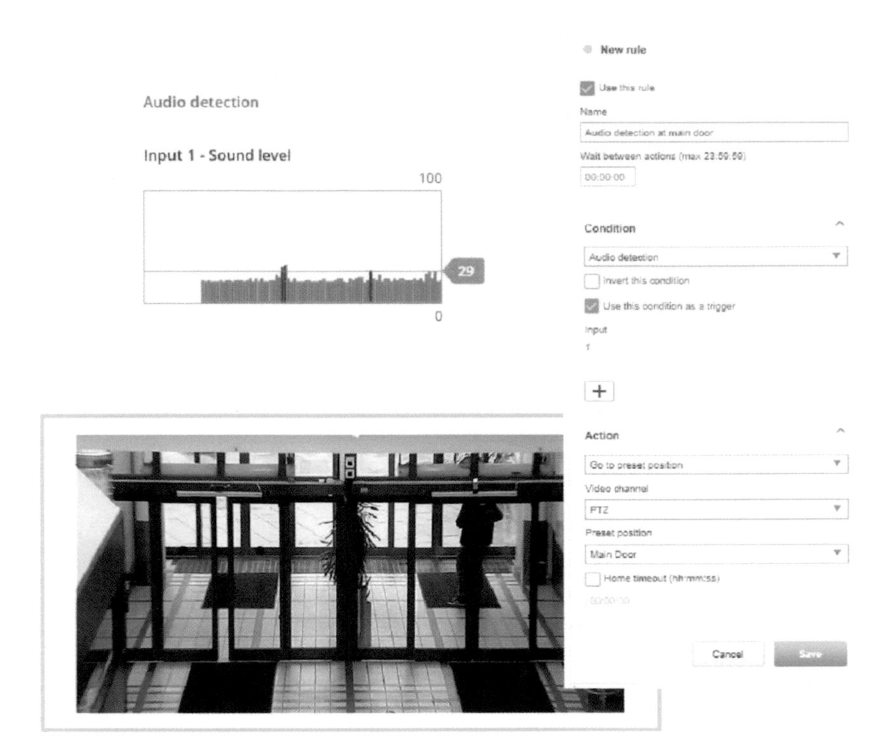

Figure 7.13 When the audio input reaches the alarm level (see "Audio detection"), it triggers the PTZ camera to move to a preset position (see "New rule"). This helps an operator both detect sound and verify whether it is cause for further investigation.

7.6.2 Visualizing sound in video

The sound captured in a video can be visualized and displayed as a sound spectrum diagram on a screen. If a set threshold is exceeded, the diagram will indicate an alarm. This can be valuable in situations where you need to monitor sounds from multiple sources at once, for example, from several patient rooms. While it is too difficult to listen to many audio sources (from many rooms) simultaneously, it is much easier to view visualizations on a monitor. The visualizations can be added as overlays to the video images.

7.7 AUDIO COMPRESSION

Analog audio signals must be converted into digital audio through a sampling process, after which it is almost always compressed for efficient transmission and storage. The sampling is often done by dedicated hardware, while the compression is done in software, using an audio codec – an algorithm that codes and decodes audio data. The following sections present some factors that can influence audio quality and file size.

7.7.1 Sampling rates

Sampling rate or frequency refers to the number of times per second a sample of an analog audio signal is taken and defined in hertz (Hz). The human ear can hear sounds up to 20 kHz, but to capture this level of sound with good quality, a sampling rate of at least 40 kHz is necessary. Music CDs, for example, use a sample rate of 44.1 kHz. Music streaming services generally use similar sampling rates.

The sampling rate (according to the Nyquist–Shannon sampling theorem) must be at least twice the maximum required frequency. If the human voice is the only sound of interest, a sampling rate

Figure 7.14 Sound visualizations overlaid on video feeds.

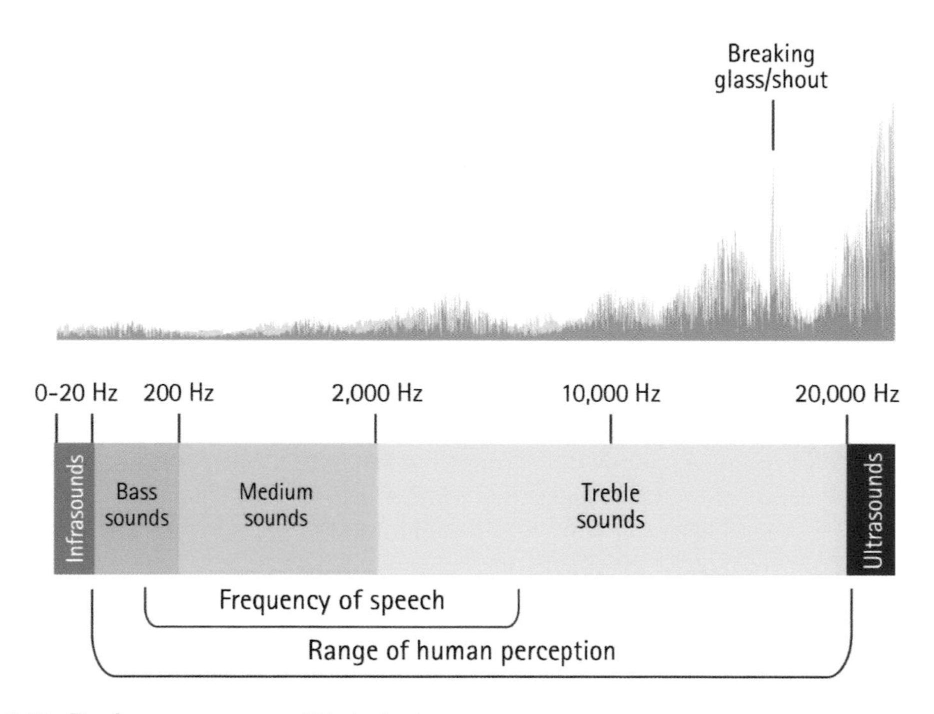

Figure 7.15 The frequency range audible to the human ear.

of at least 8 kHz is needed because the frequency of the human voice is normally below 4 kHz. In general, the higher the sampling rate, the better the audio quality and the higher the demands on bandwidth and storage.

Note that when reducing the sampling rate, it is always the high frequencies (the treble) that are lost, making the result sound muffled or dull. This is important when high frequency content is of interest, such as the sound of breaking glass.

7.7.2 Bitrate

The bitrate setting is important because it determines the level of compression and thereby the quality of the audio. In general, the higher the compression level, the lower the audio quality. The differences in the audio quality of software codecs can be particularly evident at high compression levels (low bitrates) but not at low compression levels (high bitrates).

The most common bitrates for various codecs range between 32 and 64 kbit/s. Modern codecs can produce adequate sound quality for bitrates as low as 32 kbit/s. Audio bitrates, like video bitrates, are important to consider when calculating total bandwidth and storage requirements.

Some codecs feature only a constant bitrate mode, whereas others enable both constant bitrate (CBR) and variable bitrate (VBR) modes. When using VBR, the bitrate adjusts to the complexity of the audio. This means that less demanding audio is compressed more and generates lower bitrates than more complex sounds do. This enables delivery of a higher-quality stream than a CBR file of the same size. When using VBR, a target bitrate can be set so that the levels of compression fluctuate close to the desired bitrate. The downside of using VBR is that the encoding time may be longer.

7.7.3 Software audio codecs

Some of the audio codecs used in network video today:

- *AAC-LC:* requires a license and supports either CBR or VBR.
- *G.711:* non-licensed technology, only supports CBR.
- *G.726:* non-licensed technology, only supports CBR.
- *LPCM:* non-licensed technology supporting CBR and VBR.
- *Opus:* non-licensed technology supporting CBR and VBR.
- *G.722.2:* licensed technology supporting CBR and VBR.

7.7.3.1 AAC-LC

AAC-LC (Advanced Audio Coding – Low Complexity) has the official name MPEG-4 AAC. It includes four different profiles, where LC profile is the least complicated and the most widely used. AAC-LC requires a license for encoding and decoding.

AAC offers sampling rates at 8–96 kHz and bitrates ranging from as low as 2 kbit/s for low-bitrate speech encoding to more than 300 kbit/s for high-quality audio coding. It supports constant and variable bitrate modes.

If achieving the best possible audio quality is a priority, AAC (or OPUS) is recommended, particularly at a sampling rate of 16 kHz or higher and a bitrate of 64 kbit/s. If a device does not offer a sample rate of 16 kHz or higher, then AAC with a sample rate of 8 kHz at a bitrate of 24 or 32 kbit/s is recommended.

AAC-LC was developed by a group of companies including Dolby, Fraunhofer IIS, Sony, and AT&T and has been part of the MPEG standard since 1997. It is specified as Part 7 of the MPEG-2 standard and as Part 3 of the MPEG-4 standard. AAC-LC is used widely in the video surveillance industry, although other standards are also used. For more information on the MPEG group, see Chapter 6 about compression technologies.

7.7.3.2 G.711 PCM

G.711 PCM (pulse-code modulation) is an unlicensed speech codec from the ITU's Telecommunication Standardization Sector (ITU-T). It has lower delay and requires less computing power than AAC-LC. It was developed in 1972 as a telephony standard. All IP telephony and VoIP (Voice over IP) manufacturers support this standard, which makes it very useful when integrating audio into a VoIP system. G.711 PCM has a sampling rate of 8 kHz and a bitrate of 64 kbit/s. It has a rather poor quality-to-bitrate ratio but is still used in some systems.

With G.711, it is important that the client also uses µ-law compression – a technique that takes 14-bit signed linear audio, increases the magnitude by 32, and converts it to an 8-bit value.

7.7.3.3 G.726 ADPCM

G.726 ADPCM (adaptive differential pulse-code modulation) is an unlicensed speech codec from the ITU-T. G.726 is a low-power and low-cost implementation standard with low latency. Like G.711, G-726 has lower delay and requires less computing power than AAC-LC. It has a sampling rate of 8 kHz and bitrates of 16, 24, 32, and 40 kbit/s. The most commonly used bitrate is 32 kbit/s. It is commonly used within the security industry but not widely used elsewhere. G.726 ADPCM was introduced in 1990.

7.7.3.4 LPCM

LPCM (linear pulse-code modulation) is encoded but uncompressed audio, as used on, for example, music CDs. As file sizes are generally too large for regular use in most situations, LPCM is most useful in cases that require the best possible sound or when compression is not recommended, such as when using recorded sound as forensic evidence. LPCM uses, for example, a sampling rate of 44.1 kHz for audio CDs and 48 kHz for DVD audio. The bitrate is the sampling frequency multiplied by the bit depth, so for LPCM24 at 48 kHz, it will be 24 × 48k = 1152000 bps (1152 kbps). G.722.2 (AMR-WB)

G.722.2 or AMR-WB (adaptive multi-rate wideband) is a licensed speech codec from the ITU-T. It has a sampling rate of 16 kHz and offers bitrates ranging from 7.60 to 23.85 kbit/s. This is the codec used in networks such as UMTS, a 3G mobile phone technology. G.722.2 offers good speech performance at rates of 12.65 kbit/s and higher.

7.7.3.5 Opus

The unlicensed Opus codec was developed by the Xiph.org Foundation. It is designed for efficient speech and music transmission over the internet and has been standardized by the Internet Engineering Task Force (IETF). Opus offers sampling rates from 8 kHz to 48 kHz and supports bitrates up to 510 kbit/s.

Opus combines and switches between the SILK algorithm (from Skype) and the lower-latency CELT algorithm, using them as required for maximum efficiency. Opus can be used for real-time communication links, networked music performances, and the like. By juggling quality and bitrate, delay can be reduced to 5 ms, which is considered exceptionally low.

7.8 AUDIO AND VIDEO SYNCHRONIZATION

A media player (software used for playing back multimedia files) or a multimedia framework (such as Microsoft® DirectX®, a collection of application programming interfaces that handle multimedia files) manages the synchronization of audio and video data. Audio and video are two separate packet streams that are sent over a network. For the client or player to synchronize the audio and video streams perfectly, the audio and video packets must be time-stamped.

Some network cameras may not support time-stamping of video packets when using Motion JPEG compression. If this is the case and if synchronized audio and video is critical, it is better to use H.264 or H.265 compression because such video streams are usually sent using RTP (Real-time Transport Protocol), which timestamps the video packets.

However, there are many situations where synchronized audio is less important, such as when audio is monitored but not recorded or when used for audio detection with audio analytics.

7.9 THE FUTURE OF AUDIO IN NETWORK VIDEO

Audio is already a key factor when selecting a video surveillance solution and will likely be more so in new installations. When used together, audio and video offer a more complete monitoring solution.

Many network cameras support dual audio codecs, just as multiple video compression formats (H.264/H.265 and Motion JPEG) can be supported simultaneously. This enables users to take advantage of the different strengths of the codecs and apply them for different purposes, for example, one codec for recording and another for communications.

Improved functionalities will likely lie in the following areas: audio-video synchronization, audio analytics (real-time or post-event), the ability to identify certain sounds such as speech and breaking glass, and stereo/multi-channel audio for spatial awareness.

7.10 OTHER AUDIO DEVICES IN NETWORK VIDEO SYSTEMS

As audio continues to be a more important complement to network video and physical security systems, IP-based audio solutions are now widely available to augment the systems. Some of those products are network speakers and intercoms, which are often based on the SIP protocol.

7.10.1 Network speakers

Just like a network camera, a network speaker (see Figure 7.16) connects directly to the network, is powered by Power over Ethernet, and includes local intelligence. The benefits are ease of installation, ease of integration, as well as built-in intelligence, such as automatic monitoring of functionality, the ability to play pre-recorded messages and to communicate with a phone or cell phone over the SIP protocol (see section 7.10.5).

There are several different models of network speaker available, the one to use depends greatly on your application. For deterrent purposes, a horn speaker or sound projector can be used. Other models, for announcements and music, can be wall-mounted, fitted in recessed ceilings, or fixed to pendants.

7.10.2 Audio management software

Much like the tools available for managing multiple network video devices, there are also dedicated management tools for network audio systems. Some simpler applications for managing smaller

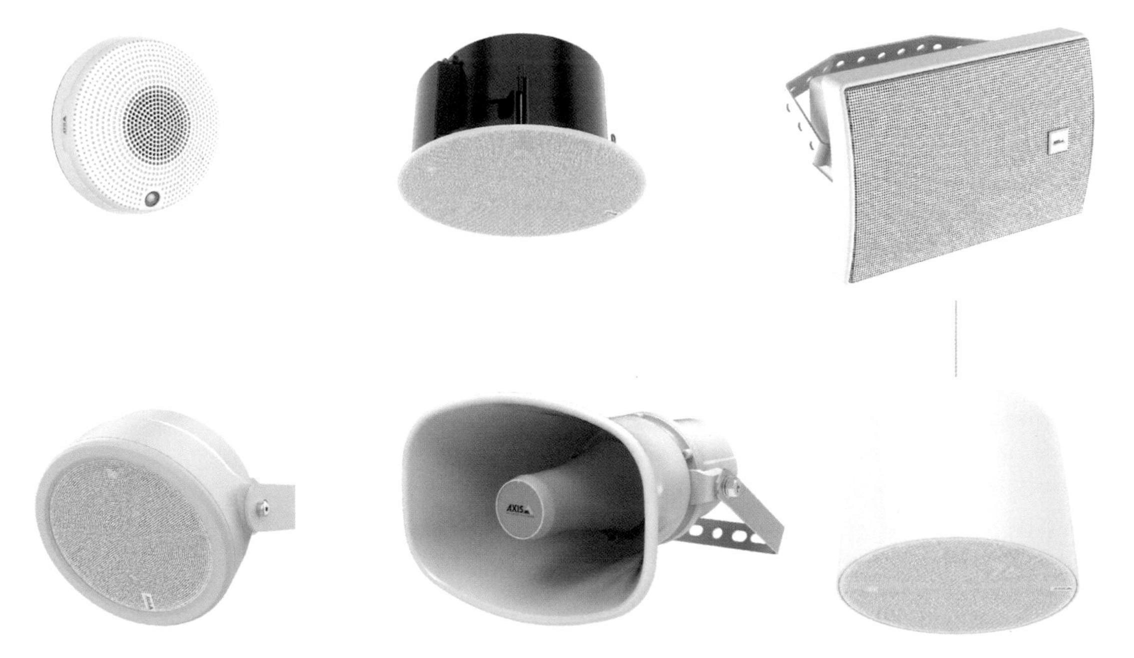

Figure 7.16 Examples of network speakers. Upper row: left: mini speaker, center: ceiling speaker, right: cabinet speaker. Lower row: left: sound projector, center: horn speaker, right: pendant speaker.

systems may be installed directly in the audio devices and be available via a standard browser. More advanced tools for larger and more complicated systems may need to be installed on a computer instead.

The functions provided by these management tools vary, but they often include the following:

- *Content management and scheduling:* This refers to the mixing and planning of live and pre-recorded announcements, commercial messages, and background music. You can make announcements, and play music in multiple so-called zones, through hundreds or thousands of speakers. Depending on the tool, you can plan and schedule the same or different content for the same or different zones. The more advanced the tool, the greater the complexity and flexibility available. It is also possible to prioritize content so that live messages always take priority over scheduled material to ensure that critical information such as emergency announcements and paging is always delivered promptly.
- *Zone management:* A zone can be made up of just a single speaker, or it might contain 5,000 speakers spread throughout an entire building. Zones can be addressed individually or all at once, meaning, for example, that the same background music can be played throughout the whole system, or alternatively, you can play different music in different zones. One of the main benefits of IP audio systems is that grouping and configuring zones is very easy, while analog systems require expensive rewiring and reprogramming.
- *Device management and health monitoring:* A central part of any audio management tool is the configuration and setup of the devices and how they interact with each other. As well as installation and initial configuration, there is also functionality to, for example, manage device firmware updates. Some tools also include system health monitoring so you can be sure that all components are working correctly. For example, a speaker can play a test signal and listen to itself via its built-in microphone to ensure the speaker is working properly, thus removing the need to send staff out to check.

If the audio management tool builds upon open protocols, it will be easy to integrate it with other systems. For example, as the audio system is connected to the network and supports SIP, this means you can use a Voice over Internet Protocol (VoIP) phone system to make live announcements in the system from any phone in your setup.

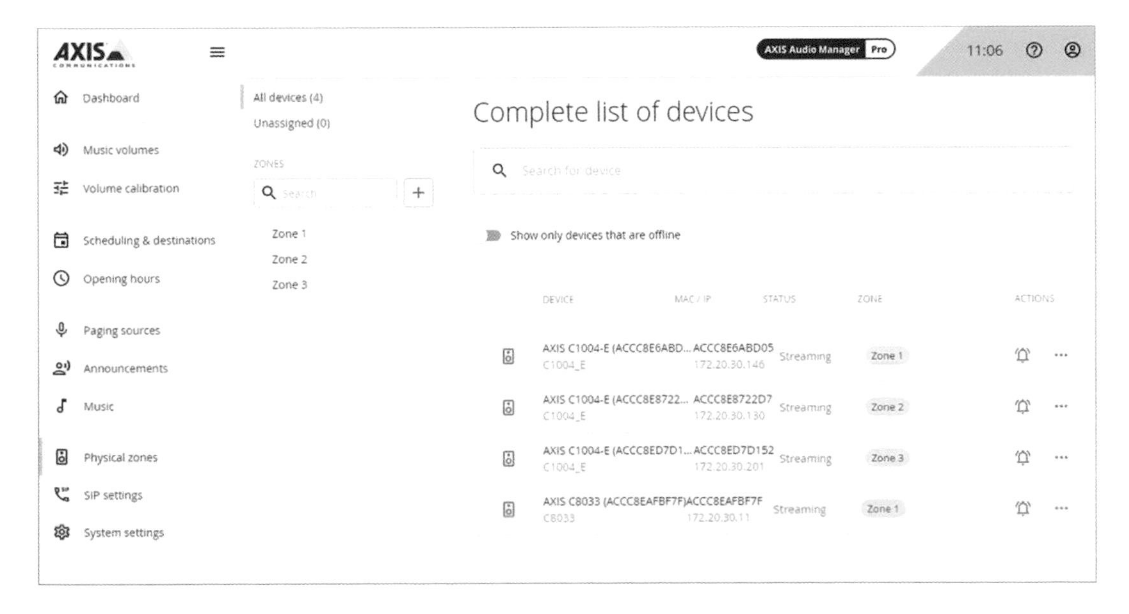

Figure 7.17 A screenshot from an audio management tool.

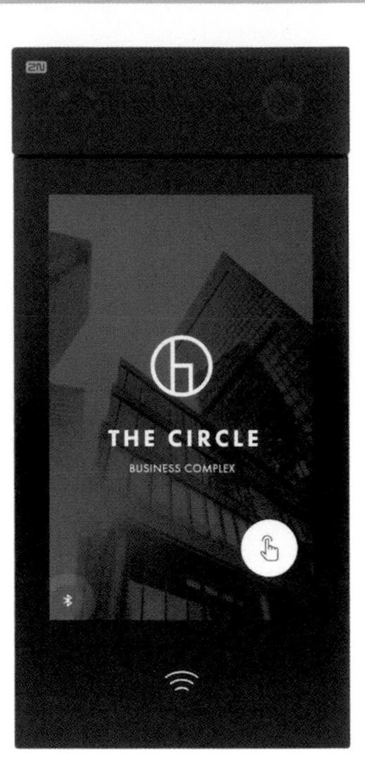

Figure 7.18 A network intercom.

7.10.3 Network intercom

A network intercom (see Figure 7.18) is basically an IP-based camera with intercom functionality. It is used for two-way communication, video identification, and remote entry control. Simply put, a network intercom is a virtual door attendant or receptionist. Persons on the inside can see and talk to others on the outside while they remain at their desk or are on the move. Then through the same interface, they can choose different actions, such as unlocking the door, starting a recording, or even locking down the whole building. This is a perfect complement to any surveillance system, as it offers effective access control and is as easy to install as any network camera.

See Figure 7.19 for an example of a surveillance system with network intercoms and Figure 7.20 for a scenario where intercoms help retail staff work more effectively.

To make calls, a network intercom uses Voice over IP (VoIP), which is a group of technologies that enable voice communication and multimedia sessions over IP networks. A number of different proprietary and open-source protocols can be used to implement VoIP. One of the most popular open-source protocols is SIP. For more information about SIP, see section 7.10.5.

The use of IP standards and the open interface makes it easy to integrate a network intercom in smaller installations as well as more advanced enterprise systems. It connects to the existing IP network, and thanks to Power over Ethernet, a single network cable is all it takes to power both the intercom itself as well as a standard door lock. If the intercom has I/O and relay connectors, it can be connected to other devices, too, for example, a request to exit (REX) device or a safety relay.

7.10.4 Audio and I/O interface module

An audio and I/O interface module enables two-way audio and I/O connectivity for network cameras that lack such functionality. The module uses the camera's IP address to provide seemingly integrated audio and I/O connectivity in the camera, with audio and video provided in a single network stream. The module is installed in-line anywhere between the PoE switch and the camera and allows the ports to be placed where required. The audio functionality is usually wanted close

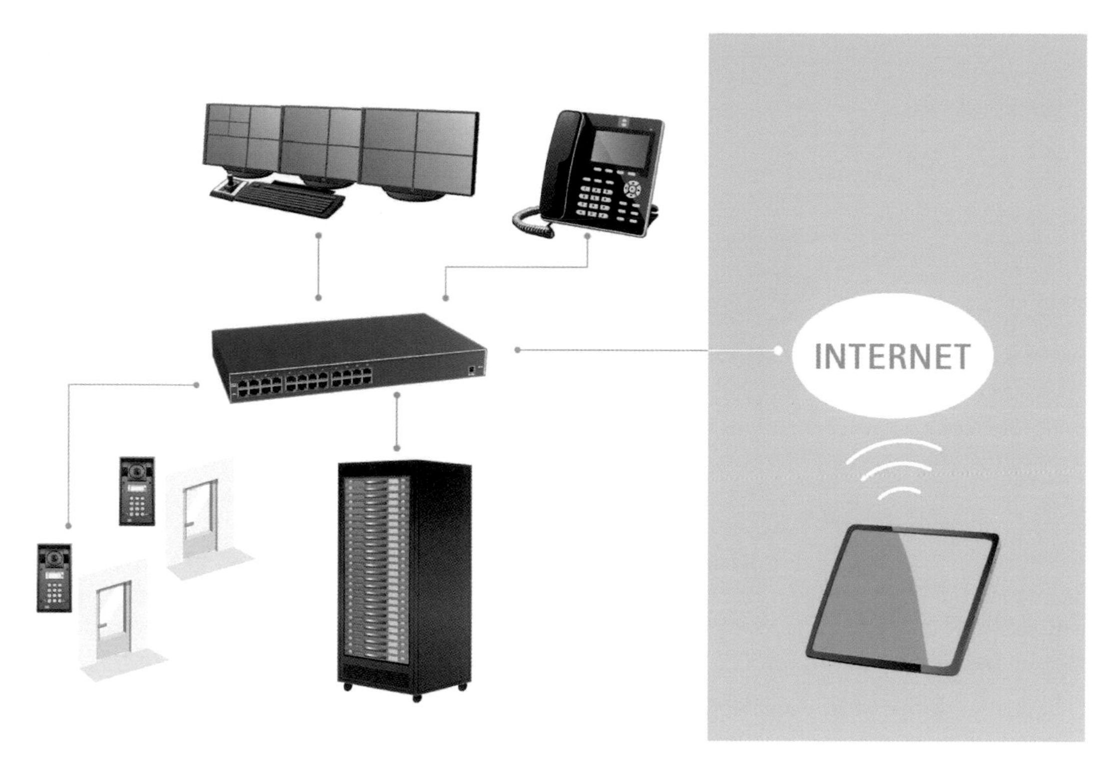

Figure 7.19 Network intercoms can communicate with monitoring stations, IP phones, and smartphones or tablets.

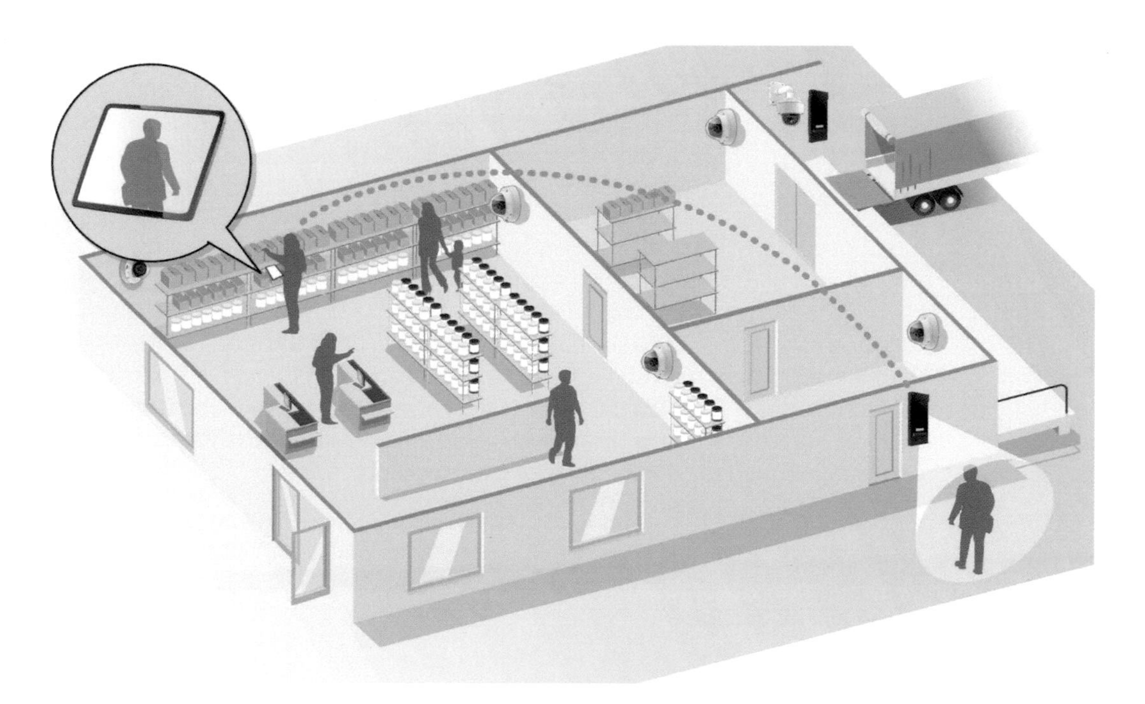

Figure 7.20 Using an IP phone with a screen, a smartphone, or a tablet, the user can identify visitors, such as sales representatives and delivery drivers, and let them in without having to go to the door.

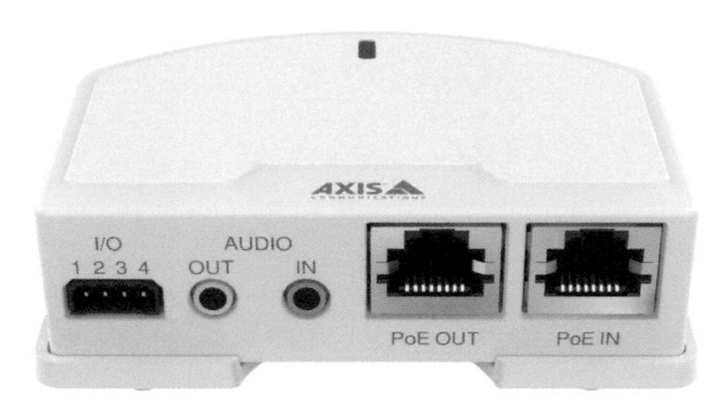

Figure 7.21 An audio and I/O interface module.

to the camera to capture the audio at that location, whereas the I/O functionality might be used for controlling other devices located elsewhere.

7.10.5 The SIP protocol

Session Initiation Protocol (SIP) is a text-based protocol similar to HTTP and SMTP, for communication over IP networks. It is used to start, change, and end media stream sessions, which can include voice and video elements. It has become a widely adopted standard protocol for IP telephony, video conferencing, call control, and instant messaging.

SIP calls can be set up in many ways, but there are three main types:

- *Peer-to-peer calls* (also called local calls)
- *SIP server calls* (also called PBX calls)
- *SIP trunk calls*

Peer-to-peer calls are made between two devices (such as computers, softphones, intercoms, cameras, IP desk phones) that belong to the same network. The call is made to the device's SIP address.

To make SIP server calls, devices must be connected to a SIP server that handles the call exchanges. A SIP server, or a private branch exchange (PBX), is a hub that works like a traditional switchboard. It can be hosted on an intranet or by a third-party service provider. The SIP-enabled devices register with the SIP server and can contact each other through their SIP addresses. A PBX can show call status, allow call transfers, handle voice mail, and redirect calls.

SIP addresses (also known as SIP URIs or SIP numbers) are used to identify users within a network just as you would use a phone number or an email address to contact a friend or colleague. Like email addresses, SIP addresses are a type of uniform resource identifier (URI) that includes two user-specific parts, a user ID or extension, and a domain name or IP address. Together with a prefix and the @ symbol, they make up a unique address. For example, if Caesar of ancient Rome had both an email address and a SIP address, these could be mailto:caesar@ancientrome.it and sip:caesar@ancientrome.it, respectively. In the case of a peer-to-peer call, the SIP address would include the IP address rather than the domain name ancientrome.it.

With a service provider that offers SIP trunking, the traditional telephone network can be used to make calls and traditional phone numbers can be assigned to the SIP devices. This way, calls can be made from a network speaker or a network intercom to a cell phone or the other way around. Often, providers charge extra for this service.

7.11 BEST PRACTICES

There are some things to consider when pursuing ideal audio performance in an installation:

- *Audio equipment and placement:* Select and set up your audio equipment based on context, needs, and environment. The microphone type and placement, polar pattern, as well as cabling and speakers all affect audio quality. Although an audio signal can be amplified later, the appropriate selection and placement of audio equipment helps to reduce noise. The microphone should be placed as close as possible to the source of the sound. In full-duplex mode, a microphone should face away from and be placed some distance from speakers to avoid feedback.

- *Amplify the signal as early as possible:* This minimizes noise in the signal chain. In addition, make sure the signal levels are close to the clipping level but not above it. The clipping level is the signal level at which audio is distorted.
- *Apply appropriate signal processing technologies to improve audio quality:* Audio quality can be improved by adjusting the input gain and using different features, such as echo cancellation and speech filters.
- *Select the right codec and bitrate:* The codec and bitrate affect audio quality. In general, the higher the bitrate, the better the audio quality.
- *Use shielded cables:* To minimize disturbance and noise, always use a shielded audio cable and avoid running the cable near power cables and cables carrying high-frequency switching signals. Keep audio cables as short as possible. If a long audio cable is required, balanced audio equipment (the cable, amplifier, and microphone) should be used to reduce noise.
- *Understand the legal implications:* Some countries restrict the use of audio and video surveillance. It is a good idea to check with the local authorities before investing in a system.

Chapter 8

Video encoders

Video encoders are key pieces of equipment that help analog closed-circuit television (CCTV) systems migrate into an open platform–based network video system. Encoders will continue to play an important role in the video surveillance market because a significant share of surveillance cameras installed worldwide are still analog. As recently as 2022, it was estimated that 50% of all cameras installed in the USA were still analog.

And analog cameras are still being installed today. The average life expectancy of an analog surveillance camera is five to seven years, but some may last even longer. In many installations, the coaxial cable is the most expensive part of the installation, so once installed, there is often limited incentive to re-cable with Ethernet cabling to enable the installation of a network camera. However, the recording device in an analog system – most often a DVR (digital video recorder) but sometimes even an older VCR (videocassette recorder) – usually fails long before the camera does. This is where the video encoder comes into play.

Video encoders allow security managers to keep their analog CCTV cameras while building a video surveillance system that provides the benefits of network video. If a video encoder is included in the system, analog cameras can be controlled and accessed over an IP network, such as a local area network (LAN) or the internet, and old video recording equipment such as DVRs and monitors can be replaced with standard computer monitors and servers (see Figure 8.1).

Analog cameras of all types, such as fixed, indoor, outdoor, fixed dome, and PTZ (pan, tilt, zoom), as well as specialty cameras, such as covert, miniature, and microscope cameras, can all be integrated and controlled in a network video system using video encoders. Many encoders provide support for analog high-definition (AHD) cameras, often referred to as HD-CCTV cameras.

The following sections give an overview of the components of a video encoder and the different types of video encoders available. The type of video encoder used depends on system configuration, camera count, camera types, and whether coax cabling is installed. The chapter concludes with a few best practices for encoders.

8.1 THE COMPONENTS OF A VIDEO ENCODER

A video encoder connects to an analog camera through a coax cable and converts analog video signals into a compressed digital video stream that is transmitted over an IP-based network. The device is called a video encoder because it encodes video using compression standards, such as H.265, H.264, or Motion JPEG. Once the video is on the network, it behaves exactly the same as a video stream from a network camera and is ready to be integrated into a network video system. A video encoder also often includes a serial port, which is commonly used for controlling the PTZ functionality of the connected analog camera.

DOI: 10.4324/9781003412205-8

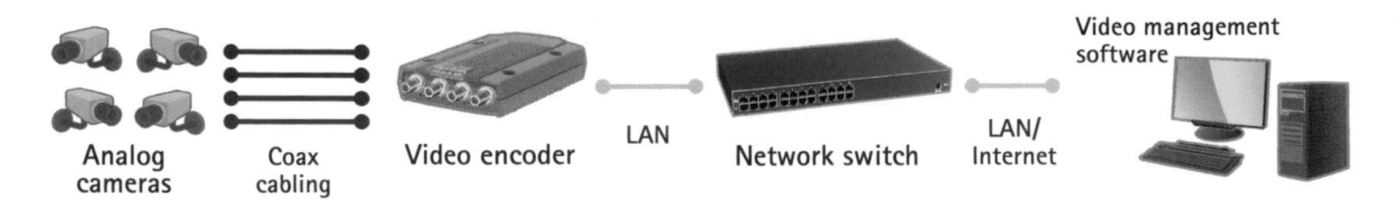

Figure 8.1 Analog cameras connected to a video encoder. This makes it possible to include the camera in a network video system.

A video encoder can also offer many advanced functionalities, such as deinterlacing, video motion detection and other video analytics, alarm handling, one- or two-way audio support, and audio alarms. Third-party applications can also be uploaded to some video encoders to further enhance the system. Encoder models that have input/output ports can connect to external devices. This allows for control of other devices, such as doors and lighting, and external sensors can be used to trigger an alarm event in the encoder.

Many video encoders also include Power over Ethernet (PoE) functionality (see Figure 8.3). This enables the encoder to receive power through the same cable used for data transmission and sometimes also to power the connected analog camera.

PoE can give substantial savings for the entire system because power cables can be omitted. In addition, if the server room is connected to an uninterruptible power supply (UPS), PoE puts the encoders on centralized backup power so they will continue to operate even in the event of a power failure.

The networking functionality is also very important, and it should include all the latest security and IP protocols. The video encoder's processor, which could be a general-purpose processor, a digital signal processor (DSP), or a purpose-built application-specific integrated circuit (ASIC), determines the performance, which is normally measured in frames per second (fps) per channel in the highest resolution. The performance also depends on whether one chip is used per channel or if multiple channels share the same chip. Today, the highest performing encoders can provide multiple individually configured video streams at full–frame rate and still have headroom for running video analytics.

Some of today's video encoders can deliver up to 60 frames per second when connected to standard analog cameras. This is achieved by using powerful processors and new methods of deinterlacing. The benefit of this higher frame rate is smoother video in high-motion scenes.

Many encoders have a memory card slot that enables video to be recorded and stored locally. Local recording can be used in several ways. It could be the primary recording location to avoid the need for a server. It could also be used for redundancy. Even if the network fails, the video is still available because it is stored in the encoder. Lastly, an operator can view live video at a lower resolution or frame rate while the encoder records full-quality video locally. This is a huge benefit in low bandwidth environments because less data needs to be sent over the network.

Video encoders use autosensing to detect the incoming video signal. Historically, these formats were PAL or NTSC signals only. The 75 Ω video termination for the video input can be enabled and disabled. In most cases, the best thing is to only enable termination in the last device in the video signal chain. For more information about PAL and NTSC, see Chapter 4. Some encoders today also support several analog high-definition signals, such as AHD, TVI, and CVI (also known as HD-CCTV). For more on analog HD formats, see section 1.4.2.

Some video encoders are specifically designed for tough conditions and can be used in harsh environments, for example, where temperatures and vibration could be a problem. For long-distance network connectivity, some encoders have SFP (small form-factor pluggable) slots. This makes it possible to plug in an SFP module and connect the encoder to the network using a fiber-optic cable.

Figure 8.2 The front panel of a typical stand-alone video encoder, with four video inputs, two audio inputs, and one audio output.

Figure 8.3 The rear panel of a video encoder, showing (left to right) connectors for PoE, external power, a microSD card, serial ports (RS485/R422), and input/output devices.

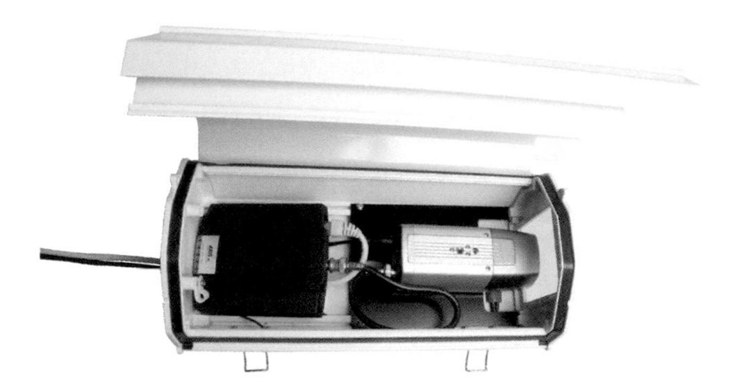

Figure 8.4 A single-port video encoder positioned next to an analog camera in a housing.

8.2 STAND-ALONE VIDEO ENCODERS

The most common video encoders are stand-alone versions that offer one or more (often 4 or 16) ports to analog cameras (Figure 8.2). Multi-port video encoders have better cost efficiency, but performance and flexibility can sometimes be limited. A multi-port encoder is ideal in situations where, for example, there are a few analog cameras located in a remote facility or far from the central monitoring room. The video signals from the remote cameras can then share the same network connection on the video encoder, which dramatically reduces cabling costs.

In situations where investments have been made in analog cameras but coaxial cabling has not yet been installed, it is best to use and position stand-alone video encoders close to the analog cameras (Figure 8.4).

Placing the camera and video encoder together like this reduces installation costs. Because the video can use existing network cabling, there is no need to run new coaxial cables to a central location. It also eliminates the loss in image quality that would occur if video had to travel long distances over coaxial cabling. A video encoder produces digital images, so the distance has no effect on the image quality. But with coaxial cables, the signal strength decreases the further the signals travel.

8.3 RACK-MOUNTED VIDEO ENCODERS

Most companies have a dedicated control room to gather equipment in one location so that operations can be efficiently monitored in a secure environment. In a building containing large numbers of analog cameras, this means that vast amounts of coax cabling run to the control room.

If all the coax cabling has already been installed and is available from the control room, the installation would benefit from using a video encoder chassis with blade video encoders, which are basically video encoders without a casing. A blade video encoder cannot function on its own. It must be mounted in the chassis.

A video encoder chassis allows a great number of video encoder blades to be mounted in a standard 19-inch rack and managed centrally. A video encoder chassis offers functionalities such as an integrated network switch and the hot-swapping of blades, which means that blades can be removed or installed without having to turn off the power. Chassis have different heights depending on the number of slots for encoder blades (see Figure 8.5).

A chassis can provide network, serial communication, and I/O connectors, as well as a common power supply. A chassis is a high-density solution that can support up to 84 channels and save valuable rack space. Power and network redundancy can be used to make a high-density chassis very robust, secure, and reliable.

There is another popular video encoder version that has 16 channels and also fits into a 19-inch rack, although its video channels are mounted in a chassis and not on blades (see Figure 8.6). While this solution lacks some of the redundancy and flexibility of a true rack-based system with blades, these encoders often have a lower cost per channel. Also, the DVRs they are replacing very often have precisely 16 channels, and therefore, the replacement process is very simple.

8.4 VIDEO ENCODERS WITH PTZ CAMERAS

The serial port (RS422/RS485) built into most video encoders is used to control the movement of analog PTZ cameras. In an analog CCTV system, separate serial wiring runs from the control

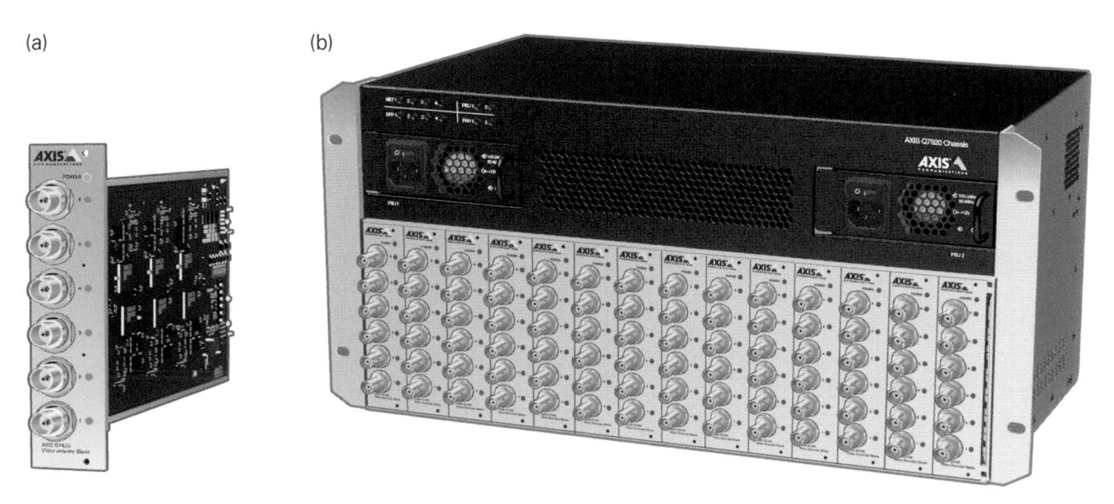

Figure 8.5 (a) A six-channel encoder blade. (b) A video encoder chassis that can support up to 84 channels.

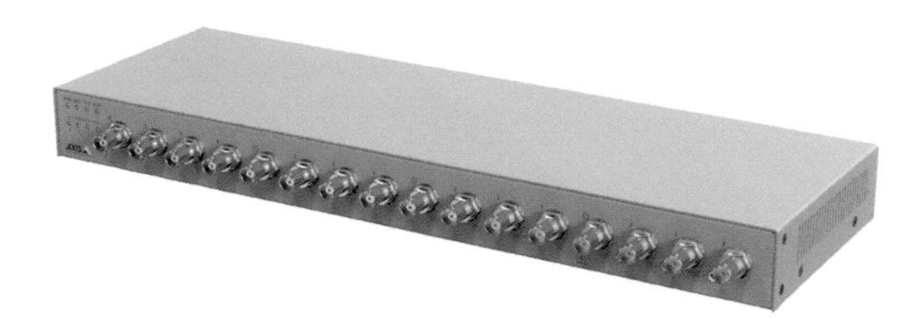

Figure 8.6 A 16-channel video encoder, which can be mounted in a 19-inch rack.

Figure 8.7 Using the encoder's serial port (RS485), an analog PTZ camera can be controlled remotely over an IP network.

board (with joystick and other control buttons) to the PTZ camera. This separate wiring is not needed if a video encoder is placed close to the camera.

In a network video system, commands from the control board are carried by the same cable as the video and are forwarded by the video encoder through the serial port to the PTZ camera. Video encoders, therefore, enable control of PTZ functions over long distances using the internet. To control a specific PTZ camera, a driver must be uploaded to the video encoder. Many manufacturers of video encoders provide PTZ drivers for PTZ cameras. A driver installed on a PC that runs video management software can also be used if the serial port is set up as a serial server that simply passes on the commands.

RS485 is most commonly used for controlling PTZ functions (see Figure 8.7). One of the benefits of RS485 is the possibility to control multiple PTZ cameras using twisted-pair cables in a daisy-chain connection from one camera to the next. The maximum distance of an RS485 cable without using a repeater is 1220 meters (4000 feet).

Some video encoders also allow control of PTZ cameras through the coaxial cable connecting the video encoder and the analog camera. This is sometimes referred to as "up the coax". If a compatible camera is used, no additional cabling is needed for PTZ control, which reduces installation costs.

8.5 VIDEO DECODER

In some installations, there is a need to watch and listen to network video and audio streams on existing equipment. When using a video decoder, the network video and audio streams are converted back to analog signals. These can then be connected to analog monitors and video switches. A typical case is in a retail environment where the user wants to have traditional monitors in public spaces to demonstrate that the area is under video surveillance. A video decoder is used to connect such monitors to a network video stream, coming from either a video encoder or a network camera.

Some video decoders can decode video from several cameras sequentially. This means that the decoder decodes video from one camera for a set period and then automatically switches to the second camera, then the third, and so on. This feature allows a guard to monitor video from

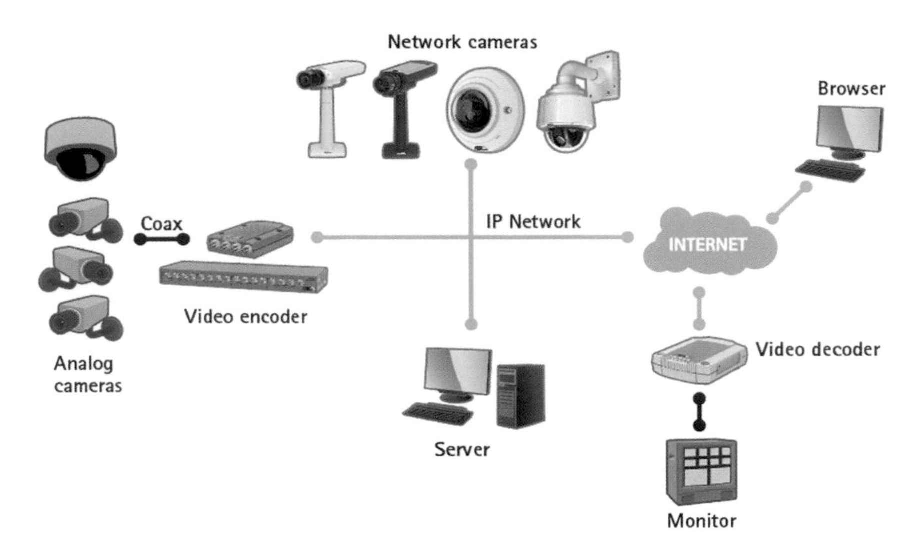

Figure 8.8 With a network video decoder, existing monitors can be used to show video and play audio from remote cameras.

Figure 8.9 An encoder and decoder can be used to transport video over long distances from an analog camera to an analog monitor.

the most important cameras (Figure 8.8). Video decoders can sometimes also show a split view from several cameras on the monitor.

Another common application for video decoders is to use them in an analog-to-digital-to-analog setup for transporting video over long distances (Figure 8.9). Distance does not affect the video quality when images are sent digitally. The downside is that there could be some latency, from 100 milliseconds to a few seconds, depending on the distance and the quality of the network between the endpoints.

8.6 BEST PRACTICES

Video encoders offer a valuable solution to the challenge of migrating analog CCTV video to network video. Video encoders play a significant role, particularly in enterprise installations where there may be large numbers of analog cameras, meaning the installation can be sustained for longer and the investment protected.

It is easy to view video encoders as straightforward pieces of technology that are little more than analog-to-digital converters. However, the demands on video encoders are very high, and there are several considerations to make when selecting a video encoder.

Considerations include the following:

- *Image quality:* Can the video encoder deliver high-quality, deinterlaced digital video at a high frame rate? For more information about deinterlacing, see Chapter 4.
- *Resolution:* What resolutions can the video encoder offer?

- *Compression:* What compression standards does the video encoder support?
- *Performance:* On how many channels can the video encoder deliver full frame rate at full resolution? Can it send multiple simultaneous streams? Can the streams be configured individually?
- *Analytics support:* Does the encoder support video analytics like video motion detection, and does it have a platform for adding analytics from a third-party vendor?
- *Rack solution:* Are there rack-mounted versions of the video encoder available?
- *Density:* How many analog channels per chassis or per Ethernet channel can the system handle?
- *PTZ control:* Does the encoder support PTZ control over the coaxial cable?
- *Onboard storage:* Does the encoder support onboard storage using SD card storage for redundancy?
- *Audio:* Does the surveillance situation require audio support?
- *External devices:* Is it possible to connect external devices to the encoder to create a more intelligent system?
- *AHD cameras:* Does the system include modern analog HD (HD-CCTV) cameras? Be sure to get an encoder that supports the correct AHD format.

Video encoders typically fall into the category of products that no one thinks about until something fails. Consequently, reliability and quality are key criteria for video encoders. These are advanced products that require careful investigation when purchasing.

Chapter 9

Networking technologies

Networking technology has experienced tremendous development over the past decades. Today, billions of people are using networking technology, tens of billions of devices are connected to networks and the internet, and the numbers are growing fast. The technology is quite complex, but most users are not exposed to this. When a user switches on their laptop, a series of networking technologies is automatically initiated to ensure that the laptop gets an IP (Internet Protocol) address and that the network communication is secure. The Internet Protocol is the common denominator in network technology and is used in home networks, enterprise local area network (LANs), and the internet for applications such as email, web browsing, telephony, and network video.

Many different protocols come into play when data is transferred securely from one networked device to another. The best way to understand how the different protocols interact is to study the OSI communication model, as explained in the first section of this chapter.

Another topic that is becoming increasingly important as network use continues to grow is network security – often referred to as cybersecurity. This is discussed in detail in Chapter 12.

9.1 THE OSI REFERENCE MODEL

Data communication between open systems is described using the open systems interconnection (OSI) reference model, which is composed of seven layers (Figure 9.1). Each layer provides specific services and makes the results available to the next layer. To provide a service, each layer utilizes the services of the layer immediately below it. Communication between layers occurs via specific interfaces. Each layer must follow certain rules, known as protocols, to perform its services. It is important to note that the OSI is not itself a protocol but rather a model used to understand the function of the included protocols.

Communication between two systems always occurs on the same layer. This is known as virtual communication, which can only occur when the same protocols are implemented in the corresponding layers of each system. A system passes data down to the lowest vertical layer (the physical) and then transfers it to the other system, in which the data is passed up again. In this way, the data reaches the corresponding layer on the other system and communication takes place.

9.1.1 Layer 1: The physical layer

The physical layer is the lowest layer, and it provides services that support the transmission of data as a bitstream over a medium, such as a wired or wireless transmission link. This layer describes the transmission medium and its physical characteristics, as well as the mechanical and electrical means that permit a physical connection to the transmission medium. This layer also defines the

DOI: 10.4324/9781003412205-9

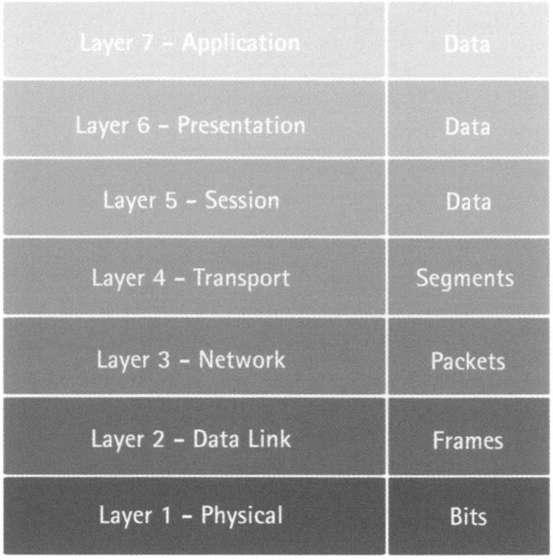

Layer	Data unit
Layer 7 - Application	Data
Layer 6 - Presentation	Data
Layer 5 - Session	Data
Layer 4 - Transport	Segments
Layer 3 - Network	Packets
Layer 2 - Data Link	Frames
Layer 1 - Physical	Bits

Figure 9.1 The OSI reference model consists of seven layers, each one performing a particular service.

form for electrical signals, optical signals, or electromagnetic waves, as well as the connectors and sockets needed for a network cable connection.

9.1.2 Layer 2: The data-link layer

The data-link layer provides data transmission and controls access to the transmission medium, by combining data into units known as frames. These frames are provided with a checksum, which the recipient uses to detect possible transmission errors. According to the Institute of Electrical and Electronics Engineers (IEEE), the second layer is divided into two sublayers, with the upper range corresponding to the Logical Link Control (LLC) and the lower part corresponding to the media access control (MAC). The LLC is used in the same way by all IEEE network technologies and simplifies data exchange. The MAC controls access to the transmission medium and depends on the network technology used.

Examples of protocols and standards are IEEE 802.2 (LLC), IEEE 802.3 (Ethernet MAC), and 802.11 (WLAN MAC).

9.1.3 Layer 3: The network layer

The network layer performs the actual data transfer by routing and forwarding data packets between systems. The most important tasks of this layer include the creation and administration of routing tables, and it provides options for communicating beyond network boundaries. The data in this layer is assigned destination and source addresses, which are used as the basis for targeted routing. Addressing in the third layer is independent of addressing at lower levels, which means that routing can extend to multiple logically structured networks.

Examples of protocols that operate in this layer are the Internet Protocol (IPv4 and IPv6), Routing Information Protocol (RIP), and Internet Protocol Security (IPSec).

9.1.4 Layer 4: The transport layer

The function of the transport layer is to provide reliable data transfer service to Layer 5 and above. The transport layer controls the reliability of a wired or wireless link through flow control and error control. Some protocols that perform in this layer are oriented towards state and connection, meaning that they can keep track of segments and re-transmit those that fail.

Examples of protocols in this layer are the Transmission Control Protocol (TCP) and the User Datagram Protocol (UDP).

9.1.5 Layer 5: The session layer

The session layer provides an application-oriented service and takes care of the process communication between two systems. Process communication begins with the establishment of a session, which provides the basis for a virtual connection between two systems.

Examples of protocols that provide Layer 5 functions include Remote Procedure Call (RPC) and Network File System (NFS).

9.1.6 Layer 6: The presentation layer

The presentation layer converts system-dependent data formats, such as ASCII, into an independent format, thus permitting syntactically correct data exchange between different systems. The tasks of this layer also include data compression and encryption. The presentation layer ensures that data sent by the application layer of a system can be read by the application layer of another system.

Examples of protocols that provide Layer 6 functions include Telnet and Apple Filing Protocol (AFP).

9.1.7 Layer 7: The application layer

The application layer is the highest layer of the OSI model. It makes functions such as web, file, and email transfer available to applications. The actual applications exist above this layer and are not covered by the OSI model.

Typical protocols in this layer are the File Transfer Protocol (FTP), Simple Mail Transfer Protocol (SMTP), and HyperText Transfer Protocol (HTTP).

9.2 THE TCP/IP REFERENCE MODEL

The TCP/IP reference model can be used to understand protocols and how communication takes place. In this model, the protocols fall into four different layers, which correspond to the seven layers in the OSI model as shown in Figure 9.2.

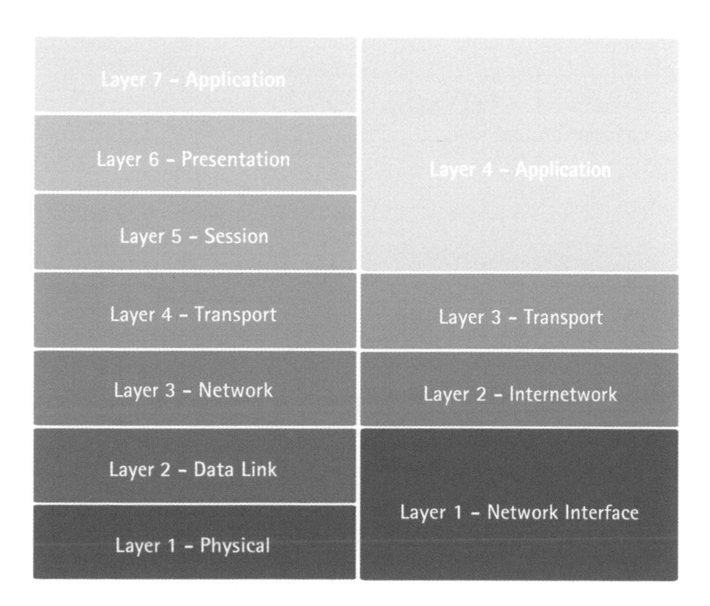

Figure 9.2 The TCP/IP reference model can be split into four layers (right), comparable to the seven layers in the OSI model (left).

9.2.1 The Internet Protocol (IP)

The Internet Protocol (IP) is a Layer 3 protocol in the OSI model and a Layer 2 protocol in the TCP/IP reference model. It is the key protocol in most of today's networked applications.

For a server to operate on the internet, it must have its own individual public IP address. The Internet Corporation of Assigned Names and Numbers (ICANN) is a non-profit global organization that coordinates the maintenance and procedures of databases related to the namespaces and numerical spaces of the internet. An IP address can be assigned by an internet service provider (ISP), which can allocate either a dynamic IP address, which can change from session to session, or a static address, which is normally provided for a monthly fee.

IP offers a connectionless service that splits data into IP datagrams before they are transmitted. *Connectionless* means that IP does not guarantee whether or in what sequence IP datagrams will arrive at a recipient. Because IP operates on a connectionless basis, individual IP datagrams can be routed to a recipient via different paths. Guaranteed data transmission and reassembly into the correct sequence are performed by protocols in the transport layer of the OSI and TCP/IP models.

There are currently two versions of the Internet Protocol: IP version 4 (IPv4) and IP version 6 (IPv6). IPv6 is slowly replacing IPv4, but this older version still provides the vast majority of the service today. The significant difference between the two IP versions is the size of the address space, which is substantially larger in IPv6.

9.2.2 IPv4 addresses

An IPv4 address is 32 bits long, meaning that 2^{32}, or 4.3 billion (4,294,967,296), unique IP addresses can be assigned. To make them more readable, IP addresses are grouped into four blocks of 1 byte (8 bits) each. The individual blocks (referred to as octets) are separated by a dot, and each block represents a decimal value between 0 and 255 (for example, 85.235.16.37). This is known as dot-decimal notation.

Each IP address consists of a network ID and a host ID. The network ID represents the logical IP network in which the host resides. The host ID represents the host itself. All devices with the same network ID will reside on the same network segment, meaning that these hosts can communicate directly with each other without the use of a router to forward the traffic to the correct network.

IP addresses can be classified in five groups, referred to as classes (Table 9.1). These classes are based on the number range of the first octet of the IP address. Each class defines how many networks and host addresses it has available. IP addresses used in LANs and WANs come from classes A, B, and C, with class C being used in most applications. Classes D and E are for multicast and experimental uses.

Note, however, that the class-based addressing model of IPv4 is not commonly used today. Instead, Classless Inter-Domain Routing (CIDR) is used, which permits a more efficient use of the limited supply of IPv4 addresses.

Table 9.1 Classes of IP addresses

Class	Value range of first byte	Bytes for net ID	Number of networks	Bytes for host ID	Number of hosts
A	1–126	1	126	3	16,777,214
B	128–191	2	16,384	2	65,534
C	192–223	3	2,097,152	1	254
D	224–239	Multicast addresses	N/A	N/A	N/A
E	240–254	Reserved	N/A	N/A	N/A

Table 9.2 Permitted values for subnet mask bits

128	64	32	16	8	4	2	1	–
1	0	0	0	0	0	0	0	128
1	1	0	0	0	0	0	0	192
1	1	1	0	0	0	0	0	224
1	1	1	1	0	0	0	0	240
1	1	1	1	1	0	0	0	248
1	1	1	1	1	1	0	0	252
1	1	1	1	1	1	1	0	254
1	1	1	1	1	1	1	1	255

Table 9.3 Examples of IP addresses broken down into network ID and host ID

	Example 1	Example 2
IP address	192.168.1.5	10.50.88.129
Subnet mask	255.255.255.0	255.255.0.0
Network ID	192.168.1.0	10.50.0.0
Host ID	0.0.0.5	0.0.88.129

9.2.3 Subnets

An IPv4 address is divided into two parts: the network ID and the host ID. The subnet mask determines the blocks of an IPv4 address that define the network and host identifiers. A subnet mask has a length of 32 bits and is also represented by dot-decimal notation (for example, 255.255.255.0).

When determining the network and host IDs within an IP address, the computer converts the IP address and subnet mask into binary. Table 9.2 shows the binary-to-decimal numbering conversions.

When viewed as a binary number, the bits of the subnet mask are a contiguous group of 1s followed by a contiguous group of 0s. The bits of an IP address that correspond to the group of ones in the subnet mask represent the network ID. The bits of an IP address that correspond to the group of zeros in the subnet mask represent the host ID. Only subnet masks with contiguous bits are allowed.

For example, a permitted subnet mask would be the following:

255.255.192.00 (11111111.11111111.11000000.00000000)

255.255.255.0 (11111111.11111111.11111111.00000000)

The following subnet mask contains non-contiguous bits and would not be permitted:

255.255.230.0 (11111111.11111111.11100110.00000000)

Table 9.3 provides some examples of IP addresses broken down into the network and host ID. In Example 1, the subnet mask indicates that the first three blocks of digits in the IP address define the network ID and the last block defines the host ID. In Example 2, the subnet mask indicates that the first two blocks define the network ID, whereas the last two blocks define the host ID.

If a host wants to transmit data, it determines the network ID of the destination IP address by means of the subnet mask. If the destination IP address has the same network ID as the host, then the host sends the data directly to the destination.

If the network ID of the destination IP address is located on another subnet, the host instead sends the data to the default gateway as defined in the host. The default gateway is the IP address of a router that forwards data to the correct network. There is an alternative notation based on IPv4-CIDR

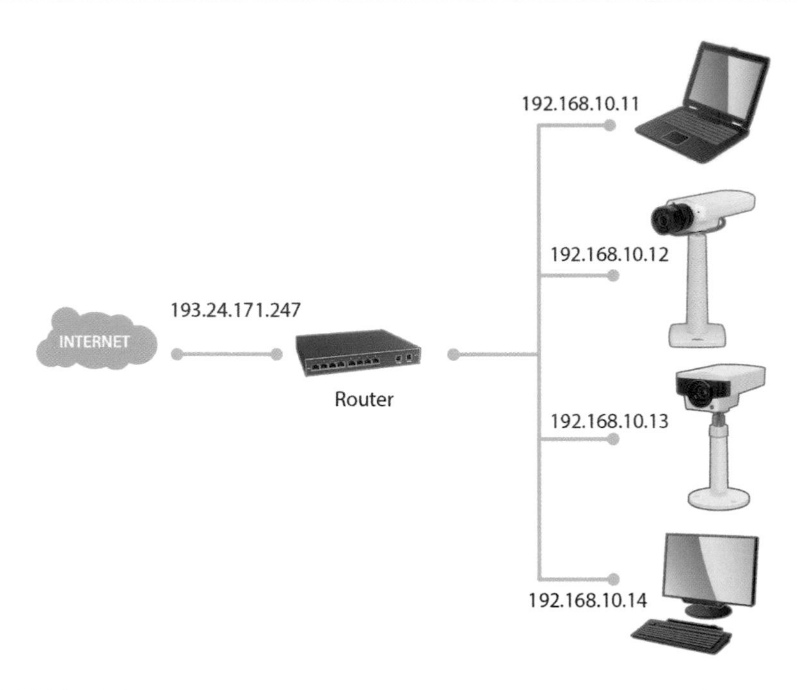

Figure 9.3 Multiple devices can connect to the internet through a unique public IP address using a NAT-enabled router.

(Classless Inter-Domain Routing) for representing the IP address and subnet mask, and it uses a suffix to indicate the length of the subnet mask (the number of 1s). The number of bits used for the subnet mask is appended to the IPv4 address as a decimal number followed by "/". For example, 192.168.12.23/24 corresponds to the IP address 192.168.12.23 with the subnet mask 255.255.255.0.

9.2.4 Network Address Translation (NAT)

Three IP address ranges are defined for use exclusively for private purposes. These are the following:

- 10.0.0.0 to 10.255.255.255
- 172.16.0.0 to 172.31.255.255
- 192.168.0.0 to 192.168.255.255

These addresses can be used only on private networks and are not permitted to traverse a router to the internet. If internet connectivity is required, this can be provided using a technique called Network Address Translation (NAT), which operates on Layers 3 and 4 of the OSI model.

In such a case, IP datagrams are forwarded by a NAT-enabled router that translates the private IP address into a public IP address, as allocated by the internet provider. Larger private networks with multiple servers also can use the same re-addressing technique. The router manages the connection between the private network and the internet (Figure 9.3).

9.2.5 Services and port numbers

Data sent from one system to another must be associated with a particular service or application so that the recipient knows how to process it. This is done by associating the data with a service or process defined by, or mapped to, a particular port number on a server. For example, a service that runs on a web server is typically mapped to port 80 on a computer (Figure 9.4).

A port number is 16 bits long, that is, it can have a value from 0 to 65535. Certain applications use port numbers preassigned to them by the Internet Assigned Numbers Authority (IANA). These numbers are in the range 0–1023 and are called well-known registered ports. The ports 1024–49151 are known

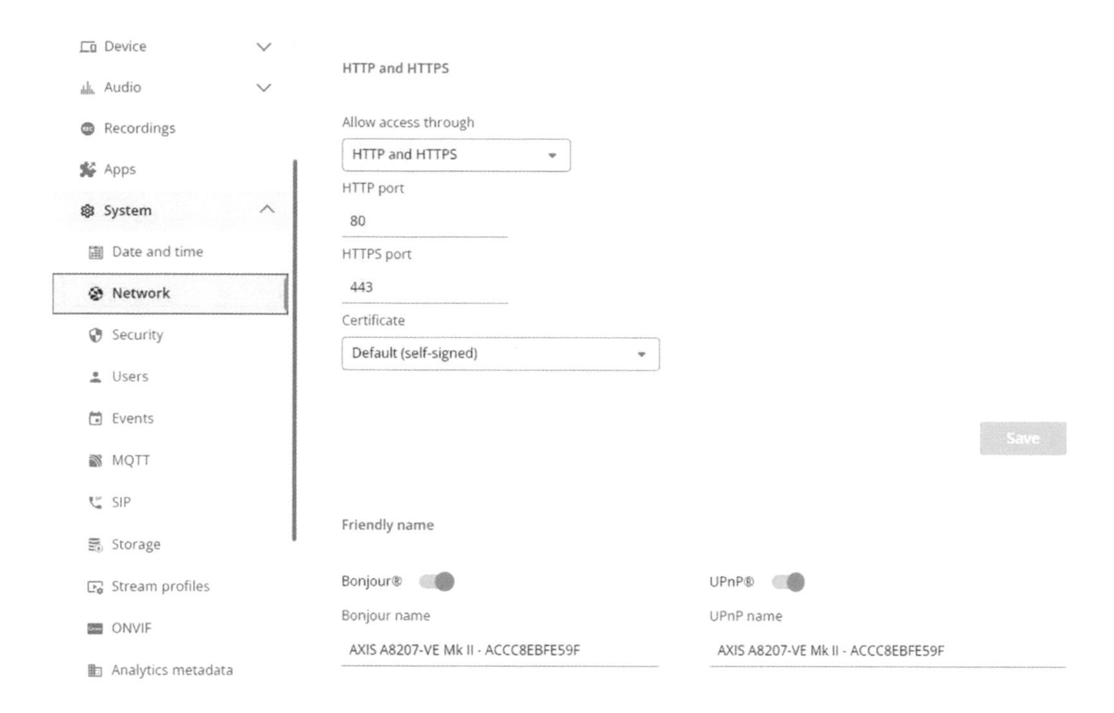

Figure 9.4 In many network cameras, the port number can be set as required. The port is normally set to port 80 if the camera is accessed as a web service over HTTP.

as registered ports. When requested to do so, manufacturers of applications can register ports from this range for their own protocols. The remaining ports (49152–65535) are called private ports and can be used freely as they are not registered and not assigned to any particular application.

Many network video products permit reconfiguration of the port numbers of individual services. For example, the port number of the web server service on network cameras can be changed from port 80 to a private port. Using a private port also means it will be more complex to access the camera through a browser, which may actually be desirable in some applications.

9.2.6 Port forwarding

To configure access from the internet to cameras located on a private LAN, one possible technique is port forwarding, which is available in most routers today.

In port forwarding (see Figure 9.5), incoming data packets reach the NAT-enabled router by way of a public IP address. The router is configured to forward any data arriving at a predefined port to a specific host on the private network side of the router. The router then replaces the address of the incoming packet with a private IP address. To the recipient, it looks like the packets originated from the router. The reverse happens with outgoing data packets. The router replaces the private IP address of the source device with a public IP address before sending the data over the internet.

To access a camera over the internet, the public IP address of the router should be used together with the corresponding port number of the device on the private network. In the system illustrated in Figure 9.5, typing the URL http://193.24.171.247:8032 in a browser would give the user access to a device with a private IP address of 192.168.10.13 Port 8032.

9.2.7 IPv6

For some time, the shortage of IPv4 addresses was managed by using techniques such as NAT. Now, however, these addresses are no longer sufficient. Fortunately, IPv6 offers a solution to the problem, as it provides a huge number of IP addresses.

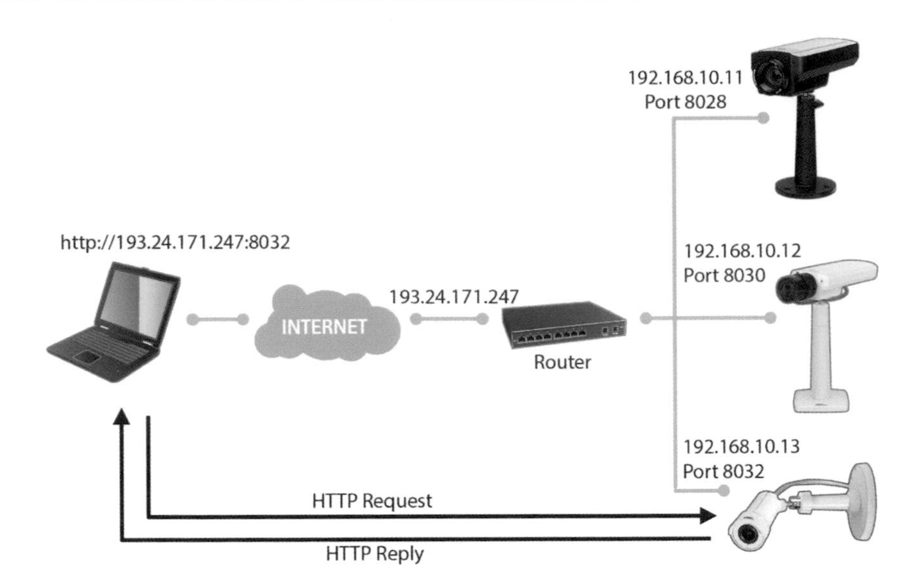

Figure 9.5 Thanks to port forwarding, cameras on a local network can be addressed individually over the internet.

Other major advantages of IPv6 include IP auto-configuration based on the MAC address, renumbering to simplify switching entire corporate networks between providers, faster routing, point-to-point encryption according to IPSec, and connectivity using the same IP address in changing networks (Mobile IPv6).

9.2.8 IPv6 addresses

An IPv6 address has a length of 128 bits, which means that it offers 2^{128}, or approximately 340.28 sextillion (3.4×10^{38}) addresses. An IPv6 address is written in hexadecimal notation with colons dividing the address into eight blocks of 16 bits each. An example of an IPv6 address:

2001:0da8:65b4:05d3:1315:7c1f:0461:7847

To simplify how the address is represented, two consecutive colons can replace one or more 16-bit groups with the value 0000. However, the resulting address may contain two consecutive colons once only.

These two addresses are equivalent to each other:

2002:0da8::1315:7c7a

2002:0da8:0000:0000:0000:0000:1315:7c7a

Leading zeroes in a 16-bit group also can be omitted; for example, 2002:0da8::0017:000c can be written as 2002:da8::17:c.

Address ranges in IPv6 are indicated by prefixes; subnets, too, are determined by the prefix. The prefix length (number of bits) is appended to the IPv6 address as a decimal number followed by "/" (forward slash). Subnet masks as used in IPv4 do not exist in IPv6 – instead, a notation similar to IPv4-CIDR is used.

In IPv6, the first 64 bits of the address are usually intended for network addressing, and the last 64 bits are for host addressing. For example, if a device has the following IPv6 address:

2002:0da8:67f3:08a4:1511:aa56:0361:7a4f

then the device comes from the subnet:

2002:0da8:67f3:08a4::/64

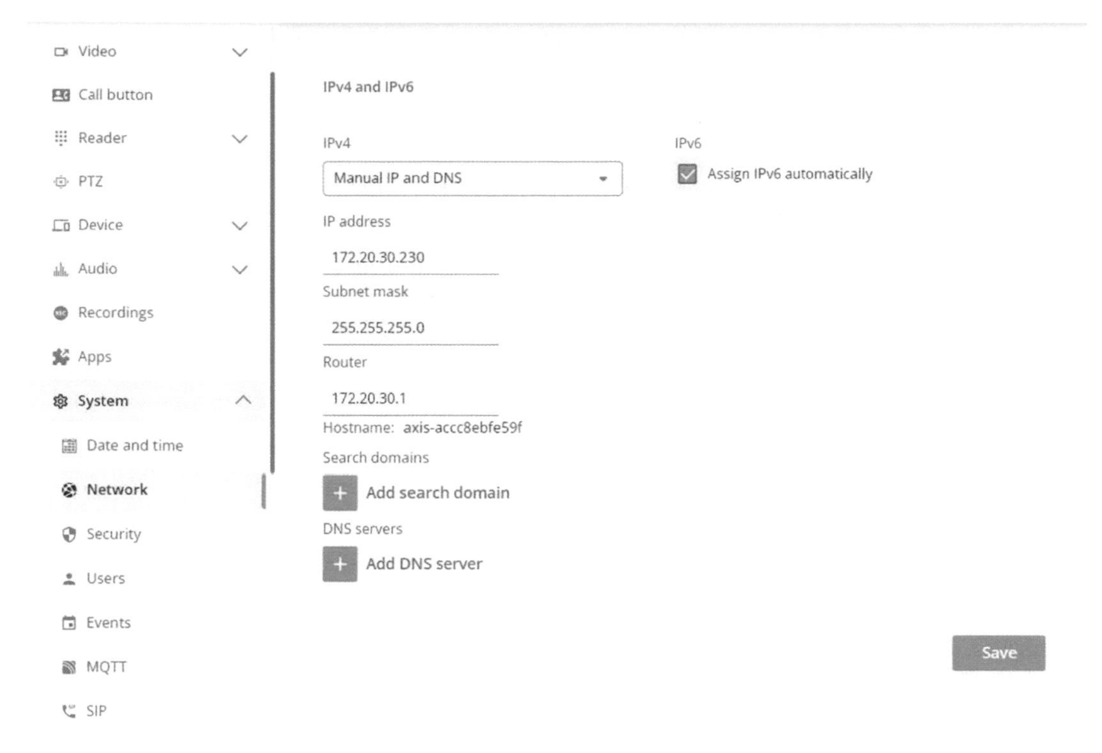

Figure 9.6 Many network cameras have support for IPv6, which is becoming increasingly important in network video applications.

IPv6 enables a device to automatically configure its IP address using the MAC address. In these cases, the prefix – the first 64 bits – is always the same; fe80 and the remaining 48 bits correspond to zeroes (fe80:0000:0000:0000). For the remaining 64 bits (suffix) of the IPv6 address, the MAC address of the system is converted into the Extended Unique Identifier-64 (EUI-64) numbering system. The result would be the following:

fe80::1511:aa56:0361:7a4f

The MAC-based address permits a networked device to communicate on the local network. However, for communication over the internet, the first 64 bits of the IPv6 address must be adapted to the network address of the router, as allocated by the ISP. To do this, a device sends the router a corresponding host request and receives the necessary prefix of the public address block and additional information from the router. Using this information, the device can create the IPv6 address from the prefix and its suffix (EUI-64 address). Services such as DHCP for IP address allocation and tasks such as the manual configuration of IP addresses are not required in IPv6.

The IPv6 address is enclosed in square brackets in a URL. An example of a correct URL is:

http://[2002:0da8:67f3:08a4:1511:aa56:0361:7a4f]/

A specific port also can be addressed by changing the address as follows:

http://[2002:0da8:67f3:08a4:1511:aa56:0361:7a4f]:8081/

9.3 MANAGING IP ADDRESSES

In the early days of networking, most devices had a fixed IP address that was set and managed manually. As networks grew, so did the demand for techniques to manage networked devices and automate the tasks of setting and tracking IP addresses.

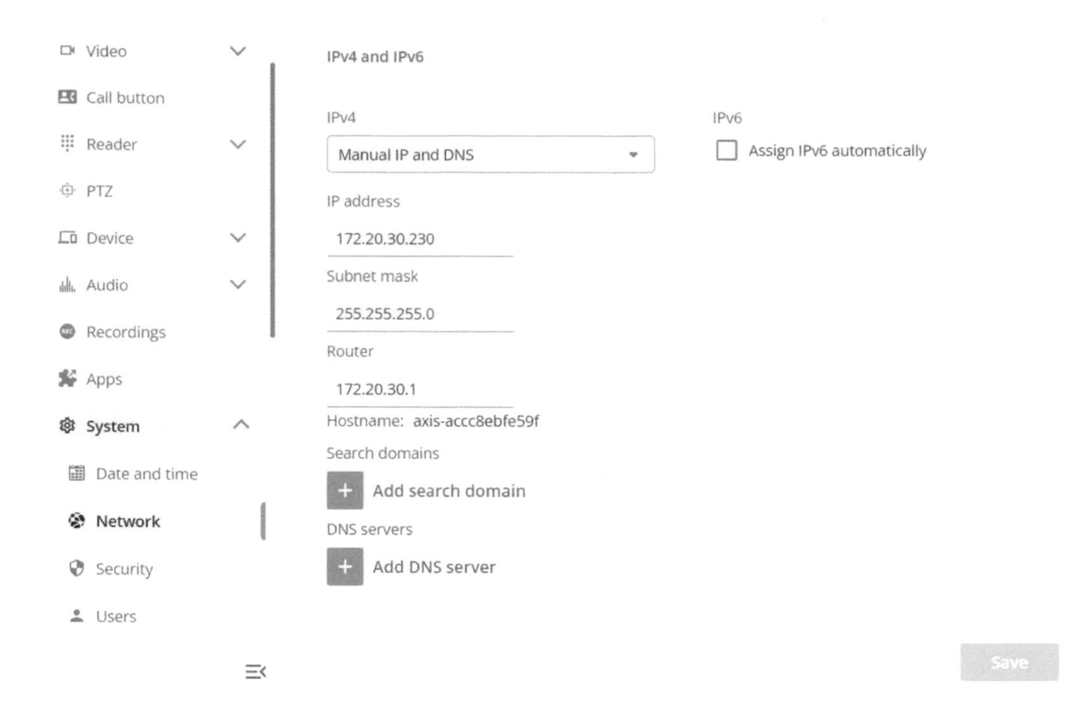

Figure 9.7 Setting a static IP address from a camera's web interface.

The following section describes the different ways of setting and managing IP addresses. The IP addressing of data packets and the use of Domain Name System are also explained.

9.3.1 Setting IP addresses

Any device on an IP-based network must have a unique and valid IP address. Setting the IP address can be done in two different ways:

- *Manually:* using a static address
- *Dynamically:* using DHCP

9.3.2 Manual address allocation

Setting IP addresses manually is labor-intensive. Static addresses are normally used only in smaller systems or for devices that need static addresses for particular reasons. One way to set a static IP address for a camera is to access its built-in web pages (Figure 9.7).

Many vendors offer tools for not only setting IP addresses but also, more importantly, for finding and managing devices on a network. In a network video solution with potentially thousands of network cameras, these tools are essential for effective management (see Figure 9.8).

9.3.2.1 Dynamic address allocation

The Dynamic Host Configuration Protocol (DHCP) is the standard protocol used for the automatic assignment and management of IP addresses. A DHCP server manages a pool of IP addresses, which it can assign dynamically to DHCP clients upon request. The DHCP server can also provide other IP configuration parameters to DHCP clients, such as host names and DNS server addresses (see section 9.3.4). A DHCP server is the central point of client configuration management; it removes the need to maintain individual network configurations for each network client.

When a client configured for DHCP comes online, it sends a query requesting configuration from the DHCP server, which replies with an IP address, subnet mask, and other configuration

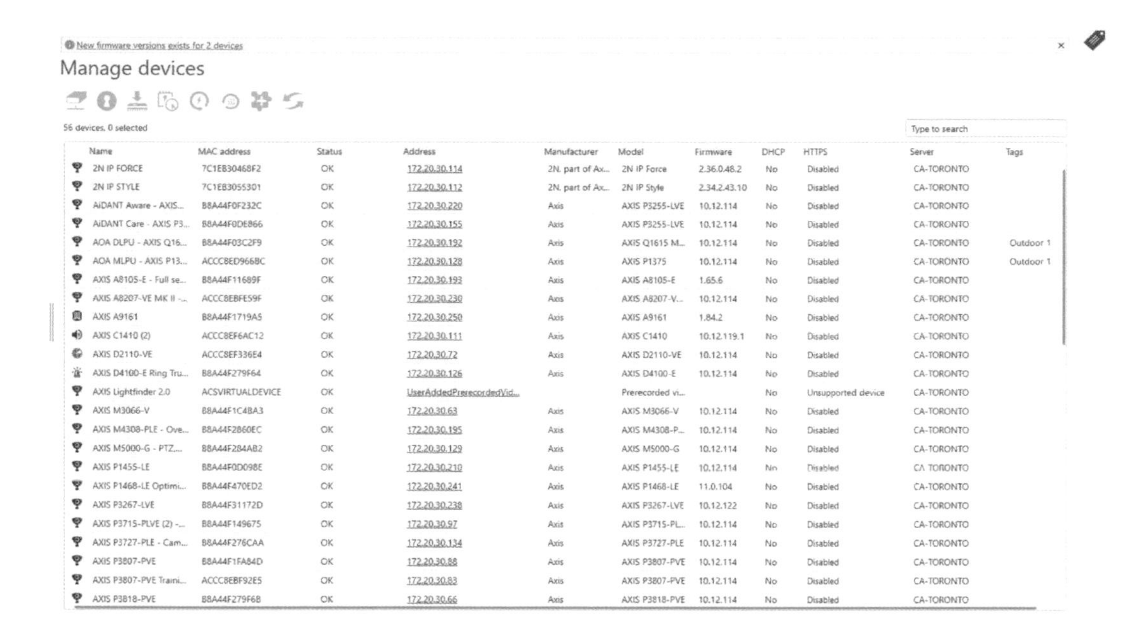

Figure 9.8 Most vendors provide tools for managing the IP addresses of networked devices.

Figure 9.9 Any networked device, for example, a camera, must have a unique IP address, which can be set dynamically by DHCP.

parameters. In a typical configuration, the IP address provided by a DHCP server is leased to the client. Once half the lease time has expired, the client will request renewal of the lease (see Figure 9.9).

9.3.3 Configuration-free networking

In many systems – even in small ones – setting and managing IP addresses is complex and cumbersome. To address these issues, several techniques have been developed, which simplify and automate IP addressing to the greatest possible extent. The two most well-known techniques are UPnP® and Zeroconf (of which Apple's Bonjour® is an implementation).

9.3.3.1 UPnP® and Zeroconf

Zeroconf is a component of Universal Plug and Play (UPnP). By using UPnP, Microsoft operating systems can automatically detect resources on a network, meaning that a network camera will automatically be listed under "Network" in the Windows operating system.

With Zeroconf, networked devices attempt to independently allocate an IP address in the range 169.254.1.0 to 169.254.254.255. If a system wishes to configure an IP address, it simply selects an address from the range using a random number generator based on system-specific information, such as the MAC address. After selecting an appropriate IP address, the system must first check to see if it is already in use by another device.

9.3.3.2 Bonjour®

Bonjour is another protocol for the announcement and discovery of services in an IP network. This protocol is based on open standards and was introduced by Apple. Using Bonjour, services can be discovered independently, and any necessary configuration can be performed automatically without user intervention. Bonjour is based on the exchange of multicast DNS packets sent via the UDP port 5353 (224.0.0.251).

In addition to multicast DNS, Bonjour is based on DNS Service Discovery (DNS-SD), which is an expansion of the Domain Name Service as used for domain management. Bonjour is comparable to UPnP and Zeroconf.

Bonjour is appropriate for use in discovering network video products using Mac computers, but it can also be used as a discovery protocol for new devices in any network.

9.3.3.3 MAC and IP address resolution

IP addressing of data packets operates on the third OSI layer. Before data can be transported over a physical network, it must be packaged into frames, a process that occurs in the data-link (second) layer of the OSI model. Addressing on the second layer uses MAC addresses, which means that in addition to knowing the destination IP address, the sending host must also know the MAC address of the destination host – so that the destination IP address can be associated with a MAC address. To discover the MAC address of a given destination host, the sending host transmits a request using a protocol called ARP.

9.3.3.4 Address Resolution Protocol (ARP)

Address Resolution Protocol (ARP) is used to discover the MAC address of the destination host. An ARP request specifies the IP address of the destination host, in addition to the IP and MAC addresses of the sending host. When a switch receives an ARP request, it broadcasts it to all devices on the local segment. The devices then compare the destination address to their own IP addresses. The device with the corresponding IP address replies by disclosing its MAC address.

The requesting host then enters the MAC address it received – together with the associated IP address – in its ARP cache. The ARP cache is a table that temporarily stores the mapping of addresses, usually between MAC and IP addresses, in the RAM memory of a host. Before transmitting the data, the sending host checks whether or not it can resolve the necessary MAC address using its ARP cache. If this cannot be done, it sends an ARP request (see Figure 9.10).

When a destination host resides outside a sending host's own LAN, the sender must instead use the MAC address of its router, as it is not able to discover the MAC address of the destination host (Figure 9.11). If the sending host does not know the MAC address of its router, an ARP request is sent to discover it. Thus, to properly address a data packet that will be sent to another LAN, the sending host uses the MAC address of its router together with the IP address of the destination host.

9.3.4 Domain Name System (DNS)

The Domain Name System (DNS) converts domain names into their associated IP addresses, and it operates on the transport layer. This can be compared to a telephone book, which links the name of a person to a phone number. The DNS-to-IP address mapping can either be done manually in a host file or automatically, which is more common.

DNS caters to the human ability to remember names better than numbers. For example, the domain name *example.com* is much easier to remember than its associated IP address *93.184.216.34*. DNS

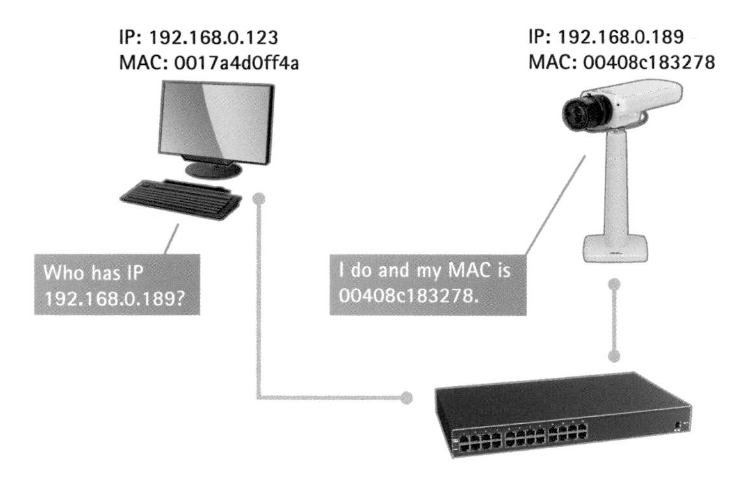

Figure 9.10 To send data over a local network, ARP is used so that the sending host can properly address the data packet with the destination host's IP and MAC addresses.

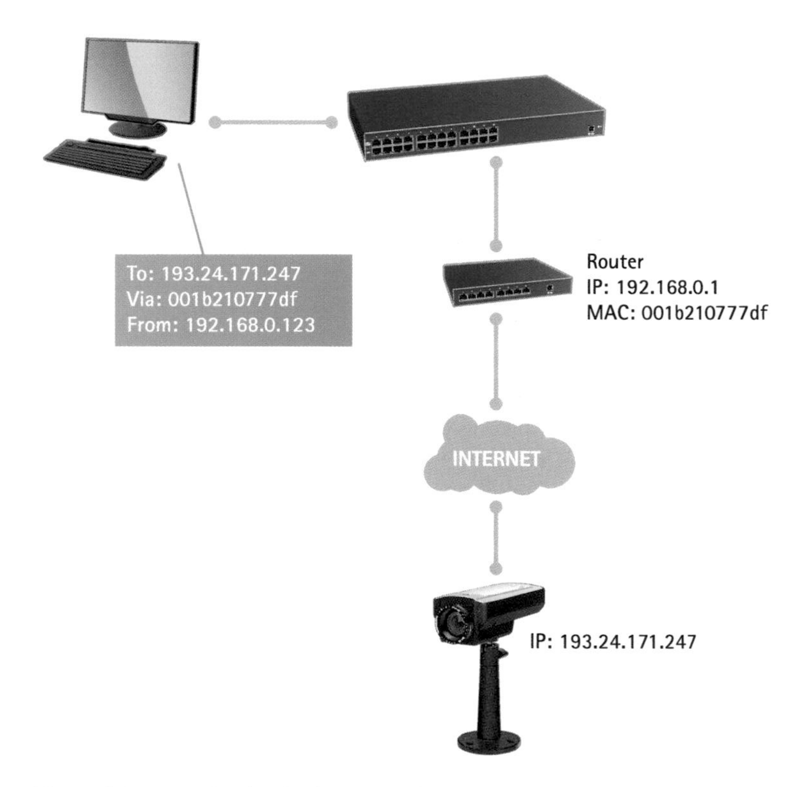

Figure 9.11 To send data to a destination host outside a local network, the sending host uses its router's MAC address, together with the destination host's IP address.

is used when a website is accessed by typing a domain name, such as www.example.com. DNS converts this name into the IP address, and communication between the systems takes place.

9.3.4.1 Dynamic DNS

When a PC or network camera connects to the internet, the internet service provider will typically assign an unused IP address dynamically through DHCP. This address can only be used for a short time, and several addresses may be used over the duration of a connection. When this happens, a Dynamic DNS (DynDNS) is used to keep track of a domain name's link to changing IPv4 addresses.

In Dynamic DNS, the host record on the DNS server is updated whenever the host's IP address is changed. Depending on the host's capabilities and network configuration, either the host itself or the DHCP server sends this update.

Using Dynamic DNS, a device's domain name (which never changes) can always be used to access the device, regardless of the IP address currently assigned to it.

9.4 DATA TRANSPORT

Internet Protocol (IP) is the most important protocol in the network layer and the core protocol for any data communication. However, IP never acts alone but always together with protocols in the same or higher layers. In the transport layer (Layer 4 in the OSI model), the most common protocols are User Datagram Protocol (UDP) and Transmission Control Protocol (TCP). These protocols, along with the application protocols described in section 9.5, are generally referred to as the IP or TCP/IP protocol suite.

9.4.1 User Datagram Protocol (UDP)

User Datagram Protocol (UDP) is a network protocol located on the fourth OSI layer, providing a connectionless transmission service. The fact that the protocol is connectionless means that no connection is established between the sender and receiver.

Messages are packaged into datagrams and transmitted. If the datagrams make it through the network, they are received by another application. If datagrams are lost, there is no strategy for re-transmission. Consequently, there is no guarantee that packets will be in order, and a UDP recipient can deliver data to the application in the wrong order.

Ports are used by UDP to allocate data to the correct application in the destination system, so the port number of the service receiving the data is embedded in the UDP header. UDP also sends a checksum in its header as a reliable integrity check option, which enables the recipient to detect transmission errors.

UDP does not apply any flow or congestion control to its sending strategy. If applications generate large amounts of data, the network can become flooded. One common scenario is that applications send more data than the network can handle, which leads to packet losses for UDP data and possibly for other flows.

From a video surveillance perspective, UDP favors timely delivery of data over reliability, but it does so without the congestion control and burst control available to TCP. UDP tends to be a better choice when trying to minimize delays and jitter. UDP may be preferable when transmitting data that requires low latency and which can tolerate some degree of loss, such as multimedia broadcast applications. On the other hand, when bandwidth is limited, or if there is a firewall or NAT in the path, then TCP tends to work better.

9.4.2 Transmission Control Protocol (TCP)

Transmission Control Protocol (TCP) is the most commonly used protocol for data transport and, when used with IP, is often referred to as TCP/IP. TCP divides data into TCP segments for data transmission, adding supplementary flow-control information in the TCP header. TCP provides a connection-oriented, reliable, and in-order delivery of data streams. In addition, it is responsive to network congestion. These characteristics make TCP suitable for applications such as file transfers or email.

In contrast to UDP, TCP is a connection-oriented protocol. This means that it establishes a connection between two communicating applications before any data exchange takes place. The connection makes sure that data flows only between these two hosts, which, however, prevents the use of TCP for broadcasting or multicasting. TCP also provides transmission reliability: The recipient confirms the incoming data, and, if necessary, the sender re-transmits if no confirmation is received.

TCP provides transmission and reception of a reliable and correctly sequenced stream of bytes for applications at the upper layers. It does this by splitting the payload into sequentially numbered segments, transmitting those segments according to its protocol rules, and finally, at the receiver, verifying and reassembling the segments to reconstruct the original stream. As data is exchanged, receivers continuously acknowledge successful receipt of segments by issuing sequence numbers so that senders can re-transmit lost data as required. If segments are lost, a TCP recipient does not pass out-of-sequence data to the application. Instead, it waits for the re-transmitted segments to arrive; that is, TCP introduces a delay in its efforts to be reliable and provide correctly sequenced delivery.

While exchanging data, TCP continuously applies flow and congestion control to its sending strategy. For example, when a recipient is slower than the sender, TCP's flow control forces the sender to slow down. Similarly, when the network path is congested, TCP's congestion control also forces the sender to slow down. These are important functions for avoiding congestion collapses, excessive packet loss, and inequalities between flows.

From a video surveillance perspective, conventional wisdom dictates that TCP is not suitable for real-time traffic because it favors reliability over the timely delivery of data. Because of this, many real-time protocols tend to use UDP and fall back on TCP only when necessary (for example, due to bandwidth, firewall, or NAT issues in the path). On the other hand, for applications where delays are not critical – such as for video storage – TCP is commonly used.

9.5 APPLICATION LAYER PROTOCOLS

At the application layer, which is the highest level of both the OSI model and the TCP/IP model, different protocols are required for data exchange between a network video system and the user, for example, when accessing a network camera and viewing video. The most common protocols are HTTP, FTP, and RTP, which are all explained in the following.

9.5.1 HyperText Transfer Protocol (HTTP)

HyperText Transfer Protocol (HTTP) is used primarily to load the text and images from a website to a browser. HTTP is a stateless protocol, meaning that a connection is not maintained between systems once data has successfully transmitted. A new connection must be established for additional data transmissions.

Network video systems provide an HTTP server service that permits access to the systems through browsers for downloading configurations or viewing live images.

9.5.2 File Transfer Protocol (FTP)

File Transfer Protocol (FTP) is a network protocol for data transmission via TCP/IP. It is used primarily to transmit files from a server to a client (download) or from a client to a server (upload). FTP also can be used to create and select directories and rename or delete directories and files.

There are two modes for establishing FTP connections: *active mode* and *passive mode*. In active mode, the FTP server establishes a connection to the client following a request from the client, whereas in passive mode, the client establishes the connection to the server. Passive mode is used if the server cannot reach a client. This is the case, for example, if the client is located behind a router that converts the client's address by means of NAT or if a firewall protects the local network against external access.

Network cameras can use FTP, for example, to transmit JPEG images to an FTP server for storage. In such a case, the camera acts as an FTP client and establishes an event-based connection to the

FTP server. It then transmits multiple JPEG images to the server and stores them to a specific directory using different file names.

9.5.3 Simple Network Management Protocol (SNMP)

Simple Network Management Protocol (SNMP) represents a set of protocols for managing complex network infrastructures. It can be used to remotely monitor and manage networked equipment, such as switches, routers, and cameras. Many cameras have support for SNMP, which means they can be managed by tools such as OpenNMS. The latest version of SNMP is version 3.

9.5.4 Simple Mail Transfer Protocol (SMTP)

Simple Mail Transfer Protocol (SMTP) is the de facto standard for transferring email over the internet. Although not normally relevant in network video, many cameras include support for SMTP to allow the sending of email alerts (for example, when motion is detected in a scene) and can even include attachments with snapshots or a video clip.

9.5.5 Real-time Transport Protocol (RTP)

Real-time Transport Protocol (RTP) permits the transfer of real-time data between system endpoints. RTP is a packet-based protocol that is usually transmitted over UDP. RTP services include identification of the transmitted user data and its sources, as well as the allocation of sequential numbers and timestamps to the data packets. Using this information, the recipient can reassemble the individual data packets in the correct sequence. Video compressed using H.264 and H.265 is often transmitted over RTP, which can be used for both unicasting and multicasting applications (see section 9.6 for more on this).

9.5.6 Real-time Transport Streaming Protocol (RTSP)

In conjunction with RTP, the complementary Real-time Streaming Protocol (RTSP) can be used for extended control over the transmission of real-time media. RTSP enables low-latency audio and video to be streamed (over RTP) directly to a client, without first being downloaded.

RTSP is similar in design to HTTP. This makes it compatible with HTTP networks, and RTSP was used for many years in streaming applications over the internet. Nowadays, other protocols have taken over streaming on the internet, but RTSP is still very common in network cameras and other related devices such as NVRs. One important reason for using RTSP in network cameras is that it assists in ONVIF compatibility. For more information on ONVIF, see section 18.2.7.

RTSP allows the client to query the server to ascertain if the required services are supported. RTSP streams can be password-protected, and the client can specify the resolution to stream.

9.5.7 Session Initiation Protocol (SIP)

The Session Initiation Protocol (SIP) is a communication protocol for signaling and controlling multimedia communications sessions. The most important applications of SIP are in internet telephony for voice and video calls, as well as instant messaging over IP networks.

SIP is an application layer protocol designed to be independent of the underlying transport layer. SIP is text-based and incorporates many elements of the HTTP and SMTP protocols.

SIP also works alongside several other application layer protocols that identify and carry the session media. Media identification and negotiation is achieved with the Session Description Protocol (SDP). For the transmission of media streams (voice, video), SIP typically employs the RTP protocol or the Secure Real-time Transport Protocol (SRTP). For the secure transmission of SIP messages, the protocol can be encrypted with Transportation Layer Security (TLS) and is then referred to as SIPS. For more information about SIP, see Chapter 7.

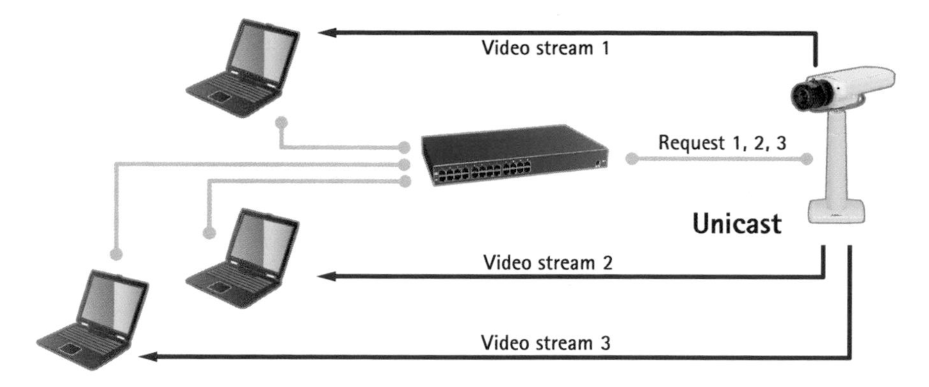

Figure 9.12 Unicast video transmission.

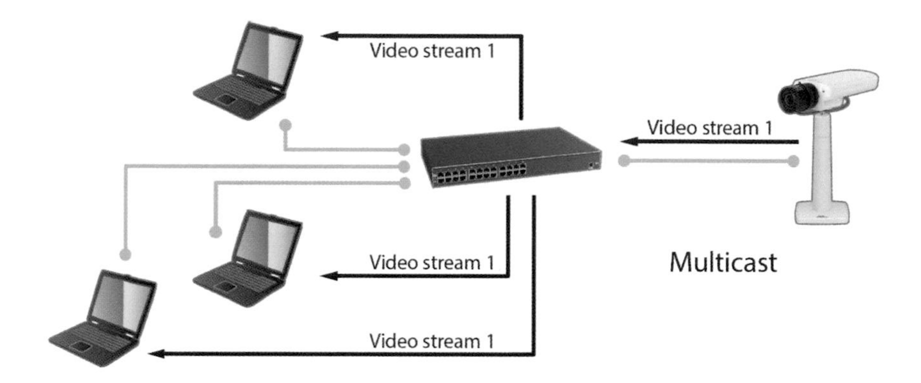

Figure 9.13 Multicast video transmission.

9.6 UNICAST, BROADCAST, AND MULTICAST

There are three different methods for transmitting data on a computer network, each catering to different needs:

- *Unicast:* This is the most common form of communication, in which the sender and the recipient communicate on a point-to-point basis. (See Figure 9.12.) Data packets are sent only to one recipient, and no other clients will receive that information.
- *Multicast:* This is the communication between a single sender and multiple receivers on a network. (See Figure 9.12.) Multicast technologies are used when many clients request the same information simultaneously, for example, live video. Multicasting reduces network traffic by delivering a single stream of information to many recipients. The biggest difference when compared to unicasting is that the video stream only needs to be sent once, whereas for unicast, a copy for each recipient is required. Multicasting is typically used when large numbers of users wish to view live video. RTP is the most common protocol used for multicasting video streams.
- *Broadcast:* This means that the sender sends the same information to all other servers on the network. When a broadcast message is sent, all hosts on the network receive the message and will process it to some extent. Too many broadcast messages will slow down a network and the hosts connected to it. Routers block broadcast messages and are used to create broadcast domains that limit broadcasts to the network segment on which the broadcast originated. In broadcast addressing, a distinction is made between a limited broadcast (address: 255.255.255.255), which is not forwarded by routers, and a direct broadcast (for example, 146.15.255.255), which will be forwarded as necessary by routers so that all hosts are reached. Broadcasts are not practical for network video transmission, and network video products only use broadcasts for specific protocols that require it, such as DHCP.

9.7 QUALITY OF SERVICE (QOS)

At present, fundamentally different applications – for example, telephone, surveillance video, and email – all use the same IP network. In such a network, it is necessary to control how network resources are shared to fulfill the requirements of each service. One solution is to let network routers and switches operate differently for different kinds of services (voice, data, and video) as traffic passes through the network. Using Quality of Service (QoS), different network applications can coexist on the same network without consuming each other's bandwidth.

9.7.1 Definition

QoS refers to several technologies that guarantee a certain quality to different services on a network. Examples of quality include a maintained level of bandwidth, low latency, or no packet losses. The main benefits of a QoS-aware network can be summarized as follows:

- The ability to prioritize traffic so that critical flows can be served before flows with less priority
- Greater reliability in a network by controlling the amount of bandwidth an application can use and, consequently, controlling bandwidth competition between applications

QoS relates to the transmission delay (latency) between systems, jitter (variation from the average latency time), packet loss rate (loss probability for individual packets), and data throughput (bandwidth).

Datagram headers in IPv4 and IPv6 contain a Differentiated Services Code Point (DSCP) flag for identifying the type of data in the relevant IP datagram. Using this flag, the data packets are divided into traffic classes and prioritized for forwarding. The DSCP flag has a length of 6 bits, which means that 64 different classes can be defined.

9.7.2 QoS in network video

To use QoS for network video, the following requirements must be met:

- All switches and routers must include support for QoS. This is important to achieve end-to-end QoS functionality.
- The video products must be QoS-enabled.

See Figure 9.14 for an ordinary (non-QoS-aware) network and Figure 9.15 for a QoS-aware network.

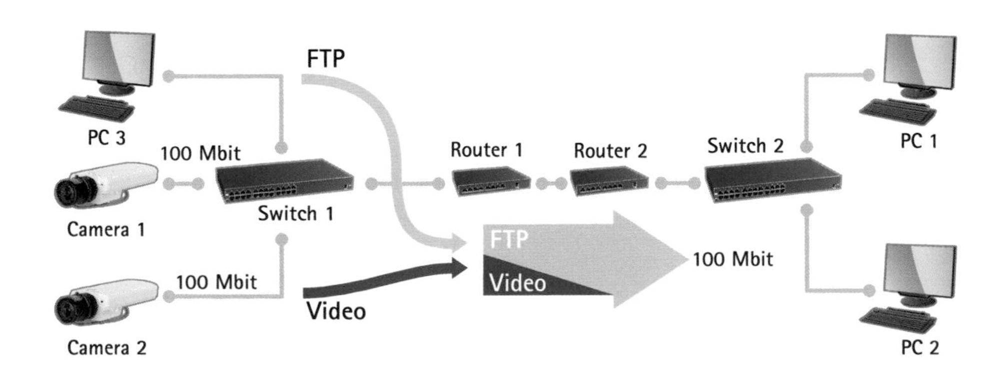

Figure 9.14 An ordinary (non-QoS-aware) network. PC1 is accessing video streams at 25 Mbit/s from camera 1 and camera 2. Suddenly, PC2 starts a file transfer from PC3. In this scenario, the file transfer will try to use the full 100 Mbit/s between routers 1 and 2, while the video streams will try to maintain their total of 50 Mbit/s. The amount of bandwidth for video can no longer be guaranteed, and the frame rate will probably be reduced. At worst, the FTP traffic will consume all the available bandwidth.

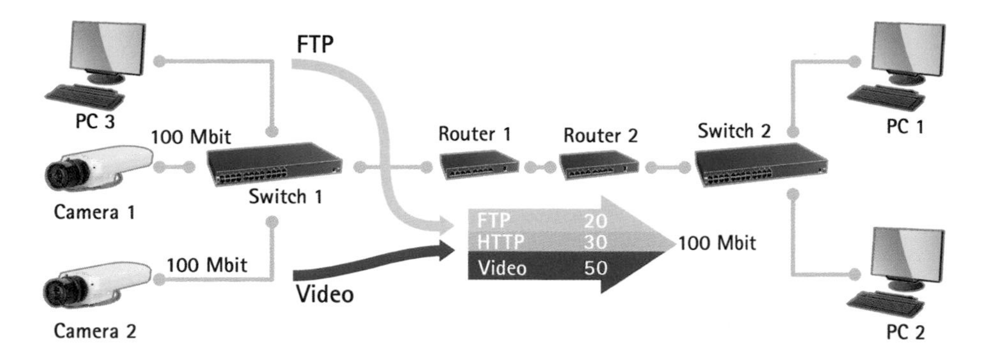

Figure 9.15 A QoS-aware network. Router 1 is configured to devote up to 50 Mbit/s of the available 100 Mbit/s to video. FTP may use 20 Mbit/s, and HTTP and all other traffic can use a maximum of 30 Mbit/s. This way, video always has the necessary bandwidth available. Note that these maximums only apply when there is congestion on the network. If there is unused bandwidth available, this can be used by any type of traffic. To guarantee fast responses, PTZ traffic, which requires low latency, is often regarded as critical. This is a typical case where QoS can be used to provide the necessary guarantees.

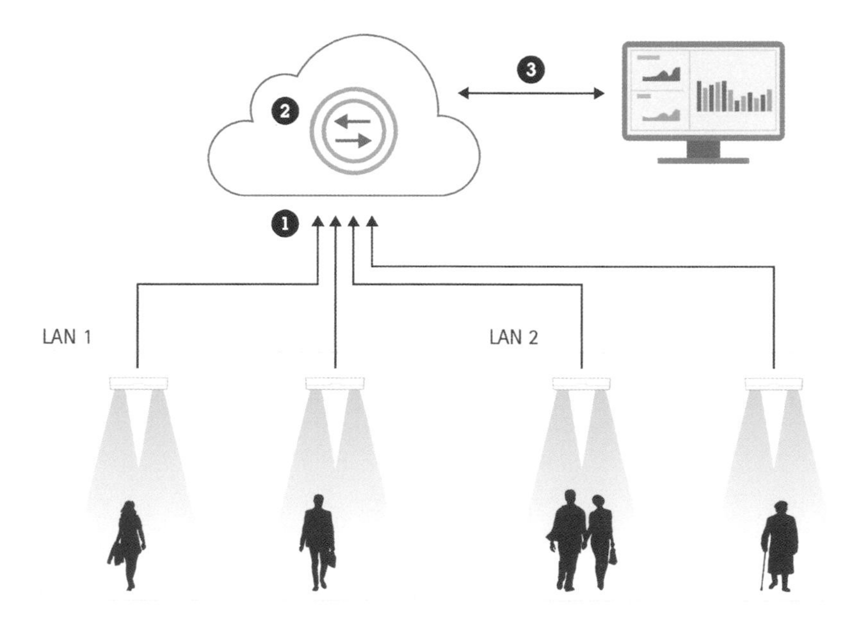

Figure 9.16 (1) Publishing clients. (2) The MQTT broker. (3) Subscribing clients.

9.8 MESSAGE QUEUING TELEMETRY TRANSPORT (MQTT)

MQTT facilitates the efficient and reliable exchange of data between network cameras, IoT devices, and cloud applications. In a typical VMS-based system, event notifications from devices are traditionally streamed to a single destination via an API interface using the RTSP streaming protocol. But the same notifications can be distributed using the MQTT protocol via the device's MQTT client, both within VMS systems but also outside them and especially over the internet. Some analytics applications use their own MQTT clients for specific use cases, such as people counting devices that send statistics over MQTT to data visualization software in the cloud.

Unlike RTSP, which uses a request/response messaging pattern, MQTT is a publish/subscribe protocol, which means that devices (clients) publish messages to an MQTT broker (a server), which keeps track of who is publishing what and who wants to see the data. Messages are published with a topic. The receiving clients (subscribers) also connect to the broker, either to get that specific topic or by using wildcards to get all subtopics. The broker forwards messages only to clients that have topic subscriptions.

Figure 9.16 shows how a device with people counting analytics generates an event notification whenever a person is detected passing in or out of a defined area. The notification is passed to the MQTT broker, which publishes the data in real time to a cloud platform. The platform uses a connection to data visualization software to display the real-time statistics from the people counters.

There are several benefits to using MQTT as compared to RTSP, including the following:

- *No passwords exposed:* There is no need for a client to access a device or server to get the data, so it does not need to know the password or know how the API works.
- *Single point of integration:* If authorized, all clients can get all other clients' published messages with a single connection to a broker. This means that the MQTT message flow can be one-to-one, one-to-many, or many-to-one, with no extra load for each client.
- *Intact firewalls:* With a public MQTT broker in-between, clients behind a firewall can publish/subscribe data without creating a hole in the firewall (firewall rules not withstanding).
- *Retained messages:* Publishers can mark a message as retained, meaning that the broker will keep a copy and send that message to newly connected clients subscribing to that topic.

Chapter 10

Wired networks

Networks provide data exchange between servers and nodes in a computer system. In the early days of office and enterprise networking, many different technologies emerged, such as Token Ring, Banyan VINES, Ethernet, and Fiber Distributed Data Interface (FDDI). Ethernet became the prevailing standardized technology for everything from home networking to large enterprise systems.

10.1 THE EVOLUTION OF ETHERNET

Bob Metcalfe at Xerox PARC documented his invention of Ethernet in 1973. During the remainder of the seventies, Metcalfe and his colleagues continued to develop prototypes, published papers, and founded 3Com to commercialize Ethernet. In 1983, the Institute of Electrical and Electronics Engineers (IEEE) approved the IEEE-802.3 standard. This basic version of Ethernet enabled a data transfer rate of 10 Mbit/s (megabits per second), with individual nodes networked through a coaxial cable. Since then, the standard has improved continuously, and new transfer media achieves increasingly higher data transfer rates.

Today, Ethernet is based mainly on twisted-pair copper cables or fiber-optic cables (often simply called fiber). Coax cables can also supply Ethernet using Ethernet-over-coax converters. In smaller systems, Ethernet over power lines is also possible. Various manufacturers offer Ethernet components for building cost-effective networks. The number of networked nodes in a single network can range from two to several thousand. The data rates available depend on the transfer media and networking equipment used in each case. Most installations today support at least 100 Mbit/s or 1 Gbit/s, but enterprise networks may even use a data rate of up to 100 Gbit/s. Today's Ethernet networks easily provide the performance level required by the most demanding network video applications.

The Ethernet standard is available in many versions, depending on the transfer medium used and the achievable data rate. The following subsections briefly describe the most important versions in terms of network video.

10.1.1 10-Mbit/s Ethernet

The 10BASE-T standard (802.3i) was released in 1990, uses a twisted-pair cable, and operates at 10 megabits per second (Mbit/s or Mbps). A twisted-pair cable is similar to an improved version of a telephone cable. It consists of four pairs of two twisted wires, which improves the electrical properties for data transfer (see Figure 10.1). A 10BASE-T system connects with RJ45 plugs and jacks and uses two of the four pairs of wires to transfer data. The maximum length of a cable segment is 100 m (328 ft).

DOI: 10.4324/9781003412205-10

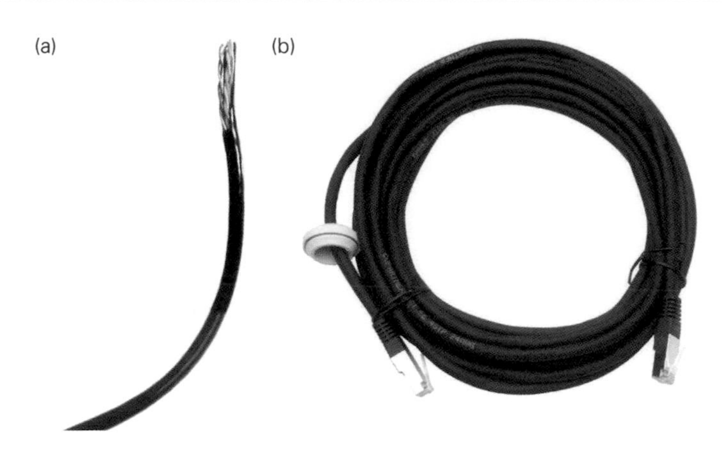

Figure 10.1 (a) Twisted-pair cable with four pairs of twisted wires. (b) Cable with RJ45 plug connectors and a protective gasket.

Although 10 Mbit/s Ethernet still exists in older installations, it is not feasible for network video. Computers, switches, and other devices with a 10/100 interface support both 10 Mbit/s and Fast Ethernet. They automatically recognize and adjust to the current network speed.

10.1.2 Fast Ethernet

Fast Ethernet refers to a 100 Mbit/s Ethernet network and was introduced with the 802.3u extension as 100BASE-T in 1995. It is described in the standard as a variation for twisted-pair cable (100BASE-TX) and glass fiber (100BASE-FX). 100BASE-TX provides backward compatibility with 10BASE-T and was, for a long time, the most popular Ethernet interface. While small businesses and residential markets may still use Fast Ethernet, most security networks have gigabit uplinks between their edge and core infrastructures.

10.1.3 Gigabit Ethernet

The third generation of Ethernet was specified in 1998 by the 802.3z extension. It includes specifications for optical fiber (1000BASE-SX, 1000BASE-LX) and short-distance copper cable (1000BASE-CX). In 1999, the IEEE updated the standard with the 802.3ab extension for twisted-pair cable (1000BASE-T). Gigabit Ethernet delivers a data rate of 1,000 Mbit/s (1 Gbit/s). The main difference, when compared with 10BASE-T and 100BASE-TX, is that 1000BASE-T uses all four pairs of twisted wires in the cable to achieve the high data rates. Most gigabit interfaces are backward compatible with 10-Mbit/s and Fast Ethernet and are known as 10/100/1000 interfaces.

Various Gigabit Ethernet versions, such as 1000BASE-SX, 1000BASE-LX, 1000BASE-LX10, and 1000BASE-BX10, are available for use with fiber (see Figure 10.2) for transmission over longer distances and different wavelengths. For example, 1000BASE-SX works with multimode glass fibers (MMF) has a wavelength of 770–860 nanometers and permits cable lengths up to 550 m (1,804 ft). 1000BASE-LX has a wavelength of 1270–1355 nanometers and permits cable lengths up to 550 m (1,804 ft) with multimode fibers or up to 5,000 m (16,404 ft) with single-mode fibers (SMF).

10.1.4 10Gigabit Ethernet

10 Gigabit Ethernet (also known as 10GE, 10GbE, or 10 Gb Ethernet) delivers a data rate of 10 Gbit/s (10,000 Mbit/s). It was first defined in 2001 by the IEEE 802.3ae standard, which specified several Ethernet-over-fiber solutions. 10GBASE-LX4 and 10GBASE-SR can bridge distances up to 10 km (6.2 miles), and 10GBASE-ER can reach up to 40 km (24.9 miles). The 802.3an specification (10GBASE-T) was published in 2006 and permits data transfer of 10 Gbit/s through twisted-pair cable. It needs all four pairs of a high-quality cable (CAT6a or CAT7).

Figure 10.2 Example of a small form-factor plug (SFP) module and a fiber-optic cable with SFP connectors. Fiber cables can bridge longer distances than twisted-pair copper cables. Backbone networks often use fiber cabling.

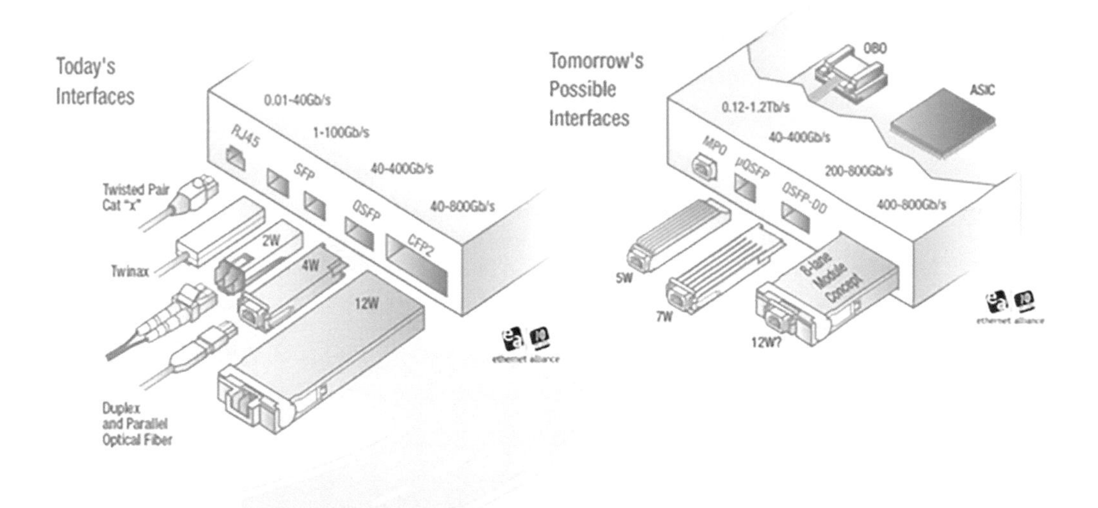

Figure 10.3 Illustration of different Ethernet connectors.

Source: Ethernet Alliance

10.1.5 The future of Ethernet

The IEEE published 802.3ba in 2010, standardizing 40-Gbit/s and 100-Gbit/s Ethernet (also known as 40GbE and 100GbE). This standard specifies single-mode (40GBASE-LR4, 100GBASE-LR4) or multimode fiber (40GBASE-SR4, 100GBASE-SR10, 100GBASE-SR4) with quad small form-factor pluggable (QSFP) connectors (see Figure 10.3). As the term indicates, QSFP connectors support four channels (4 × 10 Gbit/s, 4 × 25 Gbit/s). 40GbE is intended for local server connectivity and high-bandwidth applications, such as video-on-demand, whereas 100GbE is intended for internet backbones and data centers.

In 2016, four new standards based on the 10GBASE-T technology for twisted-pair copper cables became available. 2.5GBASE-T and 5GBASE-T is specified for 100 m of CAT5e cabling and 25GBASE-T and 40GBASE-T for 30 m of CAT8 cabling.

In the future, we will see more of terabit links (see Figure 10.4). The idea is to group 100 Gb/s lanes into 10 and 16 lanes for 1 Tb/s and 1.6 Tb/s links. However, investments still need to be made and

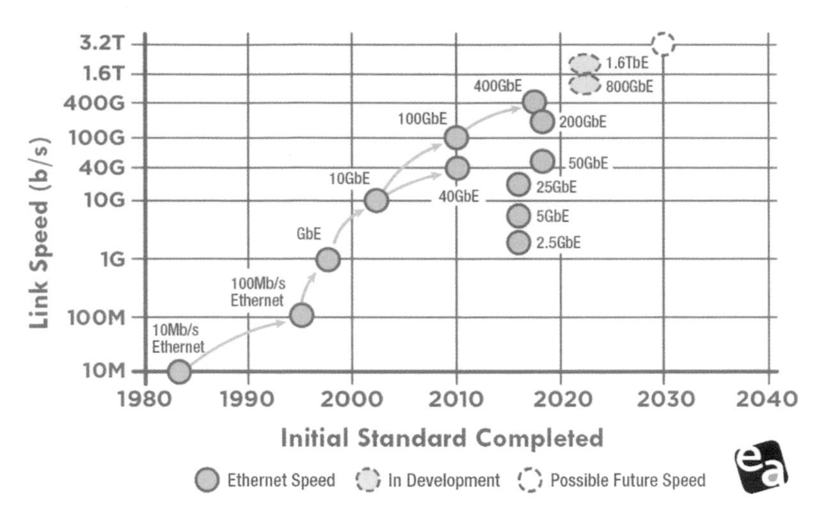

Figure 10.4 Current and projected Ethernet speeds.

Source: Ethernet Alliance

technology must advance before anyone can implement such an infrastructure in a practical and economically sustainable way. According to the Ethernet Alliance, speeds of 800 Gbit/s and 1.6 Tbit/s are expected to become IEEE standards sometime 2023–2025.

10.2 NETWORK TOPOLOGIES

Networks can be built in different ways. The network layout (also called the topology) describes how the individual nodes connect to the network. Most video surveillance networks today use a star topology.

In a star topology (see Figure 10.5), the individual nodes connect in a star formation through a central point, such as a switch. Twisted-pair or fiber cables form a point-to-point connection between the connected nodes and the central switch. Note that the central node in this topology is a single point-of-failure that will bring down the whole system if it fails.

In larger network installations, multiple star topology networks connect in a hierarchy (see Figure 10.6). An uplink is used to make connections between the network switches. The central parts of the network that connect all the local star topologies are often referred to as the network backbone.

10.3 NETWORK CABLING

In wired networks, all nodes must be connected through some type of network cable. Many different types are available, including twisted-pair or fiber cable, and each type is available in many different versions. Twisted-pair cables are separated into different categories (CAT). This section outlines the most common types and versions.

10.3.1 Twisted-pair cables and RJ45

Twisted-pair cable is still the most common type of cable used in Ethernet networks. As mentioned earlier, this cable has four pairs of twisted wires and uses RJ45 connectors. Wire pairs, which keep electromagnetic interference low, transfer complementary signals. The maximum length of a twisted-pair cable for Ethernet is normally 100 m (328 ft).

Depending on the electrical properties of the twisted-pair cable, data transfer up to 40 Gbit/s is achievable. In 10 and 100 Mbit/s Ethernet, only two of the four wire pairs are used for data

Figure 10.5 In a star topology, all nodes connect to a central point.

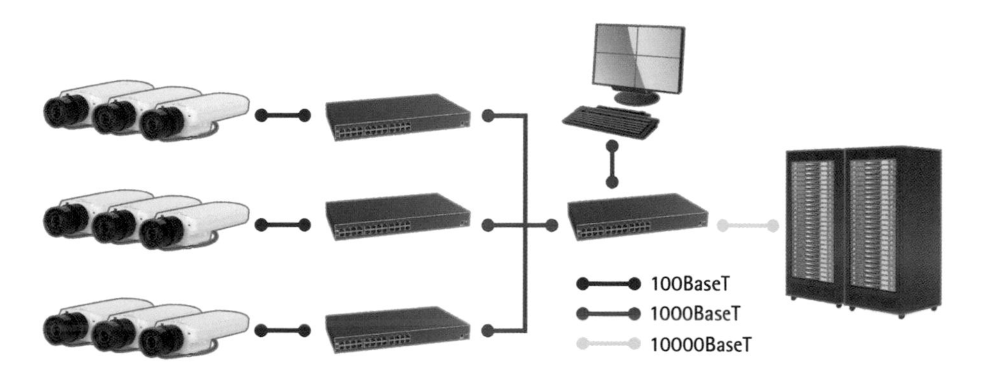

Figure 10.6 Larger networks are built in a hierarchy, with local star networks connected through a backbone.

transfer – one pair for sending data and another for receiving data. For 1,000 Mbit/s (Gigabit) and 10,000 Mbit/s (10 Gigabit) Ethernet, you need all four wire pairs. In some cases, the signals move simultaneously in both directions to reduce the frequencies.

10.3.2 Cable categories

Different data rates (measured in Mbit/s) are supported through different transfer frequencies, which are measured in megahertz (MHz). Different types of cables, referred to as categories or CAT, support the various transfer frequencies. The ISO/IEC-11801 standard defines the individual categories and specifies certain transfer properties for the twisted-pair cable, such as impedance, bandwidth, damping, and near-end crosstalk (NEXT). The categories relevant for Ethernet today are CAT6 to CAT8. In older installations where CAT3 or CAT5 cabling exists, for today's video applications, it is recommended to replace this cable with CAT6 or above.

- *CAT3:* A twisted-pair cable for transfer frequencies up to 16 MHz, which corresponds to the requirement of 10 Mbit/s Ethernet (10BASE-T). Due to the limitation of 10 MHz or 10 Mbit/s, CAT3 should no longer be installed.
- *CAT5:* A twisted-pair cable for transfer frequencies up to 100 MHz which corresponds to the requirement of Fast Ethernet (100BASE-TX). CAT5 still exists in many installations today.
- *CAT5e:* A twisted-pair cable for transfer frequencies up to 100 MHz and is built to the same specifications as a standard CAT5 cable. The *e* indicates extended inspection measurements and that the cable complies with the requirements for Gigabit Ethernet (1000BASE-T).
- *CAT6:* Originally specified for a twisted-pair cable with transfer frequencies up to 250 MHz. To reduce signal noise, crosstalk, and interference, it has thicker-gauge wire and more pair twists per inch. 10GBASE-T connections on CAT6 have a maximum distance of 55 m.
- *CAT6a:* Supports transfer frequencies up to 500 MHz or 625 MHz in special versions. It meets the requirements of 10GBASE-T and permits distances of 100 m. Unlike the majority of CAT6 cables, CAT6a cables are usually shielded. The shield reduces the risk of interference, making these cables a better choice in industrial environments. CAT6e was also introduced but has never been a ratified standard.
- *CAT7:* Also known as class F cabling, this is a twisted-pair cable for transfer frequencies up to 600 MHz. It is used in 10GBASE-T and 1000BASE-T networks. Its twisted-wire pairs are fully shielded. This is known as screened-foiled twisted-pair (S/FTP) wiring or sometimes as screen-shielded twisted-pair (S/STP). These cables successfully prevent crosstalk and provide great noise resistance.
- *CAT7a:* An augmented version of CAT7 that has a frequency of 1,000 MHz and can be used up to a maximum length of 100 m.
- *CAT8:* Meets the requirements for 25GBASE-T and 40GBASE-T. It is backward compatible with CAT6, uses RJ45 connectors, and has more or less the same diameter as CAT6a and CAT7. With a frequency of 2 GHz, it is four times faster than CAT6a but has a maximum length of 30 m. Therefore, it is more suitable for data centers than for network video systems.

10.3.3 Twisted-pair cable types

The primary difference between the different categories explained earlier lies in whether the cable is shielded and, if so, which type of shielding it uses. The shield provides electromagnetic protection for the twisted pairs within the cable, which improves the performance of the cable but also increases the manufacturing cost.

To differentiate between the levels of shielding, cables use codes that follow a specific structure: XX/XXX. The code before the slash describes the cable screen, that is, the shielding wrapped around a group of pairs. The code after the slash describes the shielding for the individual pairs.

- TP: twisted pair
- U: unshielded
- F: foil shield
- S: braid shield

Table 10.1 Twisted-pair cable types

Abbreviation	Description	Cable screen	Pair shield
U/UTP or UTP	No overall screen (U) and unshielded twisted pairs (UTP). The wires are only covered with the standard plastic cover that protects against physical damage.	None	None
F/UTP	An overall foil (F) screen and unshielded twisted pairs (UTP).	Foil	None
SF/UTP	An overall braid screen (S), an overall foil (F) screen, and unshielded twisted pairs (UTP). The overall braid screen helps prevent electromagnetic interference (EMI).	Braid, foil	None
S/UTP	An overall braid screen (S) and unshielded twisted pairs (UTP). Similar to the U/UTP and F/UTP cables, with an integrated wire that helps shield the cable.	Braid	None
F/FTP	An overall foil screen (F) and foil-screened twisted pairs (FTP).	Foil	Foil
U/FTP	No overall screen (U) but foil-screened twisted pairs (FTP).	None	Foil
S/FTP	An overall braid screen (S) and foil-shielded twisted pairs (FTP). The foil shield around each individual pair helps prevent crosstalk.	Braid	Foil
SF/FTP	An overall braid screen (S), an overall foil screen (F), and foil-shielded twisted pairs (FTP)	Braid, foil	Foil

Table 10.1 describes the main types of twisted-pair cables.

The term STP often refers to cables with an overall screen and foil shields around each individual pair of copper wires. However, most twisted-pair cables with an overall screen or individual shields have at some point claimed the STP designation. Always make sure the cables you intend to use have the correct type of shielding for the installation. The higher the transfer frequency, the better the shielding needs to be.

10.3.4 Fiber cable types

Fiber-optic cables are common in high-performance and backbone networks. Although more expensive, fiber offers several advantages. One obvious advantage is that the length of each cable segment can be much longer than a copper cable, which has a limit of 100 m (328 ft). Another advantage, especially in industrial environments, is that fiber is immune to electromagnetic interference (EMI). In a network video application, fiber makes it possible to place a network camera further away from a building, for example, in a non-adjacent parking lot.

The core of fiber-optic cables is made of glass, and light is the information carrier. Standard LEDs (light-emitting diodes) or special laser LEDs send and receive light at the two ends of the fiber. Depending on the glass fiber type, the LEDs use wavelengths of 850, 1,300, or 1,550 nm, and the light spreads through the glass fibers in different modes.

The two modes are multimode fiber (MMF) and single-mode fiber (SMF). The diameter of the cable is expressed as a ratio. Multimode fiber is typically 50/125 or 62.5/125 μm, meaning that the core diameter is 50 μm or 62.5 μm, and the outer diameter of the cladding is 125 μm. Single-mode fiber is typically 9/125 μm. As the light travels through the core, it bounces off the cladding. The greater the diameter of the core, the more reflections and the longer it takes for the light to reach the end of the cable. Single-mode fiber, with its smaller diameter, is better for long distances and high bandwidth. Multimode fiber, with its larger diameter, is good for shorter distances. As with twisted-pair cables, fiber-optic cables are divided into different categories. They also have different color jackets (see Figure 10.7). There are several standards that specify the various categories and performance categories of fiber, including EN 50173, ISO/IEC 11801, IEC 60793, and the TIA 492 series.

Table 10.2 Fiber-optic cables: Categories and characteristics

Fiber type		Applications	Wavelength (nm)	Attenuation (dB/km)	Maximum distance
Multimode	OM1	Legacy extensions	850	3.5	33 m (10G)
			1,300	1.5	
	OM2	Legacy extensions	850	3.5	82 m (10G)
			1,300	1.5	
	OM3	Large private networks	850	3.5	300 m (10G)
			1,300	1.5	700 m (100G)
	OM4	Data centers, campuses	850	3.5	400 m (10G)
			1,300	1.5	150 m (40G)
	OM5	Data centers, SWDM	850	3.0	550 m (10G)
			1,300	1.5	150 m (100G)
Single mode	OS1	Indoor, long distance	1,310	1.0	10 km (6.2 miles)
			1,550	1.0	
	OS2	Outdoor, long distance	1,310	0.4	200 km (124 miles)
			1,383	0.4	
			1,550	0.4	

1 km = approximately 3,281 feet or 1,094 yards

SWDM = Shortwave Wavelength Division Multiplexing

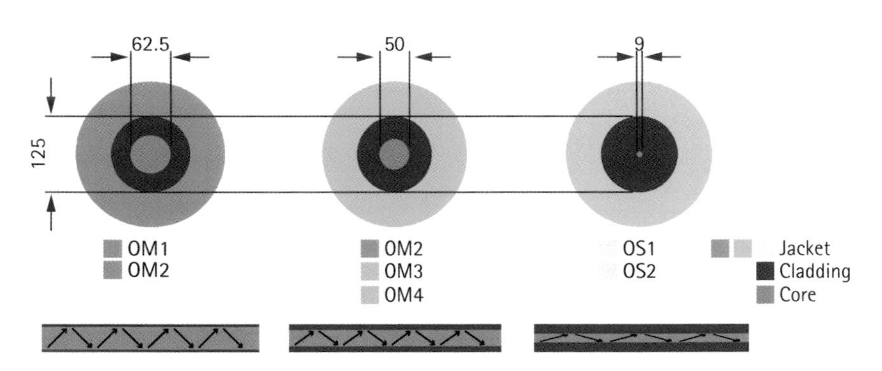

Figure 10.7 Typical dimensions of fiber-optic cables. OM1 and OM2 cables usually have orange jackets, OM3 and OM4 cables have aqua jackets, and OS1 and OS2 cables have yellow jackets. In single-mode cables, the light bounces fewer times than in multimode cables, which have larger cores.

10.3.4.1 Fiber connectors

When addressing network cabling issues such as distance or attenuation with optical fiber, it is important to match the fiber cable with the appropriate SFP modules. Common standards include SX and LX. The SX standard is for shorter distances and uses 850 nm fiber cabling. The LX standard uses 1,310 nm cabling for longer distances. There are also data communication networks that use 1,550 nm for extremely long distances.

It is important that the SFPs are compatible with the equipment in which they will be installed. Be sure to match the SFP or SFP+ modules with the connectors at the ends of the cable (see Figure 10.2). The supplier should be able to recommend modules and cables that will fit the purpose and reach the target data rate.

As the use of fiber optics in data networks has evolved, a variety of standards has emerged. Early fiber-optic networks (with speeds of 10, 100, and 1,000 Mbit/s) often used larger form-factor connectors, such as SC, ST, MTRJ, or SMA connectors, where SC was the most common in corporate networking environments. Because they are smaller and easier to use, LC connectors have replaced

Figure 10.8 Using a media converter, you can connect any device with an RJ45 jack (such as a network camera) to a fiber network. This example has two RJ45 jacks and two SFP jacks.

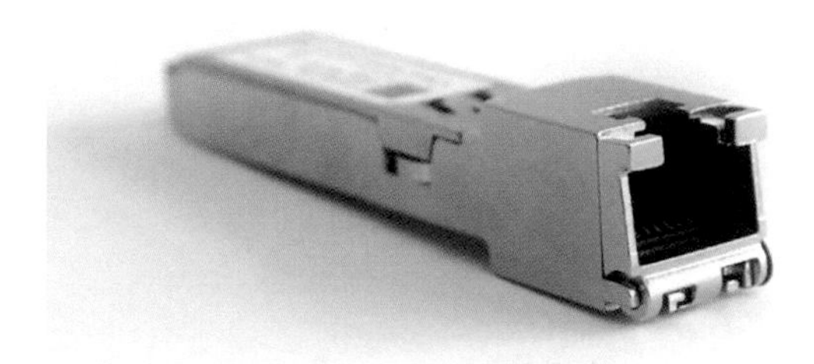

Figure 10.9 An SFP module for a twisted-pair copper cable.

SC connectors in popularity. LC connectors are often paired with SFP and SFP+ modules to connect fiber-optic cables to switches, media converters, and other network devices.

Many network video products, such as network cameras, only support a twisted-pair interface. Media converters, also known as fiber transceivers, make it possible to connect them with fiber cables anyway (see Figure 10.8).

There are also SFP and SFP+ modules for copper cables (see Figure 10.9). These provide a cost-effective way to connect devices over short distances, within racks, or to neighboring racks.

10.4 THE BASICS OF ETHERNET

The basic idea behind an Ethernet network is that the network provides a medium like the air through which all nodes can communicate. *Ether* is a word from the late fourteenth century with roots in the old French *ether*, Latin *aether*, and Greek *ather*. It means "the upper regions of space", therefore, the name *Ether-net*. An early version of Ethernet was patented in a paper in 1977. In 1980, the IEEE formed the 802 committee to develop local area network (LAN) standards, and three years later, it formally approved the IEEE 802.3 standard. Though Ethernet has undergone substantial improvements, some of the basic elements remain.

10.4.1 MAC addresses

Media access control (MAC) addresses are used as source and destination addresses. A MAC address has a 48-bit address space, which allows for potentially 2^{48}, or 281,474,976,710,656, possible MAC addresses. These are unique addresses in hexadecimal format with a length of six bytes (for example, 00-40-8C-18-32-78), where the first three bytes describe the manufacturer of the equipment and the last three correspond to the serial number of the specific device (Figure 10.10). The manufacturer sets the MAC address, and the user cannot change it. Each network device has a unique MAC address, which is often printed on the device. MAC addresses are always used when sending data

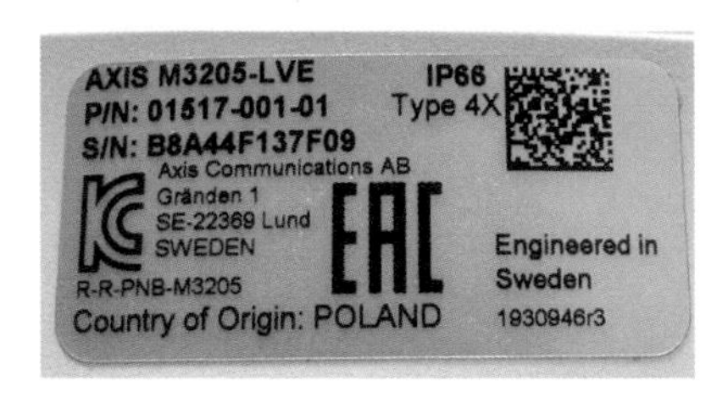

Figure 10.10 Each networking device has a unique MAC address, which is usually printed on the product label. In this example, it is shown as the serial number (S/N).

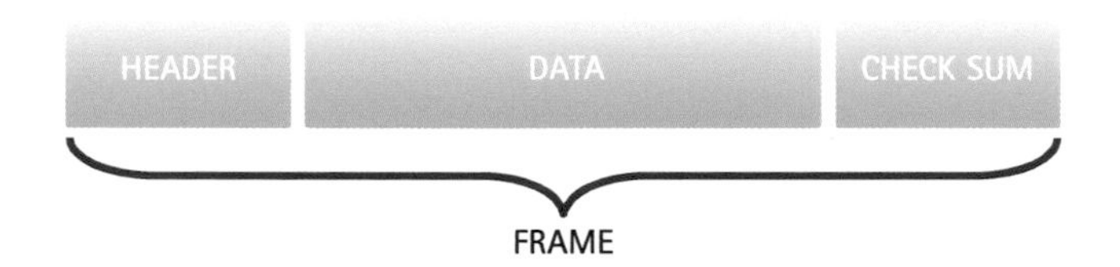

Figure 10.11 A typical frame consists of the header, the data, and the checksum. This means that only some of the bits sent over a network consist of actual data. The remainder is called overhead.

from one device to another in a network, whether in a LAN (local area network) or over the internet. For communication within a local network, MAC addressing may be all that is needed. In many cases, however, IP addresses are required in addition to MAC addresses. For more information about sending data packets, see Chapter 9.

10.4.2 Frames

When data is transferred, it is packaged into frames. A frame consists of a header, the data, and a checksum through which the recipient can recognize transmission errors (Figure 10.11). The key information in the header is the destination and source addresses of the frame. The destination address specifies the node the frame should be sent to, and the source address specifies the node that sent the frame.

10.4.3 Half-duplex and full-duplex

Simplex means that data can be sent in one direction only. Half-duplex means that data can be sent in two directions but only in one direction at a time. Full-duplex means that data can be sent and received simultaneously, which in turn means better network performance.

Coaxial cables do not provide separate transmission and receiving channels and, therefore, only support half-duplex. Twisted-pair cables and fiber-optic cables have separate transmission and receiving channels and, therefore, support full-duplex.

Ethernet products with twisted-pair interfaces normally support several data rates in half-duplex or full-duplex mode. A twisted-pair interface can adjust itself automatically to the data rate and transfer mode through auto-negotiation. Two connected nodes will automatically negotiate to find the highest common data rate and transfer mode. If the negotiation fails, many devices allow users with administrator rights to set the transfer mode manually.

10.5 NETWORKING EQUIPMENT

To network multiple nodes, you need equipment that can bridge between them. The most common network bridge is the network switch. Older networks may still use hubs. The network also uses other devices, such as a router, to connect to the internet.

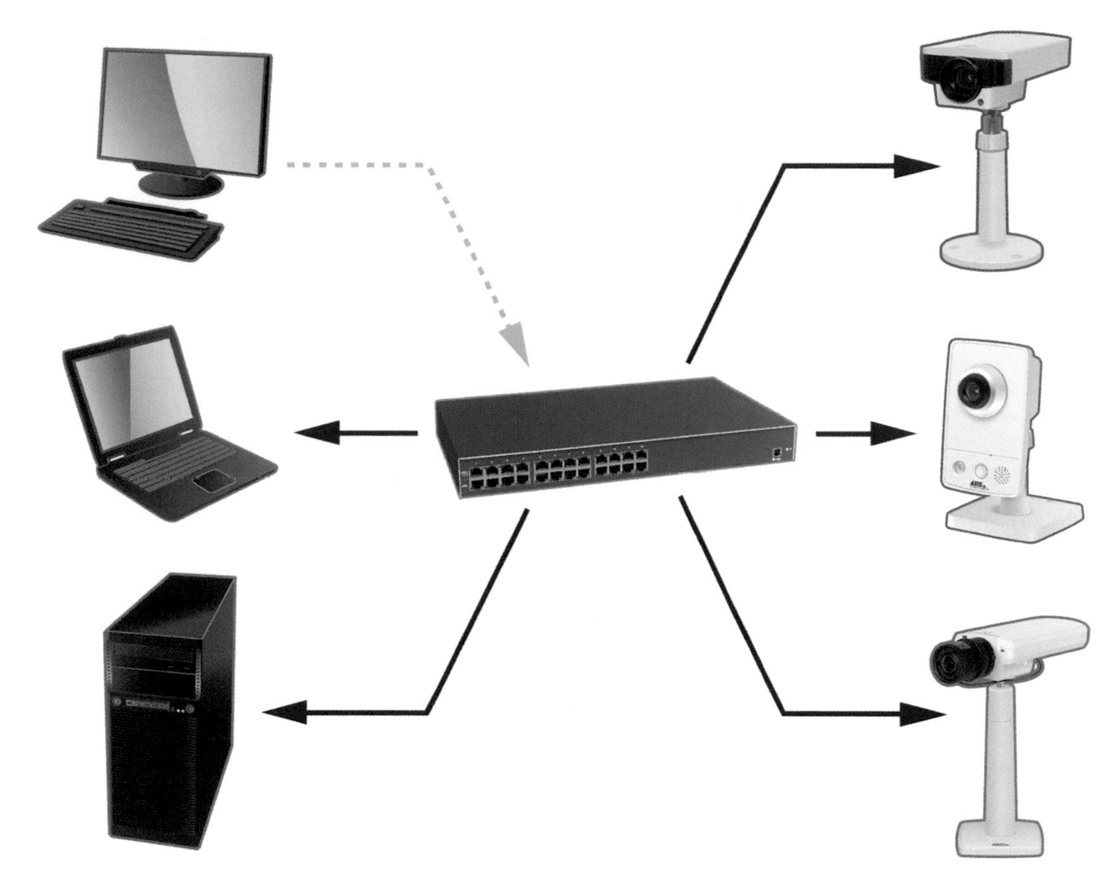

Figure 10.12 A hub is the simplest form of networking equipment. Several nodes can communicate with all other nodes connected to the hub.

10.5.1 Hubs

A hub, also called a repeater, is the simplest type of networking equipment (Figure 10.12). It works on the first layer of the OSI model (see the section about physical layers in Chapter 9). All nodes connect to the hub, forming a collision domain. Inside this collision domain, only one node can send data while all the other nodes receive data at the same time. If another hub is connected, the collision domain is extended. If a hub receives data on one connection, it sends the data to all the other connections.

In a hub environment, all nodes operate in half-duplex mode. A classic hub can support only one data rate at a time, 10 or 100 Mbit/s, meaning that all nodes in the network must support the same data rate. One exception here is the dual-speed hub, which supports two data rates. Hubs are rare in modern networks, which use switches instead.

10.5.2 Switches

Network switches are more intelligent and can forward the network traffic much more efficiently than hubs. Switches perform better and improve network security because they only send data to the devices that should receive it, rather than broadcasting it as hubs do.

Whereas a hub operates on the first layer in the OSI model, a switch manages the data also on the second layer and, in some cases, the third layer in the OSI model. This is why some switches are called Layer 3 switches. For more information about the OSI model, see Chapter 9.

There has been some confusion about what the term "Layer 3 switch" means because different vendors use it to describe different functionality. A Layer 3 switch shares some of the functionality

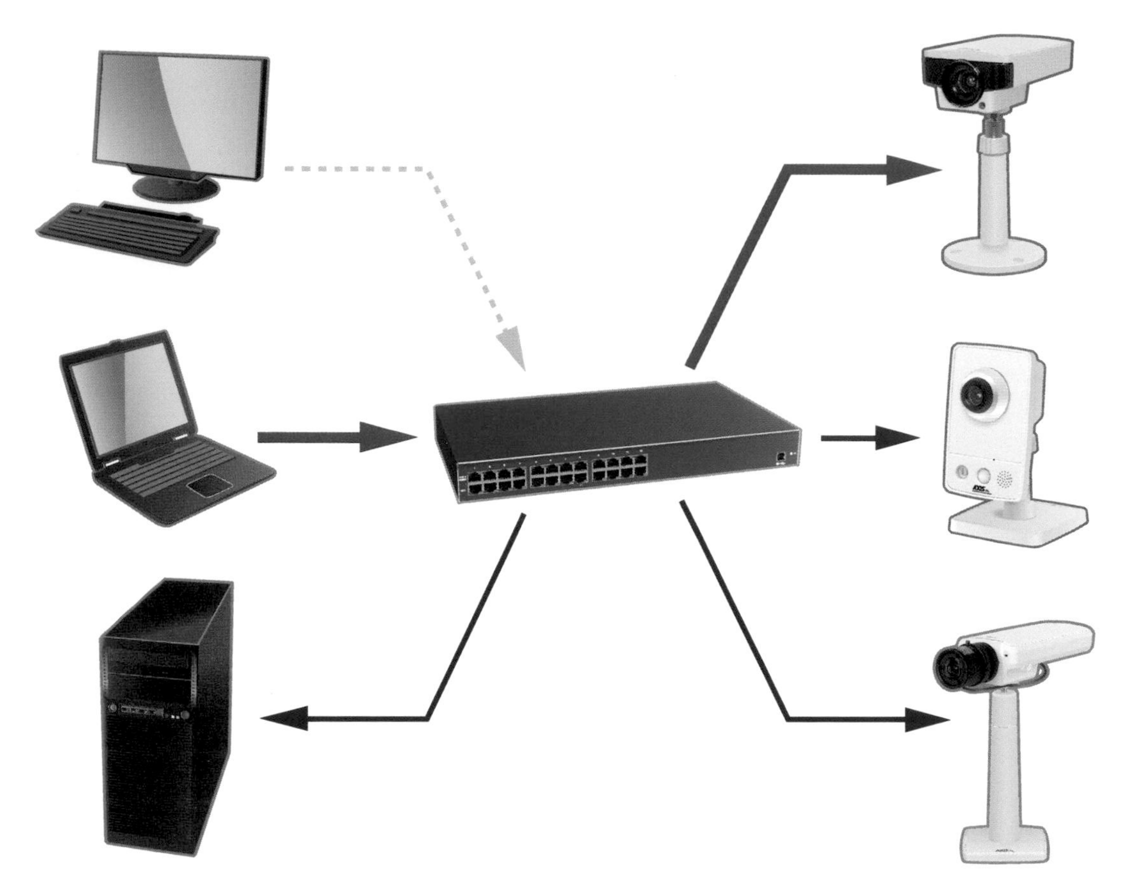

Figure 10.13 A network switch manages data transfer very efficiently because it directs data traffic between devices without affecting any other ports on the switch.

of a router (see next subsection). There are also other important Layer 3 features that are especially relevant to managing video. One is Quality of Service (QoS), which is required to manage bandwidth. Another is Internet Group Management Protocol (IGMP) snooping, which is useful in multicasting networks. Some switches with layer 3 functionality, such as IGMP snooping, are known as Layer 2+ switches. Chapter 9 provides more information about these technologies.

Data forwarding in a switch takes place through a learning process. The switch registers the address of the connected device. When the switch receives data, it forwards this data only to the port connected to the device with the correct destination address (Figure 10.13). For more information about port forwarding, see Chapter 9.

Switches typically indicate their performance in per-port rates and in backplane or internal rates (both in bitrates and in packets per second). The port rates indicate the maximum rates on specific ports. This means that the speed of a switch (such as 1,000 Mbit/s) is often the performance of each port.

A network switch normally supports different data rates simultaneously. The transfer rate and mode between a port on a switch and a connected device are normally determined through autonegotiation. This means that data always travels at the highest common rate and through the best transfer mode. A switch also allows a connected device to function in full-duplex mode, that is, to send and receive data at the same time.

10.5.3 Routers

A network router is a device that routes information from one network to another. The most common job of a router is to connect a local network to the internet. Routers only forward the data packages that are intended for another network. The main difference between switches and routers is that switches create networks, routers link networks together.

A router can forward data between completely different network technologies, thus creating a larger interconnected network (internets). Routers are sometimes referred to as gateways, as this is where they are located (i.e., where two or more networks connect). Modern integrated routers typically include a multi-port Ethernet switch, a Network Address Translator (NAT), and a Dynamic Host Configuration Protocol (DHCP) server.

10.5.4 Firewalls

Firewalls prevent unauthorized access to or from private networks. Firewalls can be implemented in both hardware and software or as a combination of both. The most typical task for a firewall is to prevent unauthorized users from accessing private networks connected to the internet. Messages entering or leaving the internet pass through the firewall, which examines each message and blocks those that do not meet the specified security criteria.

10.5.5 Bridges

A network bridge is a device or software that connects different physical networks together while filtering the data traffic between them to provide a measure of network traffic segregation. A bridge makes packet-forwarding decisions based on the MAC address of the packets' destination or destinations. It forwards packets to the network port where the particular MAC address resides.

Layer 2 switches are actually multi-port bridges that make network traffic forwarding decisions based on MAC addresses. A router, on the other hand, makes packet-forwarding decisions based on IP addresses. A Layer 3 switch is, in effect, a multi-port router. A network gateway makes forwarding decisions based on either MAC or IP addressing but adds network translation functions such as NAT, possibly provides firewall functions, and acts as a conversion point between network connectivity types such as cellular and wired.

10.5.6 Internet connections

To connect to the internet, an internet service provider (ISP) must establish the network connection.

The terms *upstream* and *downstream* are often used to describe the transfer rate between devices and the internet. Upstream describes the speed at which data can upload from the device to the internet, such as when a network camera sends video to storage in the cloud. Downstream describes the speed at which files can be downloaded or video streamed to a viewing device.

Considering that most internet users consume data rather than produce it, the download speed is usually the most important. In a network video system, however, where network cameras are located at remote sites, the upstream speed is more relevant. This is because the network cameras need to upload video to the internet.

Internet connection technologies include the following:

- *Cable:* A cable connection uses an existing cable TV connection, overlaying the data communication on the cable used for TV.
- *T1:* A T1 is a copper-based, twisted-pair connection that provides 1.544 Mbit/s for both upstream and downstream data transfer. As this bitrate is nowhere near that of other more modern solutions, T1 connections are no longer relevant for most purposes, although they can still be found in use – mostly in dedicated point-to-point connections.
- *Fiber:* A fiber-optic cable connects directly to an ISP's network, typically providing data transfer speeds of 10–20 Mbit/s up to 1 Gbit/s. Fiber supports several types of network use, for example, dedicated internet access (DIA) (as opposed to shared access), which can be beneficial for organizations with strict requirements on bandwidth and availability. Another example

is Multiprotocol Label Switching (MPLS), which routes traffic more efficiently using "labels" (paths between endpoints) instead of network addresses on private wide area networks.

The DSL, ADSL, and SDSL technologies are irrelevant in today's video surveillance networks.

10.6 POWER OVER ETHERNET

Power over Ethernet (PoE) provides the option of supplying devices with power and Ethernet through their network cable. The whole idea stems from older telephone systems where the telephone line provides both the means of communications and the required power. In the early days of IP telephony, the same functionality was mimicked using Ethernet cabling. Today, PoE is widespread and is used to power IP phones, wireless devices, and network cameras in Ethernet networks.

The primary benefit of using PoE is the inherent cost reduction. Depending on the camera location, not having to install a designated power cable can save a lot of money. It also makes it easier to move a camera to a new location or add cameras to a video surveillance system.

In addition, PoE can make a video system more secure, as the PoE system can be powered from a server room, which is often backed up with a UPS (uninterruptible power supply). This means that the video surveillance system can continue operating even during a power outage.

10.6.1 The 802.3 PoE standards

Most PoE devices today conform to the IEEE 802.3af standard or the updated version, the IEEE 802.3at standard, published in 2003 and 2009 respectively. The original 802.3af standard supports CAT3 cables for low-power devices but requires CAT5 cables for higher-power devices and to ensure stable data transfer. The 802.at standard generally requires CAT5 cables or higher, although CAT3 can be used for a Type 1 device.

The device that supplies the power is referred to as the power sourcing equipment (PSE). This functionality can be built into a network switch or provided by a midspan (see section 10.6.2). The device that receives the power is referred to as a powered device (PD). The functionality is normally built into a network device, such as a network camera, or it is provided by a stand-alone splitter (see section 10.6.2).

Backward compatibility to non-PoE-compatible network devices is guaranteed. The standard includes a method for automatically determining whether a device supports PoE, and only when this is confirmed will power be supplied to the device. This also means that if the Ethernet cable is not connected to a PoE device, it will not supply power, even if connected to a PoE switch. This reduces the risk of electrical shocks when installing or rewiring a network, as well as the risk of damaging noncompatible equipment.

In a twisted-pair cable, there are four pairs of twisted wires. Depending on the product, a PSE can either use the two unused wire pairs or it can overlay the current on the pairs used for data transmission. A PD supports both options. A Type 2 Class 4 PD supports data transmission over two pairs for 100 Mbps or four pairs for 1 Gbps.

802.3af specified a PSE to deliver a voltage of 44–57 V DC at a maximum power of 15.4 watts per port. Considering the power loss over a twisted-pair cable, only 12.95 watts are guaranteed for a PD. For 802.3.at, the specification changed to 44–57 V DC over two pairs for Type 1 PSEs and 50–57 V DC over two pairs for Type 2 PSEs.

IEEE 802.3bt was released in 2018 and is also known as "4PPoE" or "PoE++". It brings two new types, Type 3 with Classes 5 and 6 and Type 4 with Classes 7 and 8, providing greater levels of power over four pairs of wire.

Table 10.3 provides the various classes and power ranges according to the 802.3af, 802.3at, and 802.3bt standards.

Table 10.3 PoE classes and their power ranges

IEEE 802.3af PoE	IEEE 802.3at PoE+	IEE 802.3bt PoE++	Power at PSE (W)	PD power (W)
Class 0	Type 1 Class 0		15.4	12.95/13.0
Class 1	Type 1 Class 1	Type 1 Class 1	4.0	3.84
Class 2	Type 1 Class 2	Type 1 Class 2	7.0	6.49
Class 3	Type 1 Class 3	Type 1 Class 3	15.4	12.95/13.0
	Type 2 Class 4	Type 2 Class 4	30.0	25.5
		Type 3 Class 5	45.0	40.0
		Type 3 Class 6	60.0	51.0
		Type 4 Class 7	75	62
		Type 4 Class 8	90	71.3

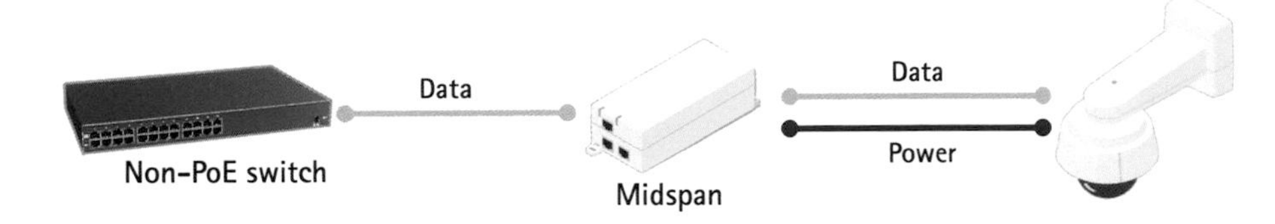

Figure 10.14 A midspan injects power into the network where it is needed but not already available.

A PSE can only deliver a certain amount of power; this is known as its power budget. For a 48-port switch with a total power budget of 500 watts, each port could potentially power one Type 1 PD with 10 watts. However, if the PD class is not known (that is, they default to Class 0), then the PSE must reserve the full 15.4 watts for each PoE port, meaning that a switch with a budget of 500 watts can only supply power on 32 of the 48 ports. If the PDs support classification, they can let the PSE know which Class they belong to. For example, for Class 2, each PD only requires 6.49 Watts, so in this case, the 500 watts will be enough to supply power to all 48 ports.

Most fixed network cameras can receive power through PoE and are normally identified as Class 1, Class 2, or Class 3 devices. PTZ (pan, tilt, zoom) and other cameras with motor control are usually Type 2 Class 4 devices (PoE+). Cameras with heaters and fans usually require even more power, typically 60 watts, which can be supplied by Type 3 Class 6 PSEs or higher.

Some PTZ camera models have a so-called low-power mode, which can be selected during the initial start-up process. Low-power mode has nothing to do with the PoE power negotiation performed during start-up. The idea here is instead to disable some or all of the heaters in the camera so that the power consumption is reduced when the ambient temperature is high enough that heaters are not needed. The aim of the low-power mode is to have the camera use less power, not to use a less powerful midspan. For example, if using a camera with IR lighting, you will still need a regular midspan to use the IR, even if the camera uses less power overall when in low-power mode.

10.6.2 Midspans and splitters

Other network devices introduced for PoE include the midspan and the splitter, also known as an active splitter (Figure 10.14), both of which add PoE to an existing network.

The midspan, which adds power to an Ethernet cable, is placed between a network switch and powered devices. Midspans with 1, 6, 12, 24, or 48 ports that support the IEEE 802.3af/802.3at Type 1 standard are readily available.

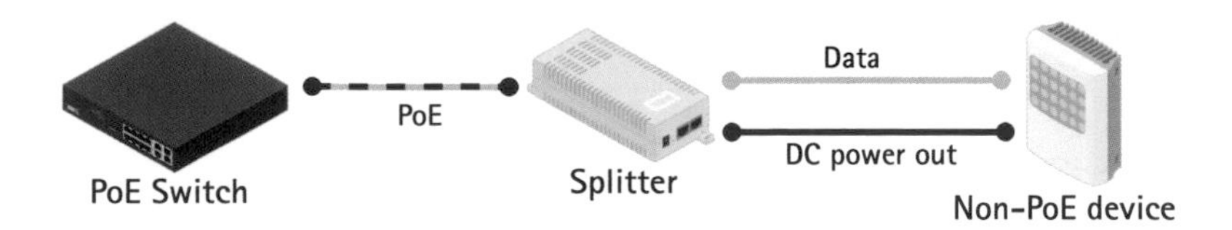

Figure 10.15 A splitter divides the data and power from the supplied PoE to provide DC power and data connectivity to non-PoE equipment.

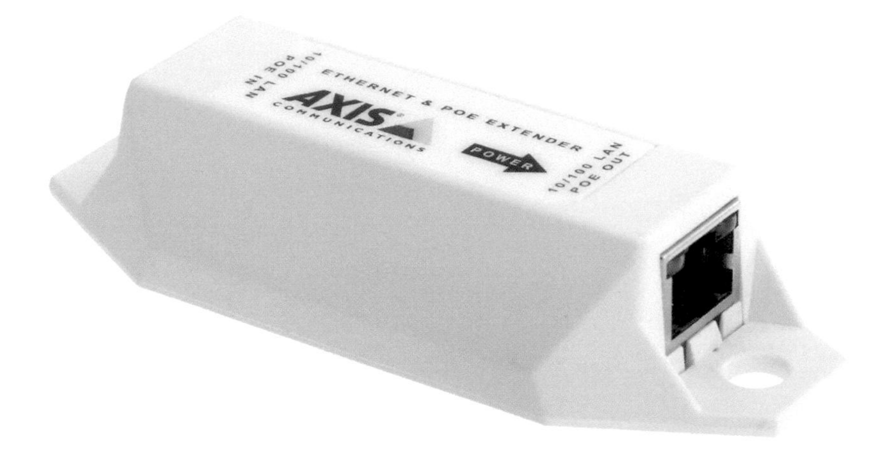

Figure 10.16 An example of a PoE extender.

A splitter splits the power and data in an Ethernet cable into two separate cables, which can then connect to a device that has no built-in support for PoE. Because a PSE delivers around 50 volts DC, another function of the splitter is to step down the voltage to the appropriate level for the device, often to 12 or 5 volts.

The maximum length of an Ethernet cable between the PSE and the powered device is 100 m (328 ft.). To cover greater distances, a switch can be used as a repeater, or optical fiber may be used. Alternatively, you can simply connect one or more PoE extenders in-line with the cabling. Depending on the midspan and the camera's power consumption type, the total connection length can be extended by up to several hundred meters.

10.7 VIRTUAL LOCAL AREA NETWORKS (VLANS)

When designing a network video system, you often want to keep the network separate from other networks for both security and performance reasons. At first glance, the straightforward approach is to build a separate network. However, the cost of purchasing, installing, and maintaining the network would probably be higher than setting up a virtual local area network (VLAN). This is a technology for virtually segmenting networks, a functionality that most network switches support. It can be achieved by dividing network users into logical groups. Only users in a specific group can exchange data or access certain resources on the network. If a network video system is segmented into a VLAN, only the servers located on that VLAN can access the cameras (Figure 10.16).

The primary protocol for VLANs is IEEE 802.1Q, which tags each frame or packet with extra bytes to indicate which virtual network the packet belongs to. Before this standard became available, several manufacturers used proprietary VLAN protocols. VLANs operate on Layer 2 of the OSI model (see Chapter 9).

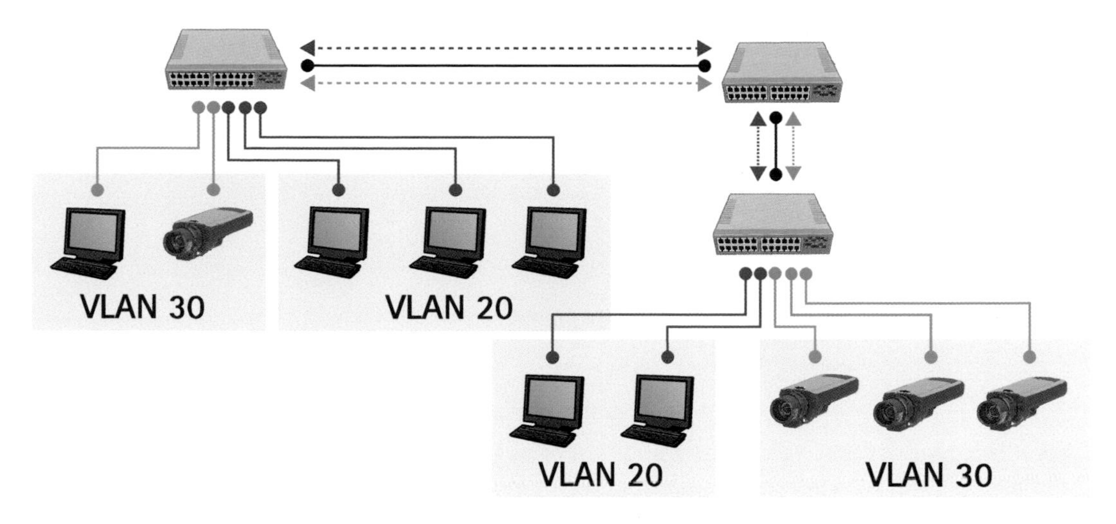

Figure 10.17 In this illustration, VLANs are set up over several switches. First, each of the two different LANs are segmented into VLAN 20 and VLAN 30. The links between the switches transport data from different VLANs. Only members of the same VLAN can exchange data, either within the same network or over different networks. VLANs can be used to separate a video network from an office network.

The four ways to assign VLAN memberships:

- *Port-based:* Each physical port on a switch is configured as either part of a VLAN or a LAN.
- *MAC-based:* Each MAC address that is part of a VLAN is listed in the switch.
- *Protocol-based:* Layer 3 data is used to determine which VLAN a frame belongs to. For example, AppleTalk® is one VLAN and IP is another.
- *Authentication-based:* Devices are placed in a VLAN based on 802.1X authentication (see Chapter 9).

10.8 CABLE INSTALLATION CONSIDERATIONS

When selecting a network cable, it is often necessary to also consider where it will be installed, as the local conditions may require particular protective properties.

10.8.1 Indoor cabling

Regular indoor Ethernet cabling has several different National Electric Code (NEC) designations, according to the type of cable jacket and its intended use:

- *CM/CMG (Communications Multipurpose General):* General purpose cabling for open areas. This cable offers no fire protection.
- *CMR (Communications Multipurpose Riser):* This type provides some fire protection and is used in riser ducts running vertically between floors and inside walls. Not for use in ventilation spaces.
- *CMP (Communications Multipurpose Plenum):* With better fire protection, this type is designed to restrict the spread of a fire and limit the amount of smoke generated, which makes it suitable for runs in ventilation spaces.

The cable types listed here are not designed to be installed outdoors and will quickly deteriorate when exposed to the elements. Nor is it enough to install indoor cabling in conduits or piping, as the plastic coating on indoor cabling affords no protection against moisture, which will be present even inside the conduit.

10.8.2 Outdoor cabling

The basic minimum rating for any cable intended for outdoors use is CMX (Communications Multipurpose, Outdoor/Residential). CMX cables have a jacket made from LLDPE (Linear Low-Density Polyethylene), which is UV-resistant and protects against sunlight, wind, rain, and extreme temperature changes. This type of cable should remain flexible and not crack due to exposure to the elements. CMX cable is not fire-resistant.

You need a CMX cable for any cable that runs up the side of a building or across a roof, whether encased in a conduit or not.

10.8.3 Underground cable runs

Besides being CMX-rated, any cable laid underground should also be of the type direct-burial cable (DBC). This type of cable requires no conduits or other protection against moisture. Depending on the exact type, DBC can be constructed using one or more of the following: multiple layers of heavy metallic sheathing, heavy-duty rubber coverings, shock-absorbing gel, waterproof tape, and stiff, heavy metal cores. Although tough and durable, remember that DBC affords no protection against mechanical damage by tools or excavation machinery.

Both regular and DBC cables may be affected by lightning strikes, even when buried. It is good practice to install surge protectors as part of any outdoor Ethernet network to guard against lightning strikes and to prevent damage to equipment.

Don't forget to test the network cable connections before burying the cable to avoid digging it up again if there should be a problem.

10.8.4 Terminating network cables

Regular twisted-pair cable is the easiest to install, as cables can be purchased pre-terminated with RJ 45 plugs. Many different lengths of complete cables are available. Twisted pair is also the easiest to troubleshoot, using tools that can indicate where in the cable a fault may lie. The tools and skills required to repair or terminate custom-length twisted-pair cables are fairly basic, and field technicians will know how to install, terminate, and repair them.

Repairing fiber is more complex, and the tools required to repair and test connections are very expensive, although pre-terminated cables are available.

10.9 BEST PRACTICES

Although today's networks generally have sufficient bandwidth available, there are several important aspects to consider when designing a surveillance video network:

- *Is there existing cabling?* If so, this can save a lot of money. Make sure the cabling is of the appropriate quality for the network speed and PoE class.
- *Does the system need new cabling?* If so, what category cable is required? Is fiber an option? When looking at the cost of installing a cable, consider the cost of labor versus the cost of the cable itself. The cable is normally a lesser part of the total cost. Select as good a cable as possible to future-proof the installation. To ensure performance and PoE compatibility, use CAT6 or better.
- *What distance does the cable need to cover?* If more than 100 m (328 ft), a fiber cable may be the best solution.
- *Is it possible to use PoE?* PoE provides huge savings, so use it for as many devices as possible. Make sure that the power available in the switch or midspan is sufficient for the devices and that the devices support power classification.

- *What are the bandwidth requirements?* Make sure that the switches provide more than the minimum required bandwidth. To future-proof a network, design it so that the original deployment only needs 50–70% of the capacity.
- *Is a WAN connection needed?* WAN (wide area network) bandwidth is normally limited. Design the system so that it does not overload the WAN bandwidth.
- *Is a VLAN or separate network needed?* VLANs normally provide better and more cost-effective solutions than separate networks.

A good rule of thumb is to build a network with greater capacity than is currently required. Whereas it is easy to replace network switches after a few years, cabling is normally much more difficult to replace.

Chapter 11

Wireless networks

Wireless networks and devices are ubiquitous in society today, and our use of them will continue to grow as billions of new devices are destined for connection to the "Internet of Things" (IoT). New products, increased data transfer speeds, better coverage, and new security technologies give us more opportunities to exchange information at lower cost and to greater advantage.

For video surveillance applications, wireless technology presents an interesting alternative to wired networks. In recent years, IP-based video applications have benefited from improved security and more plentiful bandwidth. Installing a wireless camera in a parking lot or a hard-to-reach location can often be cheaper and easier than installing cabling. Wireless also offers a way to deploy cameras over a large area quickly and efficiently, especially in urban surveillance applications.

Wireless networks also open up for mobility. A device such as a network camera can be moved around freely within the range of a wireless network. A camera in a bus or train can also be accessed live from a remote location. It is relatively simple to extend the network by using various range extenders or repeater devices. In older protected buildings, wireless networks may actually be the only alternative where the installation of network cables is prohibited.

A network camera with built-in wireless support can be integrated into an existing wireless network directly (Figure 11.1). If the camera lacks wireless support, a wireless bridge can be used to provide this functionality. (Figure 11.2).

Wireless networks come in many different forms. Wireless local area networks (WLANs) are predominantly based on the IEEE 802.11 standard. Other standards, as well as proprietary technologies are also available, some of which are of interest for video surveillance applications. Wireless networks can be designed in different ways, as point-to-point, as point-to-multipoint, and as mesh networks. Different wireless networks operate in different frequency ranges called spectra (or spectrums). One of the benefits of the 802.11 wireless standards is that they all operate in a license-free spectrum, meaning that there are no licensing fees for setting up and operating the network.

11.1 THE BASICS OF WIRELESS NETWORKS

Wireless networks involve the transmission of electromagnetic waves through the air. Ever since the Italian inventor Marconi conducted the first wireless transmission of Morse code in 1895, wireless technology has played an increasingly important role in people's lives. The basic technology remains the same, but the many developments over the years have brought about AM/FM radio, broadcast TV, mobile telephony, Bluetooth® wireless technology, and many other technologies.

DOI: 10.4324/9781003412205-11

Figure 11.1 A network camera with built-in wireless networking.

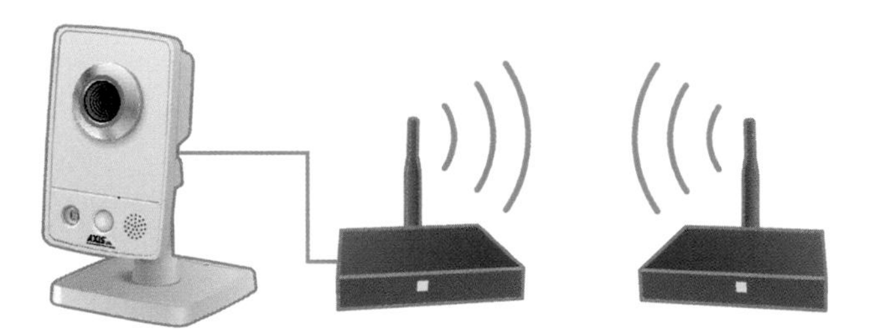

Figure 11.2 Using a wireless bridge, any network camera can connect to a wireless network.

11.1.1 The wireless spectrum

Wireless communication is transmitted at certain frequencies. These are measured in hertz (Hz), which is the unit for electromagnetic cycles per second. By using different frequencies, two or more different wireless systems can communicate at the same time, even within the same area. This is why, for example, a person can make a call on a cell phone while at the same time listening to an FM radio; these two systems are using different frequencies. If they were using the same frequency, there would be interference between them.

The risk of interference is one of the reasons we have regulatory bodies that manage radio frequencies. Some frequencies are in the licensed spectrum, which means that governmental bodies such as the Federal Communications Commission (FCC) must approve the use of certain frequency bands. Other parts of the spectrum are license-free, meaning that anyone can use them without prior approval and at no cost. See Chapter 18 for more on electromagnetic compatibility (EMC).

The lower the frequency, the greater the range of the radio signal. On the other hand, higher frequencies enable greater amounts of data throughput (bandwidth). The division and allocation of frequency ranges for different uses varies around the world. Figure 11.3 shows part of the spectra allocation in the United States. Most of the wireless spectrum from 300 MHz to 3 GHz is today already allocated. Applications such as FM radio (88–110 MHz) occupy the spectrum below 300 MHz.

11.1.2 Signal strength

There are different ways to measure signal strength in a wireless system. One way is to examine the amount of radio frequency (RF) energy transmitted. The more energy transmitted, the greater the range of the signal. The unit for energy is watts (W), and in most wireless systems, the energy is in the milliwatt (mW) range. For example, in a typical 802.11 system, the maximum energy is 100 mW at the wireless access point. At higher energy levels, the range of the radio signal will be greater but so will the risk of interference.

Although it provides an absolute value, measuring RF energy in mW is not always convenient. This is because a signal's strength does not fade in a linear fashion but instead inversely, as the square of the distance. So if the signal level at a certain distance from a wireless access point is 100, the level

Maritime radio, navigation	Maritime radio, navigation	AM radio, aviation radio, navigation	Shortwave radio	VHF television, FM radio	UHF television, mobile phones, GPS, Wi-Fi, 4G, 5G	Satellite communications, Wi-Fi	Radio, astronomy, satellite communications

100 km	10 km	1 km	100 m	10 m	1 m	10 cm	1 cm	1 mm
◀ Increasing wavelength							Increasing frequency ▶	
3 kHz	30 kHz	300 kHz	3 MHz	30 MHz	300 MHz	3 GHz	30 GHz	300 GHz

Figure 11.3 Part of the table of frequency allocations for the United States.

at twice that distance will be 25 – that is, it will have been reduced by a factor of four. The fact that exponential measurements are involved in measuring signal strengths is one reason why the use of a logarithmic scale was developed as an alternative way of representing RF power. The dBm (dB milliwatt) is a logarithmic measurement of signal strength, and dBm values convert exactly and directly to and from mW, according to the following formula:

$$p_{dBm} = 10_{log10}(P_{mw})$$

Where P is the signal strength. Therefore, a signal strength of 100 mW = 20 dBm. Halving the signal strength in mW reduces the signal strength by three dBm.

Other ways of measuring the quality of a signal are relative (i.e., not absolute). One such method is known as the receive signal strength indicator (RSSI) value, which is measured through a mechanism defined by the IEEE 802.11 standard. Here, RF energy is measured using the circuitry on a wireless network interface card (NIC) and is then represented by a value in the numerical range 0–255. Different vendors interpret the RSSI value differently, and the *RSSI_Max* value may also be selected arbitrarily within this range.

11.1.3 Antennas

The antenna is an important part of a wireless system, and a better antenna will increase the range of the wireless connection. The performance, or gain, of an antenna is measured in decibel isotropic (dBi), which describes the gain a given antenna has over a theoretical isotropic (point source) antenna, normally in the range 1–20 dBi. The effective radiated power (ERP) is defined as the effective power of the transmitter antenna. This is equal to the sum of the antenna gain (in dBi) plus the power (in dBm) into the antenna. For example, a 12 dBi gain antenna fed directly with 15 dBm of power has this ERP:

15 dBm + 12 dBi = 27 dBm (500 mW)

A longer antenna normally has a higher gain. The appropriate length of an antenna depends on the frequency. To enable the highest efficiency, the antenna length should be a multiple of the wavelength.

The two main types of antennae are omnidirectional and directional (Figure 11.4). An omnidirectional antenna spreads the electromagnetic waves in all directions, making it appropriate for multipoint networks where wide coverage is required. A directional antenna focuses the electromagnetic waves in a single direction, enabling the radio signal to reach further but only in that direction. Directional antennas are appropriate for point-to-point applications and come in different types, such as Yagi antennas, patch antennas, and parabolic grid antennas. Many have a gain between 10 and 30 dBi.

Figure 11.4 Communication towers equipped with omnidirectional antennas (stick-shaped) and directional antennas (rounded).

11.1.4 Radio wave propagation

Although open air is best for transmitting radio signals, radio signals do not require free line-of-sight for transmission. This means that radio systems are so-called non-line-of-sight systems, as opposed to optical transmission systems, where the receiving and the transmitting ends must be able to "see" each other. Some systems have better ranges, some have the ability to pass through certain materials, and some have both, as explained in the following.

A wireless signal propagates best through open air, but technically, it can pass through most other media, such as the wood and concrete present in buildings. Structures containing large amounts of metal present difficulties for radio signals, as signals reflect off metals instead of penetrating them. A building containing or covered by a metal with high conductivity, such as copper, will be virtually impenetrable for a wireless signal. Such a building or room is called a Faraday cage, after the scientist Michael Faraday.

Radio signals at lower frequencies and longer wavelengths propagate further than signals at higher wavelengths. Lower radio frequencies also have a better ability to penetrate materials such as wood or concrete. This is why a signal from an AM radio station can be received on the other side of an ocean, whereas the FM radio channels we listen to in a car must be changed every so often as we move along the road.

11.2 WIRELESS NETWORK ARCHITECTURE

Depending on the application, wireless networks can be designed in different ways. In some scenarios, such as for a single camera in a parking lot, a point-to-point wireless connection would be the best choice for relaying video to the building where the video surveillance system is located. In a surveillance application with hundreds of cameras, a point-to-multipoint or even a mesh wireless network may be more appropriate. These three types of wireless architecture are explained in more detail in the following sections.

11.2.1 Point-to-point network

A point-to-point network (sometimes abbreviated as PTP or P2P) is the simplest wireless network, as it transmits information from one point to another (Figure 11.5). Because of the nature of the system, directional antennas are used to provide the highest bandwidth for the link. The system can also be adjusted for minimal interference and the highest level of security. If there is direct line-of-sight, it is also possible to use a high-performance optical link, for example, between two buildings located on opposite sides of a highway.

11.2.2 Point-to-multipoint network

A point-to-multipoint network (sometimes abbreviated PTMP or P2MP) is the most common type of wireless network (Figure 11.6). One example of such a network is an FM radio station transmitting radio signals to many receivers. Because of the nature of the network, the central point uses an omnidirectional antenna. The surrounding points use directional antennas, unless they are mobile (for example, in a car), in which case an omnidirectional antenna is preferred. A point-to-multipoint network can be of broadcast and simplex nature, as in the case of FM radio, or it can be full-duplex, with data being sent in both directions, as in a regular wireless local area network (WLAN) application.

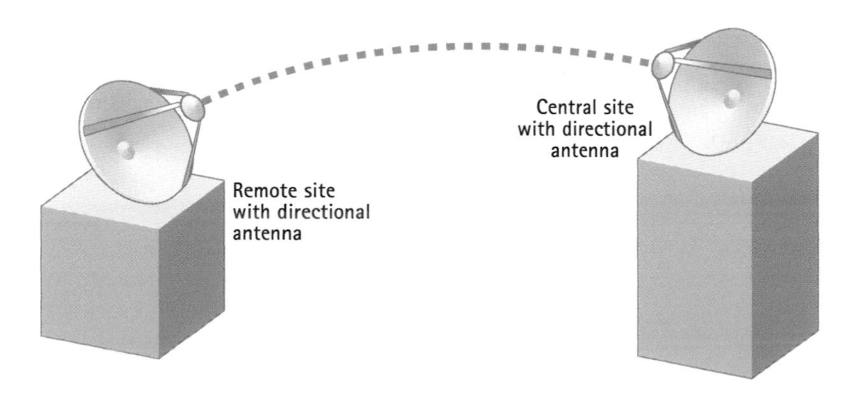

Figure 11.5 A point-to-point wireless network.

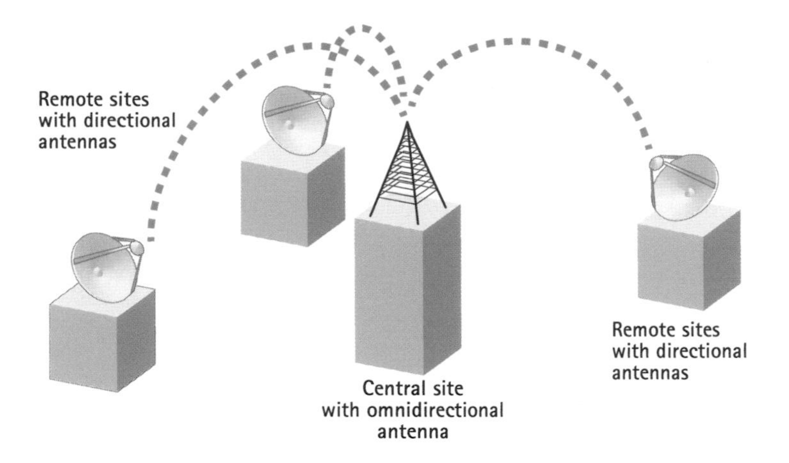

Figure 11.6 A point-to-multipoint wireless network.

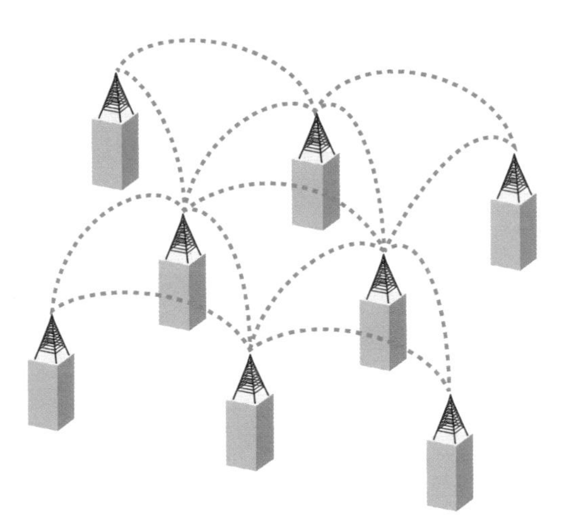

Figure 11.7 A wireless mesh network.

11.2.3 Mesh network

A wireless mesh network (Figure 11.7) is characterized by several connection nodes that provide individual and redundant connection paths from one node to another. To accommodate this, special routing protocols are used to guarantee data exchange through the most appropriate connection path. When selecting a path, factors such as bandwidth, transfer errors, and latency are considered. The number of nodes between two data points is defined as the number of *hops*. The more hops, the longer the latency. Keeping latency and the number of hops to a minimum is important in applications such as live video, particularly in cases where PTZ cameras are used.

A wireless mesh network should be capable of managing itself. This means that if a node goes down, the system will automatically set up a new path between two points. Wireless mesh networks have now developed to the point where they are a viable solution for video surveillance applications. Most current solutions are based on standard 802.11 technologies, with proprietary routing and security protocols.

11.3 802.11 WLAN STANDARDS

The most common wireless standard for data networks is 802.11, which was published in 1997 by the IEEE and specifies a media access control layer and different physical layers. The two original transfer protocols for wireless communication described by the 802.11 standard: The frequency-hopping spread spectrum (FHSS) and the direct sequence spread spectrum (DSSS), both of which provide data transfer at 1 and 2 Mbit/s. Both protocols use the 2.4 GHz frequency spectrum, which is license-free all over the world, although in some regions the signal strength is limited to 20 dBm (100 mW). The 802.11 standard has also been improved and expanded, with the most important extensions outlined in the following subsections.

11.3.1 The 802.11b extension

The 802.11b extension was approved in 1999. It uses DSSS in the 2.4 GHz range, giving data rates up to 11 Mbit/s. Until 2004, most WLAN products were based on 802.11b, and products supported data rates of 1, 2, 5.5, and 11 Mbit/s, depending on the distance.

11.3.2 The 802.11a extension

The 802.11a extension was also approved in 1999 and is based on the orthogonal frequency division multiplexing (OFDM) protocol. It operates in the 5 GHz frequency range and allows data rates of 6, 9, 12, 18, 24, 36, and 54 Mbit/s. Although higher data rates are possible, 802.11a also has some drawbacks and never became very popular.

11.3.3 The 802.11g extension

The 802.11g extension was approved in 2003 and quickly established itself as the new standard, replacing 802.11b. Wireless products that are 802.11b/g compliant use the 2.4 GHz frequency and provide data rates up to 11 Mbit/s with DSSS and up to 54 Mbit/s with OFDM.

11.3.4 The 802.11n extension

The 802.11n extension was approved in 2009 and enables data rates of up to 600 Mbit/s. These high rates are achieved with the multiple input multiple output (MIMO) protocol, in which multiple antennas and spread paths for the electromagnetic waves provide several transfer routes that can be used in parallel to make the high rates possible. An 802.11n wireless network can operate in both the 2.4 and the 5 GHz band but will provide the highest capacity when used exclusively at 5 GHz, thanks to the many non-overlapping channels and lesser degree of interference in this band.

11.3.5 The 802.11ac extension

The IEEE 802.11ac extension was approved in 2014, building on and improving the performance of the 802.11n standard, although this later standard operates exclusively in the 5 GHz band. Depending on operating conditions, data throughput can be 1.7–2.5 Gbit/s. The IEEE 802.11ac extension also includes beamforming, which can direct radio signals at specific devices, increasing overall throughput and reducing power consumption. Products referred to as 802.11b/g/n/ac are compatible with all four extensions.

11.3.6 The 802.11s extension

The 802.11s extension, approved in 2011, makes it possible to create vendor-neutral mesh networks, with interoperability between many different types of devices from different manufacturers.

11.3.7 The 802.11ax extension

Operating in both the 2.4 GHz and 5 GHz ranges, 802.11ax was published in 2021 and replaces 802.11ac. This extension supports data throughput up to 10 Gbps and is more reliable in busy networks. It also provides greater security through the use of Wi-fi Protected Access 3 (WPA3). Power consumption is reduced by the use of Target Wake Time.

11.3.8 Wi-Fi generation naming

As further extensions of the 802.11 standard are released, so does the confusion surrounding exactly what each extension supports. To rectify this situation, the branch organization Wi-Fi Alliance® launched the use of simplified generational names to give vendors and consumers an easier reference system to aid in purchasing decisions. The names of the major extensions are as follows:

- *Wi-Fi 4:* Supports 802.11n
- *Wi-Fi 5:* Supports 802.11ac
- *Wi-Fi 6:* Supports 802.11ax
- *Wi-Fi 7:* 802.11.be (under development)

11.4 THE BASICS OF 802.11 NETWORKS

Different types of networks can be set up using the 802.11 technology. The following subsections describe the basic 802.11 topologies and the frequencies in which they operate, as well as the significance of channels.

11.4.1 Infrastructure network

An access point has an Ethernet interface and at least one WLAN interface. Multiple access points can be used to cover a large area, whereby the individual access points must be positioned so that neighboring wireless cells slightly overlap one another to guarantee seamless transmission, as shown in Figure 11.8. In a case like this, it is important that the wireless cells in the immediate vicinity use different channels to avoid interference. After fulfilling these requirements, a user with a wireless device can move around the WLAN's complete coverage area without losing the connection to the network or suffering performance losses.

Moving between wireless cells is known as roaming. The access points that form a network are identified by the service set identifier (SSID). The SSID must be configured at each access point, and the nodes detect which access points belong to which network using the SSID, that is, which

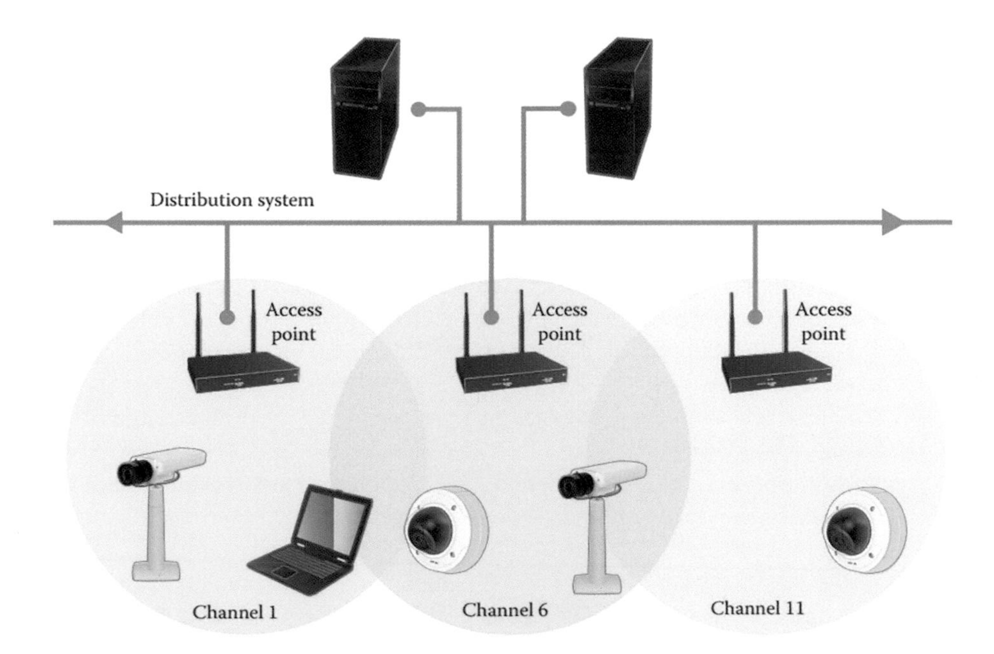

Figure 11.8 To provide good coverage, the areas of the access points should overlap slightly.

access points a node can associate with while roaming. If a device such as a network camera is to be integrated into an infrastructure network, both the infrastructure operating mode and the SSID should be selected in order to connect to the correct WLAN network.

11.4.2 802.11 frequencies

As discussed previously, an 802.11 WLAN operates in the 2.4 GHz or 5 GHz range. Both frequencies have their advantages and disadvantages. Networks that operate in the 2.4 GHz frequency are still the most common, but the greater throughput of 802.11n, 802.11ac, and 802.11ax also makes the 5 GHz band a popular choice. One of the major disadvantages of the 2.4 GHz frequency is that many other wireless technologies, such as Bluetooth, also use this frequency, with interference and reduced data rates as a result.

Although the 5 GHz frequency range offers less interference, there are also disadvantages. The higher the signal's frequency, the lower the range. This is because the damping of electromagnetic signals increases as a function of the increased frequency. This can be particularly noticeable when signals must penetrate barriers such as walls or furniture. The signal range on the latest 5 GHz access points is much improved compared to earlier models, which required more units for the same area than comparable models using the 2.4 GHz range.

11.4.3 Channels

The 2.4 GHz range is divided into 14 channels. Channels 1–11 may be used freely everywhere, whereas the use of channels 12, 13, and 14 is restricted or may only occur at low power.

Channels are divided in such a way that neighboring channels overlap. When setting up a WLAN in the 2.4 GHz range, the open channels should be separated by several clear channels to avoid interference. This means that only three independent wireless channels will be open in a single wireless cell, for example, channels 1, 6, and 11, and that these separated channels must be shared between the access points.

11.5 WLAN SECURITY

When data is transferred over a wireless link, it is important to consider that the fixed boundary available in a wired LAN does not exist in the wireless alternative. In theory, anyone within range of a WLAN could attack the network and intercept data. Consequently, security becomes even more important in preventing unauthorized access to data and the network. This section discusses some of the most common security technologies in wireless networking.

11.5.1 Wired Equivalent Privacy (WEP)

The basic 802.11 standard includes Wired Equivalent Privacy (WEP), a security protocol intended to provide a level of security comparable to data transferred by cable. WEP is used to encrypt and authorize data. Encoding is based on a symmetrical encryption protocol, which is based on RC4 Stream Encryption from RSA. All nodes and access points that exchange data must have the same secret WEP key so that the data can be encrypted and decrypted. Depending on the WEP version, various key lengths are used. In the case of WEP40, the key length is 40 bits, and in the case of WEP128, the key length is 128 bits.

Unfortunately, security loopholes in this protocol mean that WEP is not considered an adequate security method, and its use cannot be recommended.

11.5.2 Temporary Key Integrity Protocol (TKIP)

The Temporary Key Integrity Protocol (TKIP) was introduced to provide security reliably through firmware or driver updates for previously installed wireless products. TKIP addresses the security loopholes in WEP and is software implemented on top of WEP.

11.5.3 Advanced Encryption Standard (AES)

The Advanced Encryption Standard (AES) is an encryption algorithm recommended by the National Institute of Standards and Technology (NIST). This algorithm provides a high level of security and efficiency and is implemented in hardware, that is, the router chip. AES in WLAN applications uses a key length of 128 bits, although it also supports 192-bit and 256-bit keys.

11.5.4 Pre-Shared Key (PSK)

TKIP and AES both require keys from which a session key is derived. The Pre-Shared Key (PSK) protocol or authentication to 802.1X can be used to derive a session key. PSK is simpler than authentication to 802.1X. When using PSK, the key derives from a passphrase. The passphrase has a length of 8–63 characters and must be entered manually into each node and access point. A PSK passphrase is secure when based on a meaningless string of letters, numbers, and special characters, for example, 3aRs5%3?&d48fgH67, so that a so-called dictionary attack cannot take place.

11.5.5 802.1X

Authentication to 802.1X can be used as an alternative to PSK. In 802.1X, the individual nodes must identify themselves to a Remote Authentication Dial-In User Service (RADIUS) server before they can transfer data over a network. If authentication is successful, a *pairwise master key* is generated, from which the session key is derived. The use of 802.1X requires a centralized RADIUS server, which has the advantage of being simple to manage. See section 12.10.2 for more on 802.1X.

11.5.6 Wi-Fi Protected Access®

When the security risks of WEP became apparent, the Wi-Fi Alliance® was forced to react. In 2003, the alliance defined Wi-Fi Protected Access® (WPA®) as the new security standard, using TKIP as the encryption technology. WPA2 began to be used in 2004 and corresponds to the AES implementation of 802.11i. Despite claims of successful WPA2 cracking, the standard is still considered safe for most applications, as long as settings and passphrases are sufficiently strong.

The latest version of the standard is WPA3, released in 2018. When used in WPA3-Enterprise mode, it provides the equivalent of 192-bit cryptographic strength. WPA3 discontinues the use of the Pre-Shared Key (PSK) exchange and instead uses Simultaneous Authentication of Equals (SAE) exchange for a more secure initial key exchange in personal mode.

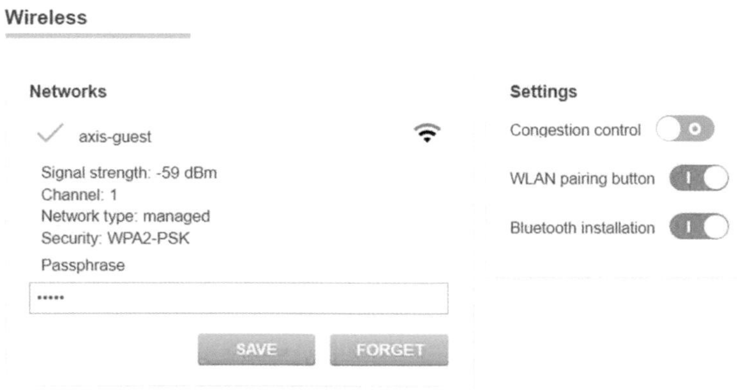

Wireless

Networks

axis-guest

Signal strength: -59 dBm
Channel: 1
Network type: managed
Security: WPA2-PSK

Passphrase

SAVE FORGET

Settings

Congestion control

WLAN pairing button

Bluetooth installation

Figure 11.9 Example of wireless settings in a network camera.

11.6 OTHER WIRELESS SOLUTIONS

In addition to the 802.11 WLAN standards, there are many other technologies available for the wireless transfer of data. Some of these are described next.

11.6.1 Bluetooth®

Bluetooth was originally a wireless solution that enabled a data transfer rate of 1 Mbit/s in the 2.4 GHz frequency range over short distances. The original key application for Bluetooth was to link peripherals wirelessly over a distance of up to 10 meters (33 feet). The standard has been continually extended so that today data rates can reach 24 Mbit/s, and distances of up to 100 meters can be bridged. Bluetooth is not commonly used in video surveillance applications.

11.6.2 Universal Mobile Telecommunications System (UMTS)

The Universal Mobile Telecommunications System (UMTS) is a mobile communications standard of the third generation and includes a multitude of technologies. 3G was the first technology generation to offer bitrates high enough for data transfer and its use is now widespread.

The 4G standard is specified by the International Mobile Telecommunications Advanced (IMT-Advanced) specification. This requires a 4G technology to provide peak bitrates up to 100 (Mbit/s) for mobile applications (for example, in moving vehicles) and up to 1 Gbit/s for more-or-less stationary use. 4G in its various forms is also in widespread use.

The successor to 4G networks currently being rolled out in many countries is generally known as 5G. The standards for 5G are set by the 3rd Generation Partnership Project (3GPP), which defines 5G as "any system using 5G NR (5G New Radio) software". Minimum standards are set by the International Telecommunication Union (ITU). As well as providing higher throughput speeds, 5G also has greater capacity and can thus connect many more devices in the same geographical area. It is predicted that 5G networks will increasingly take on the role of internet service provider (ISP) for many applications.

The data rates available now and those coming in the near future make 5G networks of great interest to video surveillance applications. Some devices such as body-worn cameras have built-in 4G or 5G technology for streaming live video over the mobile network.

11.6.3 Z-Wave®

The Z-Wave protocol was first introduced in 1999 as a consumer-level control system for lighting. Today, it has evolved into a complete home automation network mesh protocol and proprietary system that allows up to 232 smart devices to connect to each other in a single network. A maximum of four signal hops between devices provides the user with access to any device within the network. Apart from the various smart devices, a Z-Wave network also has a controller, or smart hub, which is normally the only device connected directly to the internet.

Z-Wave operates on the low-frequency 908.42 MHz band in the US and the 868.42 MHz band in Europe. It offers transmission rates up to 100 kbps for small data packets. The maximum line-of-sight signal range is 100 m (328 ft).

Z-Wave Plus is an improved version of the original. The hardware platform in Z-Wave Plus gives greater line-of-sight range – up to 150 m (492 ft) and up to 50% longer battery life.

Z-Wave is not used for video, but various Z-Wave devices can be useful when incorporated into a video surveillance system (see Figure 11.10 for examples).

11.6.4 Zigbee®

Like Z-Wave, Zigbee is a low-power, low–data rate, close-proximity network protocol. A Zigbee network can support more than 65,000 devices, with unlimited hops between devices. Zigbee was

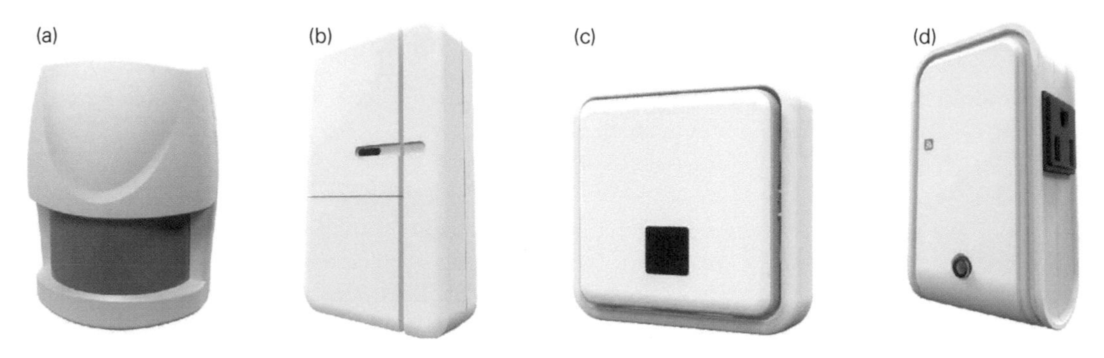

Figure 11.10 Examples of Z-Wave devices. (a) Passive IR sensor. (b) Door/window sensor. (c) Wireless alert button. (d) Wireless power on/off plug.

first released in 2005. It uses multiple protocols and open standards and is licence-free. As well as being used in home automation, it also has specifications, for example, for smart energy applications, healthcare applications, and vehicle applications. Zigbee operates mainly at the higher frequency of 2.4 GHz.

11.7 THE PERFORMANCE OF WIRELESS NETWORKS

The process of transferring data wirelessly is much more difficult than transferring it by cable, and there are significant data overheads for management, error recognition, and security. In the case of wired Ethernet, the overhead is relatively small, that is, the net data rate is close to the gross data rate.

The net data rate for a WLAN often amounts to considerably less than the gross data rate and is sometimes as low as 50% of the specified gross. When designing a wireless network for a video surveillance application and calculating the required bandwidth, the performance of the network must be taken into consideration to ensure that the desired frame rate is achievable.

11.8 BEST PRACTICES

Before selecting the technology and implementing a wireless network, there are some important things to consider:

- *How many nodes will be networked?* For two nodes only, a point-to-point connection may be sufficient, and several proprietary and standard solutions are available. If multiple nodes are involved, a multipoint or mesh network solution should be used.
- *What is the distance to be bridged?* The higher the frequency, the shorter the distance that can be covered. For very long distances, solutions at a lower frequency range, such as the 900 MHz range, may be more appropriate.
- *Are there any obstacles in the area of the wireless network?* A standard wireless network may have a range of 30–100 m indoors but up to 300 m outdoors. Buildings, trees, and other obstacles will dampen the wireless signal and limit its reach. The fewer the obstacles in the signal's path, the greater the distance that can be bridged.
- *How much bandwidth is required?* Most wireless solutions specify the maximum bandwidth. The real bandwidth depends on the distance and obstacles and may be much less than the specified maximum.
- *What is the risk of interference?* It is important to determine whether other wireless networks will be installed in the same area. Check existing equipment in the area with a spectrometer, which will be able to indicate the frequencies in use.

- *How secure does the connection need to be?* Wireless means sending data over the open air. If used for video surveillance, it is important to make the data secure by using appropriate security protocols, such as 802.1X. Remember that adding security protocols normally decreases performance.

Although a wireless network can replace a wired network, there is still a need to supply power to all nodes. Power over Ethernet (PoE) removes the need for a separate power cable but retains the data cable. For most indoor video surveillance applications, PoE is more cost-effective and more secure than wireless. For more information on Power over Ethernet, see section 10.6.

CHAPTER 12

Cybersecurity

The shift from analog to network video – in which all cameras have an IP address and stream video over standard IT networks – has had a profound impact on the capabilities and the proliferation of video surveillance. But it has also introduced new risks in the area of cybersecurity. The risks are twofold: First, there is the risk that someone outside the organization manages to get access to video from the cameras. The second and greater risk is that a hacked video surveillance system becomes the entry point to other more sensitive systems, such as those containing financial data, personal data, source code, etc. Therefore, any successful video surveillance implementation today needs to take cybersecurity measures into consideration.

12.1 WHAT IS CYBERSECURITY?

Cybersecurity focuses on the threat prevention and risk management of digital data, as well as the protection of associated physical hardware and infrastructure. The main goal is to protect digital data stored on computers, networks, mobile devices, etc., from being compromised, stolen, or held to ransom. By extension, this also includes mitigating the risk of disruption or damage to the services that use this data.

Cybersecurity is a part of "information security", which is a broader concept covering the confidentiality, integrity, and availability of data in both digital and physical form. Information security practices have been around a lot longer than cybersecurity, as there was always a need to secure paper documents, long before everything went digital.

Computers and other digital devices have long been subject to viruses and other forms of attack, with the very first computer virus being detected in the early 1970s. While many of these early attacks were not particularly malicious in their intent, they did demonstrate the need to implement security measures to prevent unauthorized access to systems.

In 1983, the Massachusetts Institute of Technology (MIT) was granted the first United States Patent for a "cryptographic communications system and method". This patented system was used to create modern protections such as the Secure Sockets Layer (SSL) encryption protocol. The granting of this patent is viewed by some as the starting point of what we today know as cybersecurity.

In today's world of constantly connected systems, attacks on digital assets have become much more frequent, ever more advanced, and they often have serious consequences for large parts of society. IT managers and security professionals need to invest considerable thought and resources in protecting their assets, and video surveillance systems are no exception.

There is no single solution to the challenges of cybersecurity, and no such thing as a 100% secure system – at least not if the system should also be usable. A system can be made more secure by

DOI: 10.4324/9781003412205-12

Figure 12.1 The five pillars of the NIST Cybersecurity Framework.

Source: National Institute of Standards and Technology

reducing exposure areas and mitigating risk, but there will always be some level of risk that needs to be accepted. Cybersecurity is more a question of trusted partnerships, where everyone, from suppliers to manufacturers, from installers and integrators to end-users, has an important role to play. This is an ongoing process and not simply a one-off undertaking.

Because cybersecurity is essentially a question of managing risk, a good starting point is to evaluate potential risks to your business or organization in terms of their probability and their potential level of harm using a risk management framework, of which there are many. One such framework is the National Institute of Standards and Technology's Cybersecurity Framework (NIST CSF).

The NIST CSF guidelines are used globally and are appropriate not just for large businesses and organizations but also for small and medium ones. The NIST CSF guidelines revolve around five pillars or functions:

- *Identify:* Understanding the context, the resources supporting critical functions, and related cybersecurity risks enables an organization to focus and prioritize its efforts.
- *Protect:* Limiting or containing the impact of potential cybersecurity events to ensure continued delivery of services.
- *Detect:* Implement appropriate activities for the timely discovery of cybersecurity events.
- *Respond:* Containing the impact of a detected cybersecurity incident.
- *Recover:* Returning to normal operations after a cybersecurity incident.

12.1.1 Other risk management frameworks and standards

The same compliance regulations met by the big data storage players also apply to cloud video providers. Data centers should be audited, and it is also possible to attain certification that guarantees compliance with a best-practice framework for secure information management. The following section outlines some of the relevant standards.

12.1.1.1 Standards for Attestation Engagements

The Statement on Standards for Attestation Engagements no. 16 (SSAE 16) was put forward by the Auditing Standards Board (ASB) of the American Institute of Certified Public Accountants (AICPA). SSAE 16 replaced the Statement on Auditing Standards no. 70 (SAS 70) in 2011. The reasons for the standard included the intention to align with the global standard for reporting on controls at service organizations, International Standard on Assurance Engagements (ISAE) No. 3402, and to overcome the limitations of SAS 70. In 2016, AICPA published SSAE 18 – a clarified and updated version of the statement.

SSAE 18 regulates how service companies report on compliance controls and, unlike SAS 70, includes a detailed description of the system, its services, processes, policies, procedures, staff, and operational activities, as well as a written statement of assertion. ISAE No. 3402, Assurance Reports on Controls at a Service Organization, was issued in 2009 by the International Auditing and Assurance Standards Board (IAASB), which is part of the International Federation of Accountants (IFAC). Although SSAE 18 and ISAE No. 3402 align well with each other, service organizations with global operations have good reason to examine both standards.

12.1.1.2 The ISO/IEC 27001 standard

ISO/IEC 27001 is part of the international ISO/IEC 27000 standard series on information technology. It was revised in 2013 and helps data centers and other organizations to identify and treat security risks and set up tools for managing information security management systems. The requirements are generic and intended for all organizations, regardless of type or size. ISO/IEC 27001 is a best-practice framework and certification is not mandatory. However, gaining an official certification by an accredited certification body can give service organizations paybacks, such as reductions in incidents, faster system recovery, greater security awareness, more credibility, and status as a preferred supplier.

12.1.1.3 The Federal Information Security Management Act

The Federal Information Security Management Act (FISMA) is a US federal law passed in 2002. Federal agencies and their contractors must follow the act, which regulates information system policies, procedures, and practices. In essence, it is a collection of standardized best practices for strengthening information security and reducing security risks. Companies dealing with federal agencies can choose to outsource to a FISMA-compliant data center or try for on-site compliance. On assignment by FISMA, the National Institute of Standards and Technology (NIST) is responsible for developing standards and guidelines for choosing, categorizing, and evaluating non-national security federal information systems. FISMA also requires annual reviews of the effectiveness of the agency's information security program and that the result is reported to the Office of Management and Budget (OMB). The OMB uses this information to prepare an annual congressional report on agency compliance with FISMA.

12.1.1.4 The European Union Agency for Network and Information Security

The European Union Agency for Network and Information Security (ENISA) is a body of expertise that helps the European Commission (EC), its member states, and the business community to address, respond to, and prevent network and information security problems. Part of this work is to assist the EC with legislative preparations in the field of network and information security. ENISA does not inspect, regulate, or enforce information security laws. Both legislation and enforcement are handled on a European and national level through EU bodies such as the European Data Protection Supervisor (EDPS), Europol, and national agencies.

12.2 RISK

Cybersecurity is about managing risks over time. While risks can be mitigated, it is very rare that they can be eliminated completely.

RFC 2828 Internet Security Glossary defines risk as "an expectation of loss expressed as the probability that a particular threat will exploit a particular vulnerability with a particular harmful result".

This can be boiled down to a formula for prioritizing risks:

Risk = Probability × Impact

Note that the RFC definition presented here uses the label "particular" for threat, vulnerability, and harmful result. This means that each threat should be considered individually, starting with the one that is most plausible and has the greatest negative impact.

The process for analyzing risk in cyberspace is the same as for physical protection. The questions to consider are as follows:

- What do you want to protect?
- From whom do you want to protect it?
- How likely is it that you will need to protect it?
- How bad will the consequences be if you fail?
- How much time and money are you willing to invest in trying to prevent those consequences? Is the cost of the preventative measures less or greater than the potential loss of the asset?

Implementing any security control measure results in some type of cost. If you do not know what the risks are, it will be very hard to estimate the budget for protection.

One challenge when discussing risk is probability – a thing might happen or it might not. The probability of a vulnerability being exploited is related to how easy it is to do, as well as to the potential benefit for the attacker.

We can plot risks graphically using the probability that a risk will occur on the one axis and the impact of the risk (if it occurs) on the other axis. This gives a clear view of the potential impact and priority that you need to give to each risk.

Applying protection measures – that is, implementing cybersecurity – will increase the cost to the attacker and thus reduce the probability. By "cost", we mean how much time and resources the attacker needs to spend to be successful. The risk of getting caught or other negative consequences is also part of the cost.

We can summarize this as follows: **Attack value = Attack benefits – Attack cost**

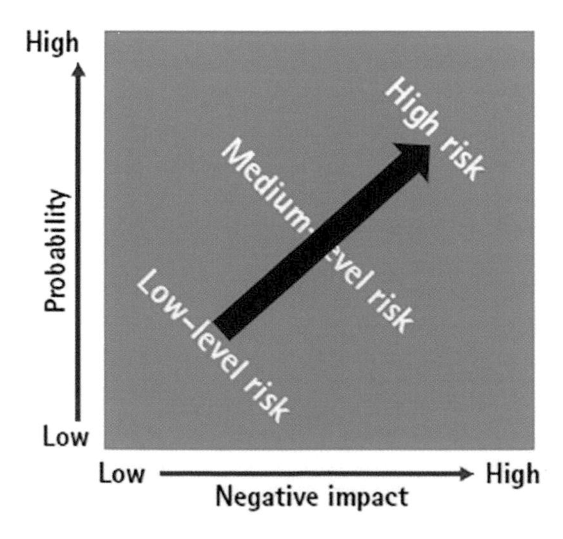

Figure 12.2 A simple graph showing how risk increases with higher negative impact and greater probability.

12.3 VULNERABILITIES

A cybersecurity vulnerability is a flaw in firmware, hardware, system interfaces, system services, software, or integration that can be exploited for malicious attacks. This does not mean that someone would easily be able to exploit such a flaw. In many cases, an attack requires various conditions to first be fulfilled, such as access to the network and its resources.

Vulnerabilities provide opportunities for adversaries to attack or gain access to a system. They can result from flaws, features, or human errors. Malicious attackers may look to exploit any known vulnerabilities, often combining several at the same time.

You need to consider both probability and potential negative impact to determine the risk associated with a vulnerability. The risk of a vulnerability may be classified as low if there is a low probability of it being exploited or if the negative impact would likely be low. For example, the risk of a vulnerability in a web server may be classified as severe on a public web server for an enterprise business portal and will need to be dealt with accordingly. The same vulnerability will have the same risk wherever the camera is located, although in some cases, an organization may be willing to accept this risk – for example, in the case of a camera deployed on a local protected network, as the exposure here is reduced.

A device application programming interface (API) and software services may have flaws that can be exploited in an attack. No vendor can ever guarantee that a product has no flaws. If the flaws are known, the risks may be mitigated though compensating security control measures, as well as updating or patching the relevant firmware or other software. On the other hand, if an attacker discovers an unknown flaw, a zero-day exploit may occur. This means that the vendor or developer has had no time to release a patch for the flaw, and the system is not protected against attempts to exploit the flaw.

In cybersecurity, the Common Vulnerability Scoring System (CVSS) is one way to classify the severity of a software vulnerability. This is a formula that examines how easy it is to exploit the

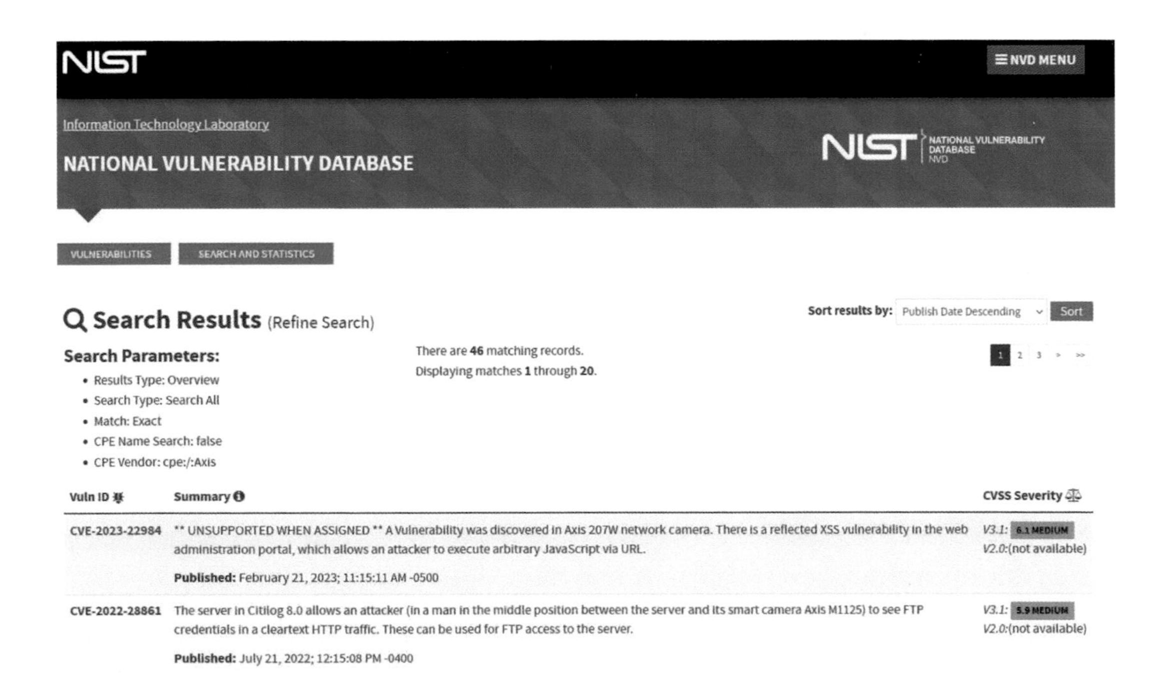

Figure 12.3 Many manufacturers publish vulnerabilities and CVSS scores that apply to their own products, but there are also independent sites that list known vulnerabilities, as in the example shown here.

Source: National Institute of Standards and Technology

vulnerability and what the negative impact might be. The score is a value between 0 and 10, with 10 representing the greatest severity. You will often find CVSS scores in published Common Vulnerability and Exposure (CVE) reports.

CVSS scores make it easier for security professionals to gauge the impact of vulnerabilities on their systems, as well as to meet the security requirements of various standards. For example, if there are unpatched vulnerabilities with a CVSS score of 4.0 or higher, this will mean noncompliance with the PCI DSS (Payment Card Industry Data Security Standard), as required by all issuers of credit card.

12.4 THREATS

A threat can be defined as anything that can compromise or cause harm to assets or resources. People tend to associate cyber-threats with malicious hackers and malware, and of course, there are innumerable instance of such attacks. However, negative impact is often also the result of accidents, unintentional misuse, or hardware failure.

In a survey published in 2022, organizations responded to the questions:

(a) "For the most significant breaches that your organization had over the last two years, what were the root causes of the attacks?"

(b) "Which causes are likely to increase over the next two years?"

50% of respondents cited human error as a root cause of an attack. Misconfigurations and poor maintenance were also cited by 44% and 43% respectively as other root causes.

These factors typically result from a failure to correctly apply security policies to all assets, undefined responsibilities, and limited organizational awareness.

12.4.1 Intentional attacks

Intentional cyberattacks can be categorized as either opportunistic or targeted. Attackers are also referred to as adversaries, whether they have malicious intent or not – so this also applies to those that cause harm to assets unintentionally.

12.4.1.1 Opportunistic attacks

Most cyberattacks today are opportunistic – that is, they occur because there is a window of opportunity, a relatively easy way to attack. In many cases, an opportunistic attacker does not even know who the victim is. These attackers often lack the determination to spend time and resources on a

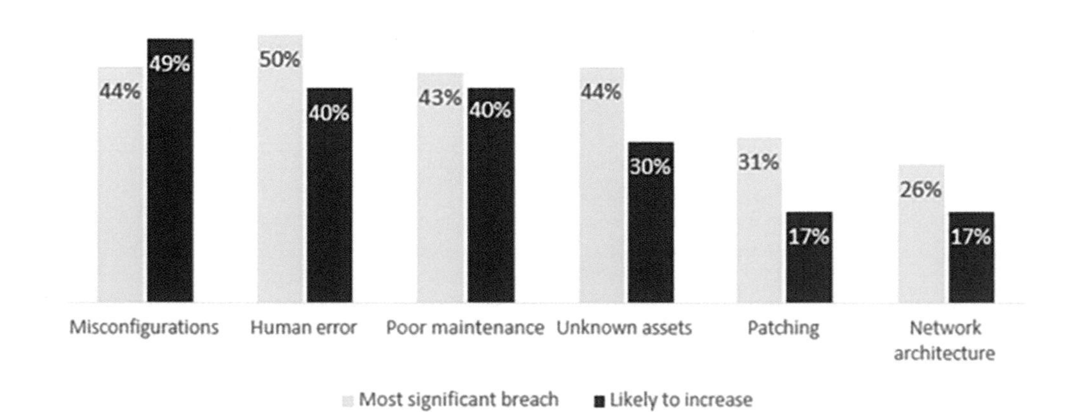

Figure 12.4 Survey responses regarding the root causes of cyberattacks.

Source: ThoughtLab survey, 2022

failed or difficult attack; they will quickly move on to their next target. Applying a standard level of protection will mitigate many risks related to opportunistic attacks.

Opportunistic attackers use readily available tools and often try to breach systems using methods that exploit weaknesses commonly associated with system entry. These tools and methods are known as "low-cost attack vectors" and can include the following:

- *Open networks, services, and ports:* As soon as a network entry point – physical or logical – is left open or poorly protected, it becomes subject to potential breach attempts.
- *Default or common credentials:* Not changing a default password or using commonly used usernames and passwords is asking for trouble. Attack tools can try out thousands of passwords in a very short time.
- *Backdoor:* This refers to a (usually secret) means of entry to a computing device or system, often bypassing authentication requirements or encryption. Many backdoor implementations are undocumented and potentially intended for nefarious purposes. Other backdoors may be more "legitimate" and commonly known, as, for example, in the case of the backdoors installed by some device manufacturers as a way to restore user passwords.
- *Unpatched services:* Software packages often contain multiple services, some of which might not even be used. These can still present a risk if left unpatched even after flaws have been found.
- *Phishing emails:* A type of "social engineering", these emails try to appear genuine and to convince the target person to reveal credentials or other information that will enable a breach.
- *DoS attacks:* Although not used specifically to gain entry to a system, a denial of service (DoS) attack will prevent legitimate users from using or accessing the service or asset being attacked, simply by overloading the target. A distributed denial of service (DDoS) attack uses large numbers of (often hijacked) computers to attack the target, making this more difficult to protect against.
- *Malware:* A general term describing malicious software designed to cause damage or disruption, usually to the computer it gets installed on but also potentially to other resources that the infected host has access to. A malware infection is often the result of a user being tricked into, for example, clicking on an email attachment or a hyperlink. One example of malware is ransomware, in which data on the infected computer is encrypted until the user pays to have the data decrypted.

12.4.1.2 Targeted attacks

Targeted attacks typically aim to extract as much intellectual property, source code, or confidential information as possible. They also will target products and attempt supply chain attacks. Attackers who target a specific system with a specific goal are, of course, harder to protect against. These attackers will start out by using the same low-cost attack vectors that opportunistic attackers use, for example, spear-phishing (sophisticated email targeting a specific recipient), to gain access to the system. If that fails and the perceived value of a successful attack remains high, these attackers will often further analyze the system, software, or processes to find alternative vulnerabilities. They are more determined and willing to spend further time and resources to achieve their goals. For these attackers, it is largely about how much value is at stake.

12.4.1.3 Actors

Attempting to understand the actors most likely to attack a system may help you gain insights into their motives, skills, and level of determination. This, in turn, may help you prioritize which countermeasures to apply. Ask yourself the following:

- *Which assets are they targeting?* Are they after trade secrets, money, and customer data, or are they simply aiming to bring down your system? See also section 12.5.1.
- *Which attack vectors* (see section 12.4.2.1) are they likely to use?

- *Which vulnerabilities will they try to exploit?*
- *How much time and resources they are willing to spend?*
- *How skilled are they?*

There are many types of actors that engage in attempts to breach computer systems. These typically include:

- *"Near and dear"*: People close to you who may want to pry into your personal business.
- *Employees:* Staff with access to the system. The breach may be accidental (through misconfiguration of user rights) or deliberate.
- *Pranksters:* Persons who find interfering with computer systems an enjoyable challenge.
- *Hacktivists:* Persons who attack organizations for political or ideological reasons.
- *Cybercriminals:* Persons aiming to make money through fraud or from the sale or ransom of stolen information.
- *Industrial spies:* Entities interested in gaining economic advantages over their competitors.
- *Cyberterrorists:* Persons or entities that carry out attacks designed to cause alarm or panic, often for ideological or political reasons.
- *Nation states:* Sometimes referred to as advanced persistent threats (APT), these are foreign intelligence services acting to gain economic and political leverage or to inflict damage on critical information systems. These attackers have very advanced capabilities and are attacking constantly.
- *Others:* Specific persons or groups with motives other than those listed earlier. This could be an investigative journalist or an "ethical" hacker. These attackers may pose a greater threat to systems that attempt to hide their flaws and vulnerabilities rather than fix them.

12.4.2 Who attacks what?

The threats, risks, and potential adversaries vary between different types of organization. In general, the size of an organization determines the likely adversary, although there are exceptions.

Small organizations: These typically include consumers, family businesses, and non-profit organizations who often want to view video of their family, home, or business from a remote location. Compared to other targets, the value for an attacker here is limited. Adversaries typically have limited skills and determination and are likely to be "near and dear", pranksters, or opportunistic cybercriminals. A video system may be exploited to pry on other individuals, or pranksters may post video clips on social media. One scenario in which larger cybercriminals may be involved in attacking small targets is when they leverage the network assets belonging to small companies and homes to attack larger specific targets by using these assets for DDOS attacks, phishing campaigns, etc.

Local business organizations: Small- or medium-sized companies, industries, or institutions, typically with outsourced physical protection and IT. The threats to these businesses include loss of money, operational downtime, and loss of trust. Adversaries include those as for small organizations but with the addition of employees and possibly hacktivists and cybercriminals.

Global business organizations (enterprises): The differences between local and global business organizations are size, exposure, and value. The potential negative impact also includes loss of competitive advantage and intellectual property. The list of potential adversaries is extended to include cybercriminals and APTs with higher levels of skill, sophistication, and determination.

Critical infrastructure: Sectors such as energy, water, transportation, telecom, public health, banking, education, police, and military are all examples of critical infrastructure organizations. The negative impact of a cyberattack here can be on a societal level, including disruption of the flow of supplies and services essential to everyday life, as well as causing widespread concern or panic.

Large enterprises may also be classified as critical infrastructure if an attack they suffer could impact the general public. The list of potential adversaries for critical infrastructure organizations also includes nation states and cyberterrorists.

12.5 MITIGATION OF RISK

As mentioned before, it is virtually impossible to eliminate risks entirely, which is why cybersecurity is largely concerned with risk mitigation (lessening the impact of attacks).

12.5.1 Assets and resources

The main purpose of cybersecurity is to protect data assets and network resources. As it is difficult to protect everything all the time, it is a good idea to do an assessment to determine which assets to prioritize. Some examples of assets and resources that can be protected in a network video system:

- *Video, live and recorded*
- *Camera operating system*
- *Camera configuration*
- *Network connectivity*
- *Passwords*
- *Interfaces*
- *User accounts on the camera and in the VMS*

Further, we can identify three main aspects of the data resources that we need to protect:

- *Confidentiality:* The assets must be protected against unauthorized access or disclosure. At the most fundamental level, an asset should only be available to those with a need-to-know.
- *Integrity:* The assets must be protected against destruction or alteration by unauthorized persons. Some resources require that changes made by authorized users can be reversed, if necessary.
- *Availability:* The assets must remain accessible to those with the proper authorization.

These areas are also referred to as the CIA triad. Trying to fulfill all three aspects can be a challenge. For example, even if all departments agree on the integrity aspect, persons working with operational technology (OT) usually prioritize availability, while their information technology (IT) colleagues will often prioritize confidentiality.

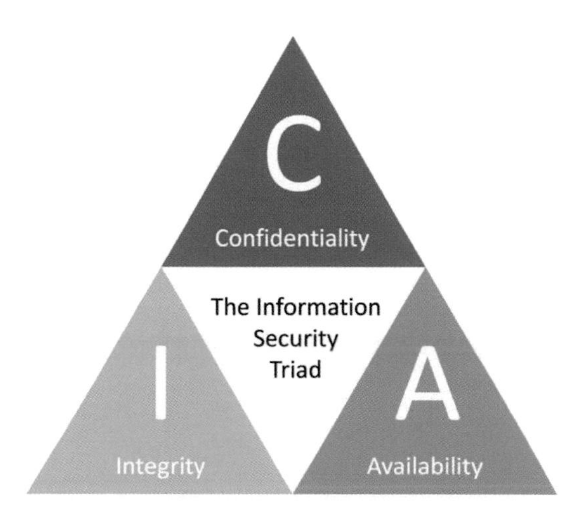

Figure 12.5 The CIA Information Security Triad provides a framework to help develop security policies.

Source: Creative Commons 4.0

Assets and resources need to be classified to determine adequate protection levels. Not all assets and resources are equal in terms of the negative impact, but they are often classified as follows:

- *Public:* The asset is for public consumption. Live video could be classified as public if it is freely available to the general public. A negative impact could be loss of availability. Live video could also be deemed public if it is available to anyone belonging to the organization that owns the video. In this case, there is also the potential for a breach of confidentiality, although the negative impact is likely to be limited.
- *Private:* The asset is privileged to a specific/selected group. Live video is often classified as private, which means it is only accessible to a specific organizational unit. Any negative impact is typically limited to the organization that owns the video.
- *Confidential:* The asset is privileged to selected individuals within an organization. In most cases, recorded video is classified as restricted, as some material could be very sensitive. Credentials and configurations for devices or systems should always be classified as restricted.

Predicting the potential negative impact on asset types is complex and estimations are subjective. The ISO 27000 impact model may help you prioritize, as it offers designated levels of impact and estimations for the time it would take to recover from an impact:

- *Limited*: from hours to days to recover
- *Serious*: from days to weeks to recover
- *Severe*: from weeks to months to recover
- *Catastrophic:* from months to years to recover – if ever

12.5.2 Security controls

Security controls are safeguards or countermeasures employed to avoid, detect, counteract, or minimize security risks to physical property, information, computer systems, or other assets. The process of deploying security controls is often referred to as *hardening* or *mitigation*.

Compensating security controls are alternative safeguards that can be used when it may not be possible to apply the preferred security control or when the preferred control may be too costly.

Security controls need to be continuously monitored and updated, as threats, value, vulnerabilities, and exposure change over time. This requires defining and following policies and processes.

There are three types of security controls:

- *Administrative controls:* policies, procedures, and processes
- *Technical and logical controls:* encryption, certificates, disabling services, and VLANs
- *Physical controls:* card readers for IT cabinets, cables that physically prevent disconnection, etc.

12.6 DATA SECURITY

Data security is top-of-mind for businesses, governments, and individuals alike. Possible risks include compromised video integrity through manipulation of the video images, the breach of information systems, and denial of service since the network camera will only support a finite number of users directly. There is also the risk of someone taking control of the device itself to destroy onboard data or to redirect the video stream elsewhere, which in turn can lead to people gaining unauthorized access to protected areas.

Businesses and governments around the world depend on the internet and cloud services. Over the years, many technologies have become available for securing IT environments. Video management systems – on-premises or in the cloud – should meet the following security standards:

- *Multi-factor authentication:* The user must provide a combination of things only the user knows, such as a username, a strong password, and an RSA SecureID token (for example, a USB dongle, smart card, or key fob). This prevents wrongful use of leaked usernames and passwords.
- *Encryption of passwords:* Usernames and passwords are always transferred encrypted between the user and the system. In addition, passwords should be stored in an encrypted format in combination with a technique known as *salted password hashing*, which ensures that nobody can use the information even if the password database is compromised.
- *Encryption of all streams:* Video and data connections are always encrypted to prevent other parties from getting access to the material.
- *Machine-to-machine authentication:* Connections between cameras and servers are authenticated using signed certificate technology to prevent man-in-the-middle attacks.
- *Digitally signed video:* A cryptographic checksum is inserted into each video frame and is then signed with a unique key from a dedicated cryptographic module that securely and permanently holds the device's unique ID. This allows video to be traced back to the camera from which it originated, making it possible to verify that the footage has not been tampered with after it left the camera.
- *Camera-owner authentication:* To avoid wrongful use of deployed cameras, it is only possible to connect a camera to a system using a unique identifier known only to the camera owner.
- *System redundancy:* This can be provided by inexpensive storage devices such as NAS or SD cards so that recordings are protected even if the network goes down.

12.6.1 Hardening guides

Many manufacturers of network video products release recommendations on how to ensure that appropriate network security is deployed in a network video system. Such recommendations are often referred to as hardening guides, and they refer to different levels of protection, examples of which are given here.

- *Default protection (Level 0):* Cameras are shipped with preconfigured default settings and a default password. Default settings should not be used for daily operations. Settings must be adjusted depending on the environment and risk analysis.
- *Standard protection (Level 1):* This is the minimum recommended level and is sufficient for home and small office use, typically when the system administrator is also the operator/user. Other actions include checking for the latest firmware, setting the time and date, and disabling audio if not in use.
- *Enterprise protection (Level 2):* Corporate and enterprise systems have dedicated system administrators, and the network has multiple users/clients and servers. Video systems must adhere to the corporate IT infrastructure and policies. Security capabilities and features must be enabled and configured. Unused camera services should be disabled. Other actions include enabling IP address filtering and encryption.
- *Managed enterprise (Level 3):* Managed enterprise network systems will typically have additional management tools and services that cameras must align with:
  - *802.1X Network Access Protection*
  - *VLAN segmentation*
  - *SNMP monitoring*
  - *Remote SysLog*
  - *Network Time Protocol (NTP)* (used in forensic investigations)

12.6.2 Vulnerability scanning

Vulnerability scanning is an automated or manual audit of a product or software. Several tools on the market are available for performing vulnerability scanning. The audit determines if the product includes services that have known vulnerabilities associated with a specific software or service version. The scan will raise a flag for the affected service, even if it is not installed or in use. Vulnerability scanning can only help to identify known vulnerabilities. It is not a good way to determine how secure a product is, as a new critical vulnerability may be discovered at any time. The result from a vulnerability scanning needs to be assessed.

12.6.3 Penetration testing

Penetration tests (sometimes truncated to "*pentests*") are simulated cyberattacks and physical attacks to discover vulnerabilities in devices, software and hardware. Tests are typically divided into two major groups – internal and external (or third-party) penetration tests. Mature manufacturers will employ internal penetration testing on new software and devices as part of their development process before release into production.

As well as testing their products, it is also important for manufacturing organizations to periodically test their own IT infrastructure, using third-party penetration tests, as the internal IT teams "don't know what they don't know". Third-party penetration is broken down into three main categories: *white hat, black hat*, and *gray hat* testing. All these tests are typically limited to two to four weeks, aiming to find vulnerabilities in the device or network service. White hat testing is where all knowledge of the organization is given to the team performing the test or attack. This type of testing is helpful for devices and software, and the scope is usually limited to specific areas. In black hat testing, no information is given to the team performing the test, which best simulates an actual attack against the organization's infrastructure or new product or service being introduced. A gray hat test is somewhere in between; some information is provided but not all. The intent of this type of test is to mix the black and the white testing, although sometimes with a reduced scope.

The purpose of any penetration test is to provide a report of any vulnerabilities found. In many cases, the test is on products or services being used in the real world, so patching software and systems in a timely manner is critical. In third-party testing, these reports are often provided to potential new or existing customers to help them evaluate the security of a product or service they are procuring, as well as the cyber-maturity of the organization they are buying from and how responsive the organization is to patching or correcting vulnerabilities.

12.6.4 Device inventory and management

A video system may have large numbers of cameras. If you do not have control over the device inventory, you may face challenges with device monitoring and maintenance. A good idea is to use a dedicated tool to manage your network devices. Such a tool lets you perform installation, security, and maintenance tasks in batches, instead of doing so one device at a time.

An entry-level tool will help you with the basic network settings for your devices, such as configuring IP addresses and setting the network's default router. A more advanced management tool is suitable for larger installations with thousands of cameras. These tools can also install certificates, configure specific camera settings, perform firmware updates, push templates to multiple cameras at once, and apply recommendations related to cybersecurity.

12.6.5 Updated firmware

Any software-based product can potentially be released with unknown vulnerabilities that may one day be discovered. In many cases, these vulnerabilities will pose a limited risk, as they may be either very hard to exploit or the impact may be limited. Occasionally, however, a critical vulnerability is discovered, one that escalates risk levels.

The manufacturer's vulnerability management process should include the patching and announcement of identified vulnerabilities.

Running up-to-date firmware versions will mitigate common risks for devices. This is because the latest firmware versions will include patches for known vulnerabilities that attackers may try to

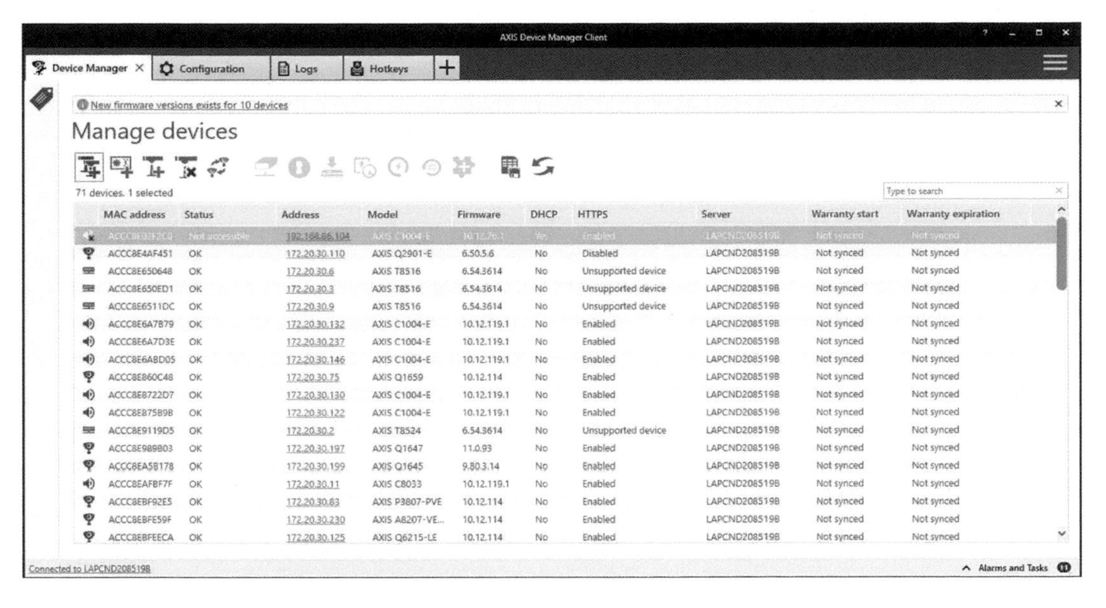

Figure 12.6 A dedicated management tool helps manage large numbers of network devices, individually or en masse, as required.

exploit. Firmware updating and patching should preferably occur according to a regular schedule, that is, weekly, monthly, quarterly, or similar.

12.6.6 Limit exposure to the internet

Exposing computers or other network devices directly to the internet is risky, as their IP addresses (or ports) are then accessible by others. Cameras or other devices behind a firewall are not internet-facing but can still access services on the internet, just like a computer.

The sheer volume of malicious computers on the internet that continuously probe public IP addresses to find known, exploitable vulnerabilities is enormous. Specialized web crawler search engines seek out potential victims by probing public IP addresses and then listing which devices or services are available on the internet. This means that all external IP addresses will eventually be indexed, and the service or interface behind the address will be identified.

The most common mistake made by a small organization is to expose a camera as a public web server to view video from their business, house, or store over the internet. A common yet unsafe solution is to use port forwarding, in which the router forwards external requests to the camera, thus poking a hole in the firewall. The risk increases when an internet-facing device is combined with a weak (or default) password.

If you require this type of access to video from cameras in your system, you should use some kind of solution for secure remote video access. Larger organizations that require remote video should use a proxy solution that does not expose the camera at all – only the video service. Consult your VMS vendor or IT security specialist to find the best solution.

12.7 DEVICE SECURITY

There is no way to guarantee that any network product or service is free from flaws that can be exploited for malicious attacks. What can be done is to make concerted efforts to ensure that the level of risk for a product is as low as possible.

Features such as signed firmware and secure boot can be employed to prevent firmware tampering and supply-chain tampering. A Trusted Platform Module (TPM) or Secure Element (SE) can be used for cryptographic operations.

12.7.1 Signed firmware

One possible attack vector an adversary may try is to get a system or device owner to install altered applications, firmware, or other software modules. Altered software may include malicious code with a specific purpose. A common recommendation is to never install any software from a source that you do not fully trust. Altering a device's firmware and tricking users into installing it is not an easy exercise and requires a detailed understanding of firmware design and operation. Even so, these types of attacks may still occur if the perceived value of attacking a system is high enough.

The countermeasure here is to use signed firmware, which is implemented by the software vendor signing the firmware image with a private key, which is kept secret. When a firmware has this signature attached to it, the device will validate the firmware before accepting it for installation. If the device detects that the firmware integrity is compromised, the firmware upgrade will be rejected.

12.7.2 Secure boot

Firmware signing protects a device from installing a compromised firmware, but what if a third party alters the device on its way between vendor and user? If an adversary has physical access to the device in transit, there is the potential that they could compromise the device's boot partition, thus bypassing the firmware integrity check and then installing altered, malicious firmware before the device is even deployed.

Secure boot is a boot process that consists of an unbroken chain of cryptographically validated software, starting in immutable memory (boot ROM). Secure boot verifies, in real time, the embedded signatures for each block of firmware loaded from the flash memory. The boot ROM serves as the "root of trust", and the boot process continues only as long as each signature is verified. Each part of the chain authenticates the next part, ultimately resulting in a verified file system. Once the file system is verified, the device will start the boot process.

12.7.3 Secure identification

But how do you know your system is communicating with an authorized device and that the device has not been tampered with? One way is to use a dedicated, built-in security component such as a Secure Element that enables automatic and secure identification and authorization of new devices, following the international standard IEEE 802.1AR. The device ID is securely stored within the Secure Element, which is a collection of certificates, including a digitally signed version of the device's unique serial number.

Trusted Platform Modules and Secure Elements are components that provide a certain set of cryptographic features suitable for protecting information from unauthorized access. The private key is stored inside these modules and never leaves them. All cryptographic operations requiring the use of the private key are sent to these modules to be processed. This ensures that the secret part of the certificate never leaves the secure environment and remains safe, even in the event of a security breach.

12.8 NETWORK SECURITY

Although early IP networks had some security flaws, huge investments in research and development have now led to extremely secure networks. There are several ways to provide security both within a network and between different networks and clients. Everything from the data sent over a network to the actual use and accessibility of the network can be controlled and secured.

Secure communication can be divided into two different types:

- *Authentication and authorization:* This initial step involves the user or device identifying itself to the network. This is done by providing some type of identity – for example, a username and password. This authentication is then authorized and accepted – that is, it is verified if the device is allowed to operate as requested (or not.) Once authorization is complete, the device is fully connected and operational in the system.

- *Privacy:* This is accomplished by encrypting communication to prevent others from using or reading the data. The use of encryption can cause communications to slow down, depending on the kind of implementation and encryption used. Privacy can be achieved in several ways. Two of the more commonly used methods are virtual private networks (VPN) and SSL/TLS (also known as HTTPS).

The main technologies for securing data transmissions are explained next.

12.8.1 Authentication

The most basic method of protecting data on an IP network is to employ username and password authentication. Data is protected from access until a user submits a valid username and password. Passwords can be sent encrypted or unencrypted, the former providing the better security.

Username and password protection may be appropriate in an installation where high levels of security are not required.

"Use a strong password" is an oft-repeated phrase, and this is, of course, important. However, a common challenge with device passwords is that they tend to spread within the organization. For example, during device maintenance, a new employee might request a device password in order to adjust something. A couple days or weeks later, another new person has the same request, and before long, multiple users now have passwords to your devices, and you lose control over them. The strength of the password makes no difference in this case. Devices should instead use multiple role-based accounts, and temporary accounts should be used for occasional maintenance/troubleshooting.

Privileged access management (PAM) covers the strategies for controlling privileged (elevated) access and permissions in an IT environment. PAM means knowing who is using accounts such as *Root*, *Admin*, etc. Passwords for these should be changed at least annually and immediately if, for example, a network administrator or security manager leaves the organization.

12.9 ZERO TRUST NETWORKING

As the name suggests, the default position in a zero trust network is that no entity connecting to it or within it – apparently human or machine – can be trusted wherever and however it connects.

The overriding philosophy here is "never trust, always verify". This requires that the identity of any entity accessing or within the network is verified multiple times in several different ways, depending on the specifics of the network being accessed. Entities are granted the minimum amount of access required to complete their task.

Zero trust can also be applied to physical security and video data. There are a number of steps that should be followed to plan your implementation:

1) *Define the assets you wish to protect:* This depends on your type of organization and field. Critical infrastructure, government, and banking typically have video identified as sensitive. In a small retail business, this might not be the case.
2) *Locate and track your assets:* Knowing where your assets are and how they move around the network allows you to protect them in transit, in use, and at rest. Data may be stored on-premises or in the cloud or both. Each organization's network topology is unique, and mapping where data flows in and out will vary.
3) *Design your perimeters:* Zero trust employs techniques such as network microsegmentation, which involves applying varying levels of security to specific parts of the network where more critical data resides. Granular network perimeter security based on users and devices, their

physical locations, and other identifying data is used to determine whether their credentials can be trusted to access the network or the data. In some cases, adding security gateways or firewalls specifically around the segments of video data will help protect and provide the means to authenticate once more, for access to sensitive video data.

4) *Monitor your system*: Giving individuals access only to the parts of the network and the data required for their roles brings obvious security benefits. Anomalies in the behavior associated with these identities brings an additional level of security. For example, a network administrator may have extensive network access for maintenance, for example, to R&D or finance servers, but if these same credentials were used to access specific data in the middle of the night, then this should probably raise a flag.

   Zero trust networks use policy engines – software that, broadly speaking, allows an organization to create, monitor, and enforce rules about how network resources and data can be accessed. The policy engine compares every request for network access and its context to policy and informs the enforcer whether the request will be permitted or not.

5) *Automate wherever possible*: The larger goal of any zero trust implementation is to get to a point where devices can be granted access to a network automatically based on the type of device and the resources it needs. For example, when a video device is booted up, the network would recognize the device's manufacturer, the model number, type, etc. While this is happening, the device may be sent to a provisional VLAN to protect against attack while its network certificate is signed and prepared. Once the certificate is fully authenticated, the device can be moved into the production network in the physical security or video system VLAN.

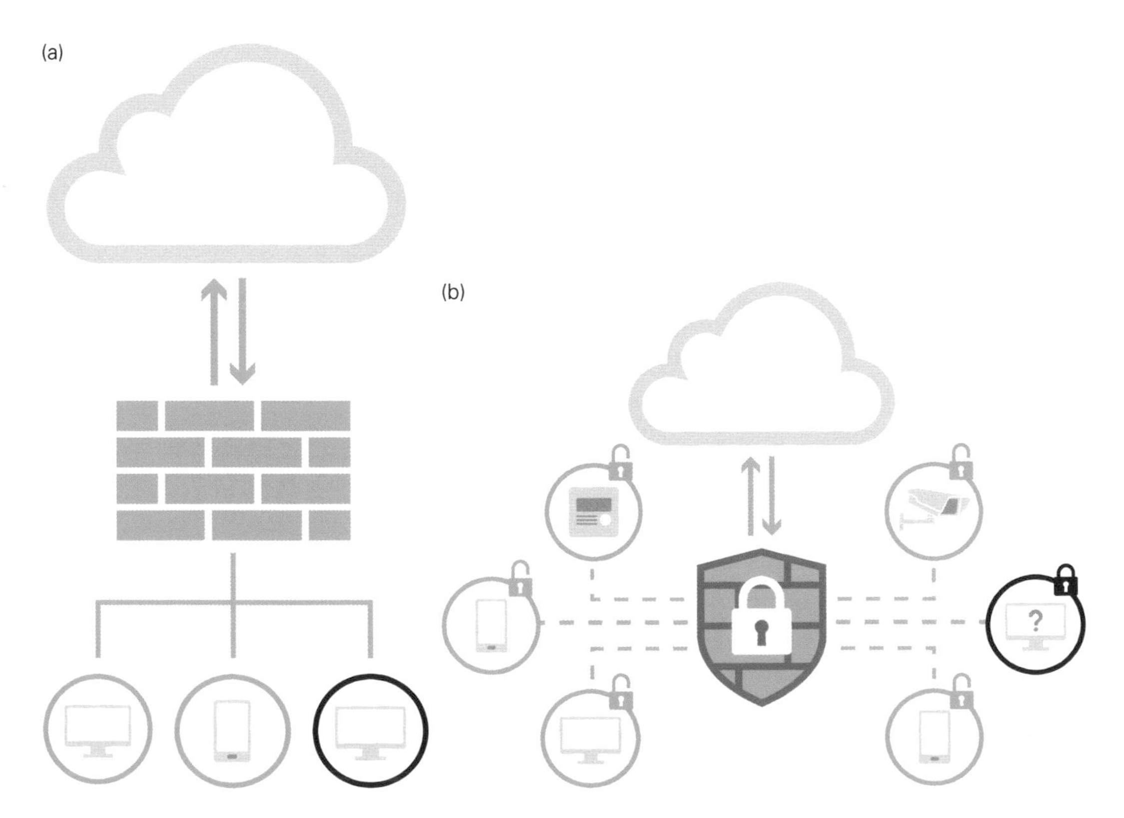

Figure 12.7 (a) In traditional network security, once an entity has been challenged and authenticated at the main network perimeter, there is little or no further authentication or controls applied. (b) In zero trust, entities are instead challenged every time they request access to individual resources (assets, services, accounts, etc.) and not necessarily to network segments. Access is determined by the roles and policies at the heart of the system.

Figure 12.8 IP address filtering in a network camera restricts access to specified IP addresses only.

12.10 OTHER TECHNIQUES AND PROTOCOLS FOR CYBERSECURITY

12.10.1 Limiting network exposure

Devices on a local network are protected by a firewall that reduces the risk of threats from the internet. Reducing the exposure on a LAN will help mitigate risks from any adversaries on the inside, as well as help mitigate risks related to compromised passwords, unpatched devices, and brute-force login attacks. The following sections describe some common security controls that can be applied.

12.10.1.1 Network isolation

Network isolation is a way to separate critical network resources from each other to reduce the risk of them having a negative impact on each other, especially if these resources do not need to interact with each other. Network segmentation can be virtual (VLAN), which requires an infrastructure of managed switches, or the networks can use separate cabling and network equipment. The decision on which type of segmentation to use depends on cost, infrastructure, and policies. Another option is to implement extra firewalls or gateways to protect certain networks, instead of separating these completely.

12.10.1.2 IP filtering (IP tables)

Many network cameras include IP filtering, which prevents all but one or a few IP addresses from accessing the product (Figure 12.8). This function is like having a built-in firewall at the device level. When a VMS is used as the core in a professional video system, clients will access all live and recorded video via the VMS and not directly from the camera. This means that the only computer or server that should be accessing cameras during normal operations is the VMS server. The cameras can be configured with an IP filter to only respond to whitelisted IP addresses, which typically will be the VMS server and administration clients. IP filtering helps mitigate risks from compromised camera passwords, from unpatched cameras and from brute-force attacks. Approved IP addresses can easily be included in a configuration file that is pushed out to devices during installation or later during an update.

12.10.2 Virtual Private Network (VPN)

A VPN uses an encryption protocol to provide a secure tunnel from one network to another through which data can be securely transferred. This allows for secure communications across a public network, such as the internet. Only devices with the correct certificate will be able to operate within the VPN.

A VPN typically encrypts the packets on the IP or TCP/UDP layers and above. The IP Security (IPSec) Protocol is the most common VPN encryption protocol, and it can use various encryption

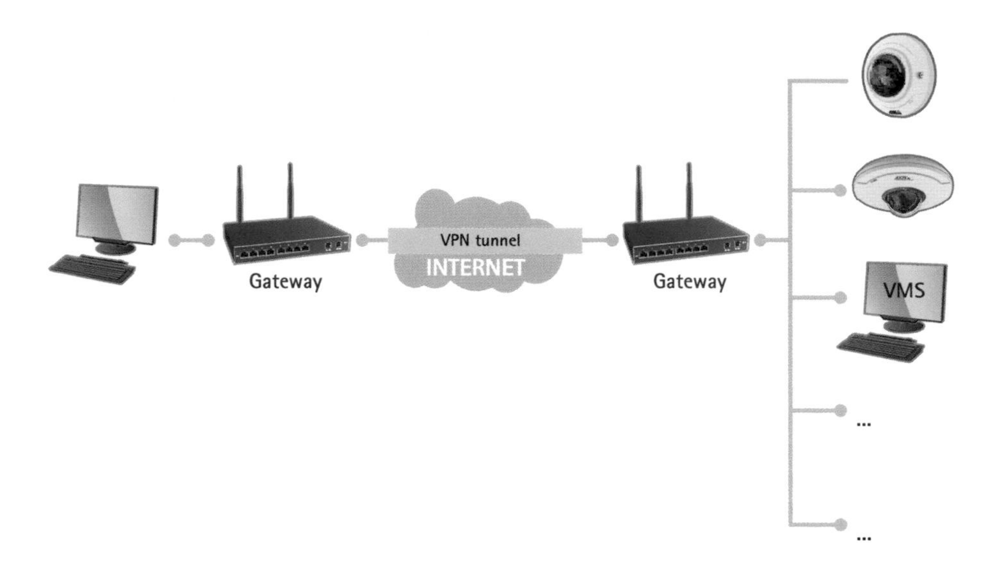

Figure 12.9 Virtual private network security on a video network.

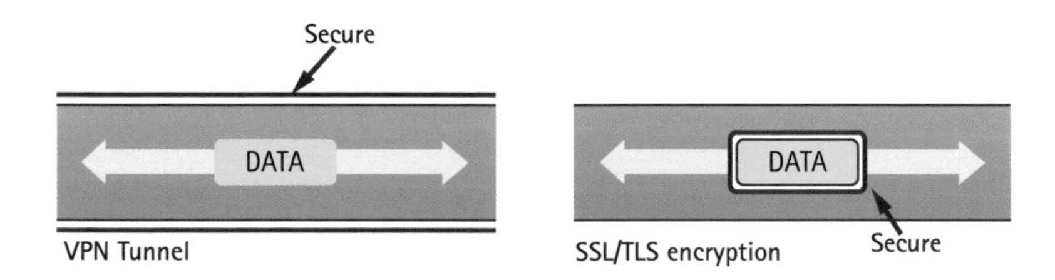

Figure 12.10 The difference between virtual private network (VPN) and Secure Sockets Layer (SSL) is that in a VPN, the tunnel itself is encrypted, whereas in SSL, the data are encrypted. Although both technologies can be used in parallel, this is not recommended because each will add an overhead and affect the performance of the system.

algorithms. Today, either the Triple Data Encryption Standard (3DES) or the Advanced Encryption Standard (AES) is used. AES offers greater security and requires considerably less computing power than 3DES to encrypt and decrypt data. AES uses either 128-bit or 256-bit key lengths.

12.10.3 HyperText Transfer Protocol Secure (HTTPS)

Another way to provide privacy is by applying encryption on a higher level – that is, the data, not the transport protocol, is encrypted. HTTPS is the most common data encryption protocol and is used, for example, in online banking to provide the requisite security for banking transactions performed over the internet. HTTPS is identical to HTTP but with a key difference: The transferred data is encrypted using *Secure Sockets Layer (SSL)* or *Transport Layer Security (TLS).*

The security offered by SSL/TLS is based on three main elements:

- *Authentication of the communication partner*
- *Symmetrical data encryption*
- *Protection against the manipulation of transferred data*

When making an SSL/TLS connection, negotiation first takes place by using a handshake protocol that determines which cryptographic methods will be used, as well as the secure identification or authentication of the communication partner. This last element is achieved by the web server identifying itself to the browser using a *certificate* (see section 12.11). A premaster secret is then exchanged between the communication partners over an *asymmetrical encryption* or *Diffie-Hellman key exchange.*

A certificate can be compared to an ID card that a person uses to prove their identity. This is a binary document usually issued by a certificate authority (CA), such as VeriSign. Certificates normally used only within closed user groups, such as for a company's private web server, can be issued by anyone.

Many network video products have built-in support for HTTPS, which makes it possible to view video securely using a browser.

12.10.4 Secure Real-time Transport Protocol (SRTP)

Although HTTPS encrypts the control data from a camera, it does not encrypt the video data. This is where SRTP is useful. SRTP is Real-time Transport Protocol (RTP) with enhanced security, providing encryption, message authentication, and replay protection for audio and video data. SRTP needs its own certificate (see section 12.10); it cannot use the HTTPS certificate. Many video devices today support SRTP.

Video and audio data can be affected negatively by latency. Although SRTP takes this into account, any encryption process will add some delay, in both encryption and decryption. Video often produces large and continuous streams of data, requiring the device to have plenty of processing power for encryption. Using SRTP on a video stream at 1080p is not a problem for today's network video devices, but testing a device's ability to use SRTP to encrypt video streams greater than 5 megapixels is paramount, as this could impact the number of frames per second (fps) the device can handle.

12.11 CERTIFICATES

There are several different kinds of digital certificates. An *SSL/TLS certificate* on your web server enables your server to identify itself to web browsers. This is one-way authentication, as it is only the server that identifies itself to the client, not vice-versa. A *digital device certificate* on the client, however, authenticates the client to the web server. This is much like when certificates enable authorized users to connect to web apps or drives on internal networks.

A digital device certificate allows devices to authenticate each other and create secure connections between them, using a *public key infrastructure (PKI)* to do so. These certificates provide secure authentication, device identity, and the encryption of data. Cryptographic keys are used to secure communications between devices.

A public *certificate authority* (CA) is a third-party organization that issues publicly trusted certificates, which are available for other organizations to purchase as required. A device certificate can also be issued directly by an organization, which means they have an internal certificate authority (a private CA). The organization's certificates are used on its own network.

12.11.1 Certificate management

All digital certificates have expiry dates and need replacing at regular intervals. To manage your certificates correctly, you will need to know the following:

- *All the devices on your network that use certificates*
- *If your devices have PKI digital certificates*
- *When your certificates were issued and when they will expire*
- *How they were issued*

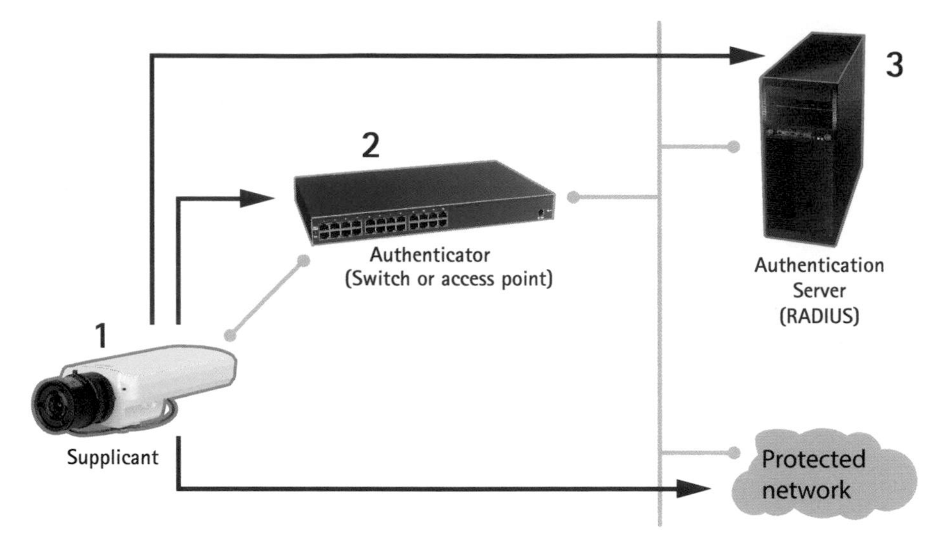

Figure 12.11 IEEE 802.1X enables port-based security and involves a supplicant (1), an authenticator (2), and an authentication server (3).

12.11.2 802.1X

The IEEE 802.1X standard is among the most popular authentication methods in use today. It provides authentication to devices attached to a LAN port, establishing a point-to-point connection or preventing access from that port if authentication fails. 802.1X is often referred to as port-based network access control and prevents port hijacking – that is, when an unauthorized computer obtains access to a network by getting to a network socket inside or outside a building.

Authentication requires three components:

- *The supplicant*
- *The authenticator*
- *The authenticating server*

The supplicant is a device such as a network camera that requests access to the network. The authenticator can be a switch or a wireless access point. Logical ports on the authenticator allow user data from the supplicant to pass once the supplicant is authenticated. The authenticating server is a (dedicated) server on the LAN to which supplicants must identify themselves in the authentication process. Some switches include the functionality of the authenticating server.

The authenticating server is called a *Remote Authentication Dial-In User Service (RADIUS)*. If a device wishes to access a network, it asks the authenticator, which forwards the query to an authentication server. If authentication is successful, the server instructs the authenticator to authorize access to the network for the querying supplicant (Figure 12.11).

802.1X is often built into network cameras, as they are often located in public spaces (such as receptions, hallways, meeting rooms, or even outside a building), where they need the security layer that 802.1X provides. Without 802.1X, a network socket that is openly accessible poses a substantial security risk. In today's enterprise networks, 802.1X is a basic requirement for anything connected to the network.

12.11.3 802.1AR

Networks that provide publicly accessible services or which cannot be completely physically secured have their own challenges. Authentication of a human user is often used to authenticate devices such as personal computers, but many network devices are designed for unattended autonomous operation and do not support user authentication.

802.1AR is a standardized device identity that facilitates secure device authentication and simplifies deployment and management.

802.1AR specifies *Secure Device Identifiers (DevIDs)*, which are designed for use as secure device authentication credentials with *Extensible Authentication Protocol (EAP)* and other standard authentication and provisioning protocols. During production, the manufacturer installs in the device a unique *Initial Device Identifier (IDevID)* and stores it so that it cannot be modified.

Devices may also support the creation of *Locally Significant Device Identifiers (LDevIDs)* by a network administrator, for the purpose of including and protecting additional information for supporting local authorization routines.

Each LDevID is bound to the device in a way that makes it impossible to forge or transfer without knowing the private key used for the cryptographic binding.

12.12 RECOVERING FROM A CYBERSECURITY EVENT

Cyberattacks are constant occurrences, and statistically speaking, any organization is going to get attacked at some point, no matter the sector or type of business. Apart from the risk of intentional interference and attempts at sabotage, there is always the risk of hardware failure, accidental damage, weather events, power outages, and other circumstances beyond your control.

Depending on your organization's use of video surveillance, this should be included in your business continuity plans, which should answer questions such as the following:

- How will your cameras continue to record if the network is compromised? You may want to configure backup recording to onboard SD cards to cover yourself against network outages.
- Are your cameras and network switches connected to an uninterruptable power supply (UPS)? How long will the UPS battery keep your equipment running?
- Do you have ready access to replacement cameras and other equipment if needed?
- Do you have a backup file for each camera's configuration? See section 12.11.1 for more on this.
- Do you have a record of each camera's field of view and viewing direction? It is one thing to configure a camera's settings correctly, but do you know exactly what it should be covering?

12.12.1 Backups

Restoring a network video system after an attack or hardware failure will often result in the need to reconfigure existing cameras or to configure replacements. As you have already taken the time to harden your cameras correctly for your environment, you will not want to do this all over again, camera by camera. Getting your cameras back online will be much easier if you have made regular backups of their configurations, which, after an event, can easily be pushed out again using a device management tool. These tools can store and push out individual configurations per device, as well as general templates that can be applied to multiple devices when restoring from backups.

Creating backups of your cameras and other devices should occur on a regular basis, according to a schedule. It is also recommended to make regular backups of the VMS installation, as well as the recordings created in the system.

Equally important as creating the backup is testing it. Does the backup file and restore process work correctly and get the camera back online and functioning as before? If the backup is out of date or, for some reason, it fails, you don't want to have to discover this when dealing with a cybersecurity incident.

12.13 CYBERSECURITY BEST PRACTICES

- *Risk analysis:* Try to determine potential threats and the possible damage or cost if the system is attacked. This will indicate the level of protection required, as well as the precautions that must be taken to reduce the risks.
  - *Evaluation and assessment:* To identify assets and evaluate their properties and characteristics.
  - *Risk assessment:* To discover threats and vulnerabilities that pose a risk to assets.
  - Risk mitigation: To address a risk by transferring, avoiding, mitigating, or accepting it.
- *Educate yourself:* Learn about system protection and possible threats. Work closely with resellers, system integrators, consultants, and product vendors. The internet is a fantastic resource.
- *Secure the network:* If network protection is breached, this increases the risk of snooping for sensitive information and attacks on individual servers and network devices.
- *Passwords:* Use strong, unique passwords and change them on a regular basis. Where possible, use multi-factor authentication.
- *Factory default settings:* Do not rely solely on a network device's default settings.
- *Encrypted connections:* Use whenever possible, even on local networks.
- *Reduce exposure:* Clients should not access cameras directly, unless necessary. Clients should normally only access video through a VMS or a media proxy. It is also recommended to use an extra firewall or gateway to control direct access to the VMS and cameras.
- *Check logs:* Check access logs on a regular basis to detect attempts at unauthorized access.
- *Monitor devices:* Monitor devices on a regular basis. Enable system notifications where applicable and supported, and consider using SysLog and remote SysLog. Devices can also be monitored using SNMP.
- *Firmware:* Use the latest available version, which may include security patches. A good management tool, usually available from the device vendor, will help you manage this process.
- *Recovery:* Plan ahead so as to lessen the impact of an event.

While the digital aspect of cybersecurity is of great importance, remember that the greatest threats to a camera are physical, including sabotage, vandalism, and tampering. Selecting a vandal-resistant model or casing and mounting it correctly is important, as is protecting cables.

CHAPTER 13

Servers and storage

Servers and storage are important components of a network video system, as they are used for monitoring, recording, managing, and archiving video. A major benefit of IP-based video surveillance is the ability to deploy standard IT server and storage components. Components can range from a low-price PC to a blade server system attached to a storage area network (SAN) with petabytes of storage that can cost millions of dollars. As they are easy to use and install, purpose-built network video recorders (NVRs) are popular in small and some midsize systems, especially if the NVRs have built-in PoE ports to power the cameras. Some of these NVRs are proprietary in nature and are not based on standard servers and components.

Another benefit of using standard components is that they can be serviced and maintained along with the other IT equipment within an organization. This not only reduces costs but also increases the availability and reliability of the equipment.

Because all the equipment is standard IT hardware, this chapter merely provides an overview of server and storage technologies, along with some best practices relating to network video.

13.1 SERVERS

Depending on the performance required, there are various aspects to consider when selecting the appropriate server platform. For a network video system that uses video management software on the server, the benefit is that servers are easily upgraded. If, for example, greater performance and capabilities are required, then the video management software can be installed on a new server.

13.1.1 Hardware platforms

Most hardware platforms are based on Intel® or AMD (Advanced Micro Devices, Inc.) processors. There are, however, some differences between a desktop platform used for word processing and email and the enterprise-class servers normally used in network video applications. Enterprise servers are designed for handling huge amounts of data and for continuous reading and writing. Thus, they differ to a standard PC in the following areas:

- *Processors:* Compared to a PC with a single processor, an enterprise server is sometimes equipped with multiple processors.
- *Internal memory:* The internal random-access memory (RAM) is the memory used for processing and buffering data. Because a server in a video system must handle large amounts of data, large RAM buffers are required. Even the smallest enterprise server will have 8–16 GB (gigabytes) RAM, and a top-of-the-line server may have several hundred GB. Servers with several RAM sockets can have many TB of RAM.

DOI: 10.4324/9781003412205-13

- *Storage:* Compared with a standard PC designed for occasional reading and writing, enterprise solutions require reliable storage solutions with high throughput and continuous reading and writing so that they can serve multiple clients. Specific hard drives rated for video surveillance data (with continuous writing) are also available. Note also that neither the operating system nor the video management system should be run on the same disks as those used for video storage.

There are also other server parameters to consider. Servers are typically available in various designs, such as stand-alone towers, rack-mounted servers, or blade servers. Blade servers are popular because of their high density and suitability for larger video management systems. Virtualization is another popular technology, where one physical server runs several virtualized servers on the same hardware, although this type of setup is not generally used for video storage or display.

13.1.2 Operating systems

Making the server hardware operational requires an operating system (OS). In theory, any operating system can be used as a platform for recording and managing video, as long as the video management software supports the operating system and meets the technical requirements. Some commonly used platforms:

- *Windows®*: This is the most common platform for video management applications. Using Active Directory® service, it is possible to provide centralized authentication and authorization services for Windows-based computers. The Windows file system is New Technology File System (NTFS), which supports a file system up to 256 TB (terabytes) and file sizes of 16 TB. With most recent versions of Windows Server, file systems of up to 8 PB (petabytes) are possible.
- *Linux®:* This is a popular UNIX™-like operating system that comes in a variety of versions, also called distributions. With low, or no, licensing fees, Linux-based servers are often used for cost reasons. The latest file system used is ext4 (fourth extended), which supports file systems up to 1EB (Exobyte) and file sizes up to 16 TB. While Linux is not commonly used in video management applications, it is very popular as an embedded platform for network video devices, such as network cameras and video encoders.
- *Android and IOS:* Many applications now support mobile applications running on these operating systems, although these are mainly for administration purposes or for client access and are not used for running the server.

13.1.3 Video file systems

As discussed earlier, different file systems have different capabilities. Regardless of the OS or file system, it is important to have a well-organized structure when storing recorded video on a hard disk. Video can be stored as standard files (JPEG, MPEG, H.264, H.265, AVI, ASF, etc.) or as raw indexed data. The structure can be a simple directory tree (with subdirectories containing files for each camera, week, day, and hour), or it can be a complex database (with or without multiple references) together with metadata information.

There is no standardized method of organizing large amounts of recordings. Video management software vendors use different technologies; some use standard databases, such as Structured Query Language (SQL) or Oracle or file systems, while others use proprietary formats and structures. Even two versions of the same application can differ in terms of how video is stored. The structure can be optimized for high recording performance, fast searches, tampering detection, stability, and recovery, among others. Unless an application has an open interface for exporting, accessing, or searching stored video, it can be extremely difficult for other applications that use a different system to read and interpret the recordings.

Before video is even stored to disk, it is important to consider the properties of the video stream being saved. Spending some time and resources on the video quality, the bitrate, etc., can have a huge impact on the amount of physical storage that a network video system requires. For more on video compression, see Chapter 6.

13.2 HARD DISKS

The hard disks used today in IT servers and desktops range from 1 TB up to 20 TB and every year sees increases in capacity. In addition to size, there are several other factors that differentiate hard disks. One is the spin speed, measured in rotations per minute (RPM), which for consumer-grade products is normally 5,400 or 7,200 RPM. Another is the non-sequential (random) read-write performance of the disk system, which indicates the speed of reading and writing to non-sequential blocks of data.

Of particular interest to network video systems are surveillance-grade hard disks. In contrast to a standard desktop or server hard disk, a disk of this type is designed for almost constant write operations. As there is usually far more video stored than ever viewed, the disk's read-write head can move a little slower and more smoothly. This radically decreases the level of mechanical wear. These disks are usually also much better at power management and are better equipped to withstand the extremes of temperature associated with constant write operations.

Hard disks are supplied in various form factors. For servers, a 3.5 inch standard casing is the norm. Another major differentiator is the interface, with some of the most common interfaces described next.

13.2.1 The SCSI interface

The SCSI (Small Computer System Interface) is a standard for physically connecting computers and peripheral devices, such as hard disks, tape drives, and optical drives. SCSI hard drives are good choices for demanding video surveillance systems, as they are designed and optimized for server applications requiring high performance and durability for continuous reading and writing 24/7. Table 13.1 provides an overview of some SCSI variants.

13.2.2 The ATA and SATA interfaces

ATA (Advanced Technology Attachment), also known as IDE, ATAPI, and PATA (Parallel ATA), is another standard for physically connecting computers and peripheral devices such as hard disks, tape drives, and optical drives. Up to two devices can connect to each controller. ATA is superseded by SATA, which is now the primary architecture in the PC and laptop market.

SAS (Serial-Attached SCSI) is a serial communication protocol that allows for much higher speed data transfers, with up to 128 direct point-to-point connections. SAS devices use serial communication, which means they are not exposed to the interference (crosstalk) that occurs over parallel communication.

Table 13.1 Some of the many SCSI variants developed over the years

Interface	Type	MB/s	Range	Devices
SCSI-1	Parallel	5	20 ft (6 m)	8
Fast SCSI	Parallel	10	10 ft (3 m)	8
Fast-Wide SCSI	Parallel	20	20 ft (6 m)	16
Ultra-320 SCSI	Parallel	320	39 ft (12 m)	16
Serial Storage Architecture (400 MHz)	Serial	80	82 ft (25 m)	96
Fiber Channel 8 Gbit	Serial	788	6.2 miles (10 km)	224 (*)
SAS 1.1 (Serial-Attached SCSI)	Serial	300	20 ft (6 m)	16,256
USB-Attached SCSI	Serial	~1,200	10 ft (3 m)	127
iSCSI	Serial	Implementation- and network-dependent		2128 (**)

* Theoretical limit

** For IPv6

13.2.3 Hard disk failure

According to research (reviews of drive vendors' MTBFs and third-party hard disk drive reliability studies), the most common failures in the current generation of hard drives appear to be related to the head-disk interface. Regarding the timing associated with an actual hard drive failure, research suggests a failure is likely to occur in either the first 60 days of service or in the third, fourth, or fifth year of operation, given the high duty cycle and operational characteristics expected in a video surveillance storage application.

Because data loss can be very costly, using RAID configurations for redundancy in video management system is recommended. See section 13.4.1 for more information about RAID. A new server in a video surveillance system should be subjected to a "burn-in" test to ensure the drives do not fail within the first few days.

13.2.4 Solid-state drives

Unlike the mechanical disks discussed earlier, a solid-state drive (SSD) is a storage device with no moving parts or disks. Instead, it uses non-volatile flash memory for its data storage, thus endowing the SSD with various desirable properties, including silent operation, fast access, low latency, and good resistance to shocks and vibrations.

As mentioned earlier, the ATA interface and its variations in speed and connector type also applies to SSDs. However, SSDs mainly use SATA or NVM Express (NVMe), the latter being an open, logical device interface specification for accessing non-volatile storage media, usually attached via PCI Express (PCIe) bus. NVMe has a superior speed interface, and many servers run their operating systems on NVMe SSDs for fast booting and OS and application operation.

Solid-state drives get hot when constantly accessed – and they get hotter than their mechanical (HDD) counterparts. Although there are no moving parts in a solid-state drive, the memory chips get hot during heavy use.

Comparing the life spans of SSDs and mechanical disks is difficult. The HDD can suffer mechanical wear, causing data to degrade. In an SSD, it is instead the memory cells that degrade, which also will also affect the data.

Data degradation and failure in these disks does not necessarily need to be an issue. Most drives of both types operate for well over five years, which is probably longer than the entire server is used.

Although fast and reliable, SSDs are considerably more expensive per GB of storage than mechanical disks. Device capacity can also be a limiting factor when considering storage for a network video system, as there are few SSD drives that offer the same storage capacities as mechanical drives.

13.3 STORAGE ARCHITECTURE

The demand for more and more storage for all kinds of applications is driving the industry to develop ever-better performance and greater capacity for storage systems. High-quality surveillance video puts very high demands on a storage system, and storage is often a considerable part of the cost of a network video system. Virtually any size of storage system can be accommodated, which means that any frame rate, number of cameras, and retention time can be handled.

There are a few basic types of architecture for storage systems, all with different complexity, performance, cost, redundancy, and scalability. The most common are described in the following sections.

13.3.1 Edge storage

In some applications, storing the video onboard – that is, in the network camera itself – can be a good choice. Many cameras provide a slot for a standard memory card such as an SD card, which can store many days of video, the total depending on multiple factors such as the frame rate, the resolution, and the capacity of the SD card. In small systems, using SD cards as the only type of video recording is today feasible and provides for a very cost-efficient system. In addition to being the primary storage for video, an SD card can also function as backup storage when primarily recording to storage located elsewhere.

As the saving of video to a physical storage medium is a very write-intensive process, it is important to always use surveillance-grade SD cards for edge storage applications. These SD cards are specially designed to handle the large number of write operations typical when saving video and will last much longer than a standard SD card.

Another scenario for a distributed storage model is a mission-critical one, where loss of video is not acceptable when the network goes down or when taken offline for maintenance. This scenario is more hypothetical in nature because today's networks have a very high degree of reliability, provided they are designed correctly.

The term SD card covers three different formats of SD card: SDSC, SDHC, SDXC, and SDUC. The original format (SDSC) supported up to 4 GB of storage, the SDHC up to 32 GB, the SDXC up to 2 TB, and the SDUC format is expected to allow storage capacities of up to 128 TB.

Also of interest is the speed class of the SD card, which specifies the minimum write speed. In a network video system, this factor is very important, as selecting the correct speed class guarantees that video can be written to the SD card quickly enough. Write speeds range from 2 MB/s for Class 2, to 10 MB/s for Class 10 sufficient for 1080p video, all the way up to 30 MB/s for UHS Speed Class 3, which is sufficient for saving video at 4K resolution.

The latest speed class is UHS III, which allows write speeds up to 624 MB/s. The use of these cards requires other hardware than for earlier generation cards.

In parallel with the advances in storage capacity and write speed, the form factor of SD cards has also changed. Two steps down in size from the original SD card have produced the miniSD and the microSD, as shown in the next figure.

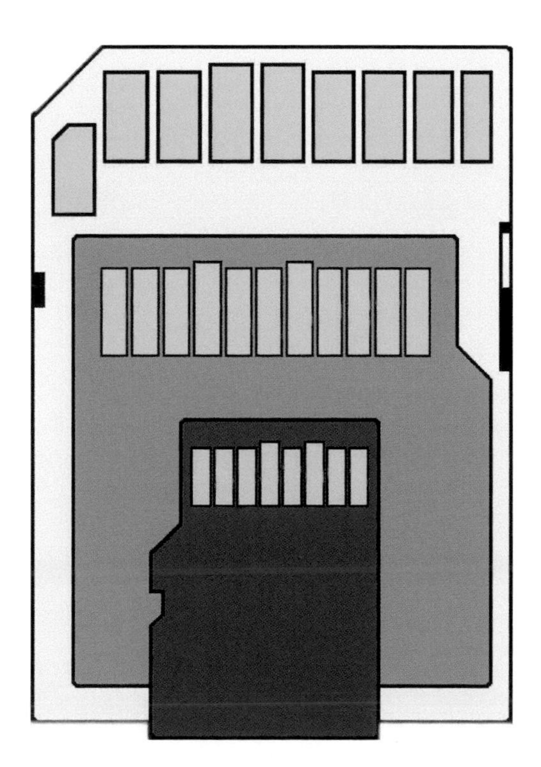

Figure 13.1 A comparison of the sizes of the original SD card (largest), the miniSD card, and the microSD card (smallest).

13.3.2 Single-server storage

Although single-server solutions are used for video storage in small installations of up to 16 cameras (Figure 13.2), many of these solutions can today handle over 100 cameras. The hard disks are in the same server (a PC in this case) that runs the video management software (application server). The server and the number of disks it can hold will determine the available space. Most modern servers provide support for multiple hard drives, which with terabytes of space can be more than enough for the most demanding application, even with long retention times.

In applications where the amount of stored data and management requirements exceed the limitations of direct attached storage, a separate storage system must be implemented. These systems are network-attached storage (NAS) and storage area network (SAN).

13.3.3 Network-attached storage (NAS)

NAS provides a single storage device directly attached to a local area network, and it offers shared storage to all clients on the network. A NAS device (Figure 13.3) is simple to install and easy to administer. It provides a low-cost solution for storage requirements and a good solution for smaller systems.

13.3.4 Storage area network (SAN)

A SAN is a high-speed, special-purpose network for storage devices. It is connected to one or more servers via a fiber channel (Figure 13.4). Users can access any of the storage devices on the SAN through the servers, and storage is scalable to hundreds of terabytes or even petabytes (1,000

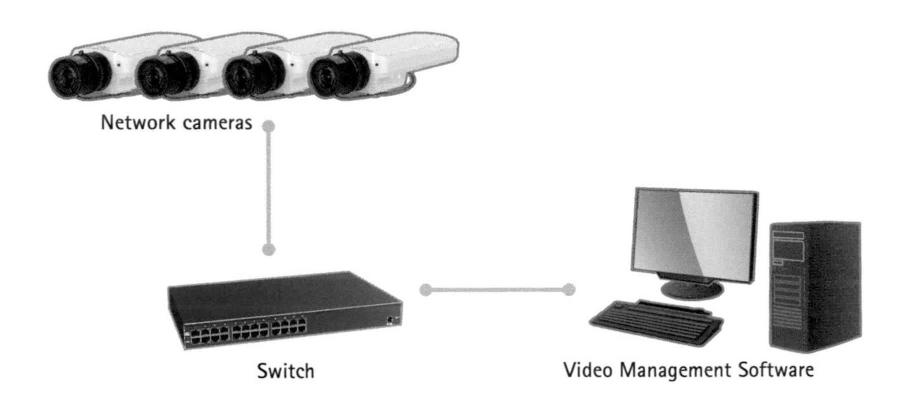

Figure 13.2 Example of a single-server storage setup.

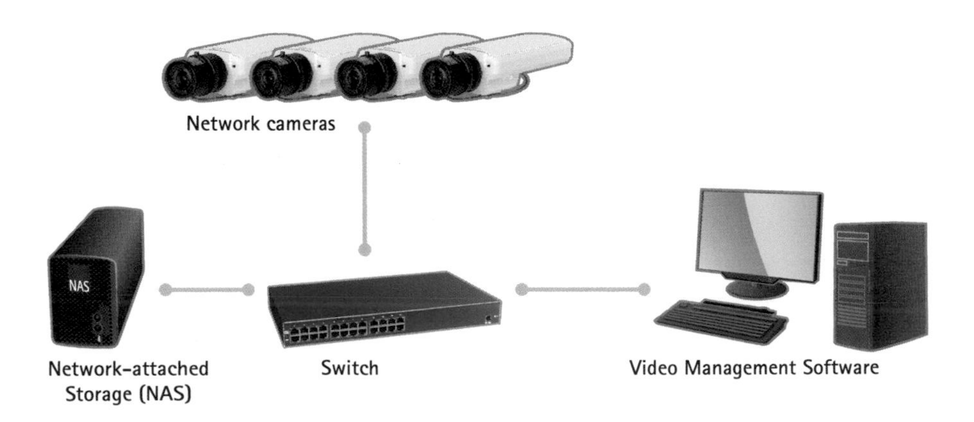

Figure 13.3 Example of a network-attached storage setup.

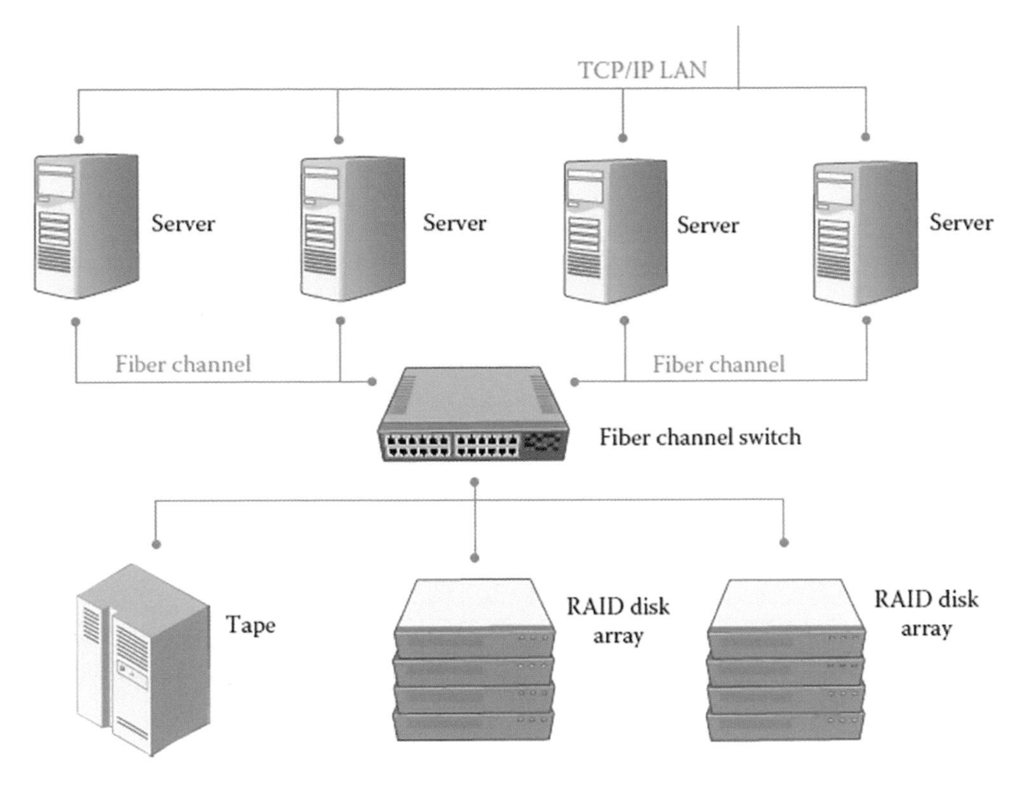

Figure 13.4 Typical SAN architecture, where a fiber channel network ties all storage devices together and lets the servers share the storage capacity.

terabytes). Centralized data storage reduces the administration required and provides a flexible, high-performance storage pool for use in multi-server environments.

The difference between NAS and SAN is that NAS is a storage device in which the entire file is stored on a single hard disk, whereas a SAN consists of several devices on which the file can be stored, block by block, over multiple hard disks. This type of configuration allows for a very large and scalable hard disk solution, in which vast amounts of data can be stored with a high level of redundancy.

13.3.5 Internet Small Computer System Interface (iSCSI)

iSCSI is a network protocol for sending SCSI commands over a TCP/IP network instead of an SCSI cable. This makes it possible to connect a storage device and address it as a local drive on the server. Unlike NAS, iSCSI devices do not support file system protocols. The main difference between iSCSI and NAS is that iSCSI provides block-level disk access, whereas NAS only provides file-level access. The host server manages the file system as it would on a local drive. iSCSI is commonly used in large installations and usually in RAID configurations, as it is a cost-efficient way of quickly adding more storage to a server.

13.3.6 Cloud-based storage and hybrid systems

As systems get larger and bandwidth costs remain stable, more and more storage is placed in locations other than the site containing the cameras. Cloud-based storage, as well as hybrid systems that combine local and cloud storage, are becoming ever more popular due to their opportunities for painless expansion and for the option to backup recordings to achieve full redundancy. Also, many organizations do not have IT operations as their main area of focus, so it is very convenient to outsource some or all the related configuration and maintenance to companies that specialize in these services. For an overview of cloud and hybrid systems, see Chapter 15.

13.4 REDUNDANCY

The server and storage components are essential in any IT system, and an IP-based video surveillance system is no different. Many different technologies are available to make the system more reliable by increasing redundancy. The following subsections discuss the most common techniques.

13.4.1 RAID systems

Redundant array of independent disks (RAID) is a method of arranging standard, off-the-shelf hard drives in such a way that the operating system sees them as a single large logical hard disk. Benefits include increased throughput and better reliability. Using hardware RAID controllers instead of software controllers is recommended to reduce performance issues.

There are different levels of RAID that offer different levels of redundancy – from practically no redundancy at all to a fully "hot swappable" mirrored solution, in which there is no disruption to operations and no loss of data in the event of a hard disk failure.

The most common RAID levels include the following:

- *RAID-0:* Also called striping. Information is spread over two or more disks to increase performance. There is no redundancy because the array is ruined if one disk fails.
- *RAID-1:* Also known as disk mirroring. The information on one disk is duplicated on one or more other disks. This increases reliability but may reduce performance, as data needs to be written to multiple disks.
- *RAID-5:* Also known as striping with parity. Data and parity are spread over three or more disks and require at least three disks in the array. Read performance is the same as for a single disk. Write performance can be lower because data must be written to two disks. RAID-5 can tolerate a single disk failure and still recover all data. Additionally, the disks can be made hot swappable. RAID-5 has become popular because it provides redundancy and maximizes disk space for data instead of backup.
- *RAID-6:* Similar to RAID-5 but with dual parity bits. This requires at least four disks, and the configuration can tolerate two disk failures.

13.4.2 Data replication

Data replication is a common feature of many network operating systems; file servers are configured to replicate data from other servers (Figure 13.5).

13.4.3 Tape backup

Tape backup is an alternative or complementary method. There is a variety of software and hardware equipment available on the market. In case of fire or theft, backup policies normally include taking tapes off-site. However, due to the relatively slow write speed of tape drives, these devices are seldom used for backing up video material.

13.4.4 Server clustering

Many server clustering methods exist. A common scenario for database servers and mail servers is to have two servers in the same storage solution, commonly in a RAID system. In such a case, when one server fails, the other (configured identically) takes over the application (Figure 13.6). These servers usually even share the same IP address, making the so-called failover completely transparent for the user.

13.4.5 Multiple servers

A common method to ensure disaster recovery and off-site storage of network video is to simultaneously transmit video to two different servers located in separate locations. In turn, these servers can be equipped with RAID, work in clusters, or they can replicate their data to other servers even further away.

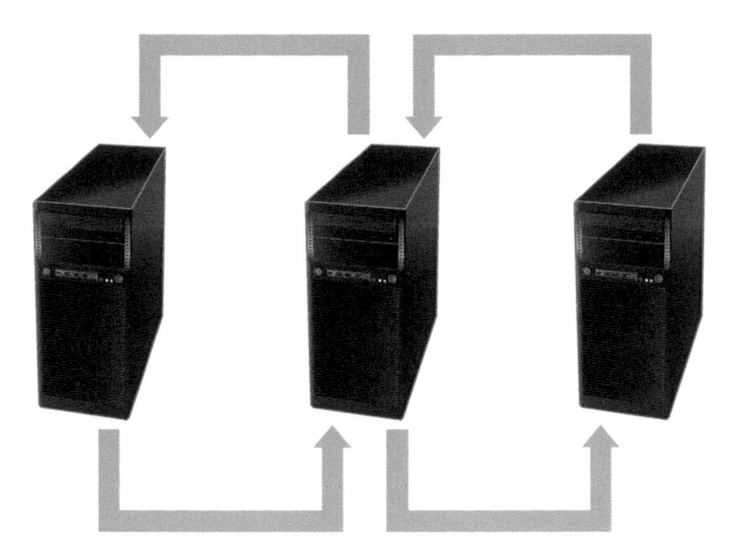

Figure 13.5 Replicating data ensures a high level of redundancy.

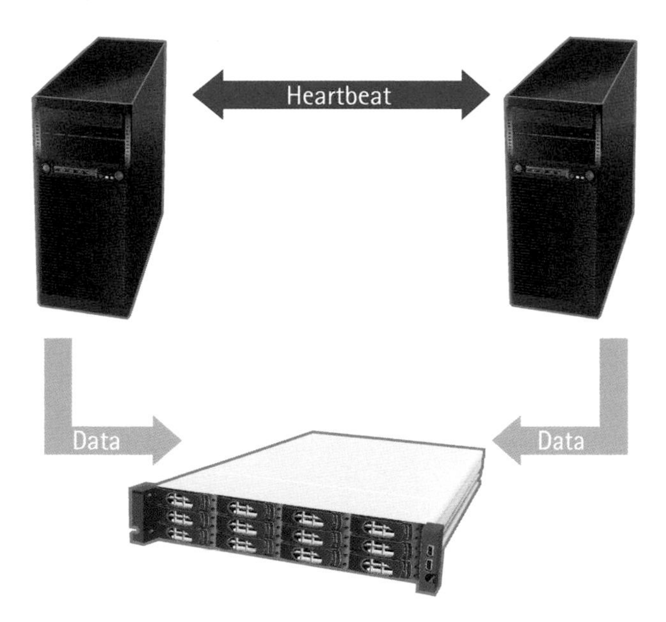

Figure 13.6 Server clustering is a popular way to ensure redundancy on the server part of the system.

13.5 BEST PRACTICES

Selecting the correct server and storage platform for a video surveillance system requires consideration. Some of the questions that need addressing include the following:

- *Is a centralized or decentralized system preferred?* This depends on the size of the system and the available bandwidth between the different locations.
- *What is the required system reliability?* The higher the requirements, the better the grade of disk and server required. Consider investing in surveillance-grade drives.
- *What is the required system redundancy?* A RAID-based recording system or redundant servers might be a smart investment that ensures no video is lost if a hard disk or server fails.

- *What is the size and scalability?* How many cameras will be managed and how scalable does the system need to be? Always plan for growth. The server should not be used at maximum performance, and it should not be the case that 90% of the storage space is used right from day one. There will always be a need to add to the system down the road, so plan for that today.
- *What is actually being recorded?* With the correct configuration, it does not necessarily need to be the case that doubling the frame rate also doubles the required disk space. Other methods can also lower requirements, for example, using event-based recording at night and reducing the resolution where possible. See Chapter 6 for more on video compression.
- *How long will video be retained?* Remember that more than 99% of all video is never watched. Having different retention times for different cameras is recommended, that is, a few days' retention for some cameras and longer retention times for other more important camera locations. In some applications, legislation or company mandates require a certain minimum retention time, often 7 or 30 days but sometimes 180 days or longer.
- *Is there a disaster, backup, and recovery plan?* Planning for trouble can make recovery faster. It is recommended that system operators understand the process in the event of failure. Having a strategy in place, together with IT teams or the parties responsible for infrastructure, while also including the system integrators is important. Ensure there is a plan to restore lost data, connectivity, or failed hardware.

Chapter 14

Video management

Video management is one of the keys to a cost-effective and successful video surveillance system. Selecting the right system requires careful consideration, including strategies for scalability, flexibility, and functionality. Along with cost, bandwidth, and storage, you need to consider the different management architectures and how they serve different user types. A well-designed system should also be easy to set up and use.

A video management system (VMS) has recording and viewing functionalities as well as functionalities for finding and configuring network video devices and other associated devices, such as audio devices, radar units, etc. A VMS also includes event handling and system security. More advanced functions include object detection, smart search, and other video analytics, as well as integration with other systems such as physical access control and building management.

Video management software is available for all sizes of systems:

- *Large systems:* more than a hundred cameras, sometimes thousands
- *Midsize systems:* ten to a hundred cameras
- *Small systems:* one to ten cameras

In some systems, there are many sites (e.g., retail stores) with small or midsize systems that are aggregated (sometimes also called federated) to a very large system. However, system size is only one aspect to consider when selecting a video management system. On a higher level, video surveillance needs to meet certain levels of comfort, forensic evidence, and preventive ability. The security demands and monitoring activity levels are in many ways more important than the number of cameras when evaluating and choosing software, system setup, and platform. Various departments and businesses will use the video management system differently depending on their purposes and needs. For example, in a grocery store, one person might use the video surveillance system for security purposes, whereas another might use it for studying customer traffic patterns.

Activity can be divided into the following levels of monitoring:

- *Passive:* Surveillance to satisfy curiosity or insecurity. Monitoring is casual and typically only happens randomly or when following up on alarms. This monitoring level is common for homeowners and small offices.
- *Occasionally active:* Surveillance to satisfy curiosity or insecurity. Monitoring activities are more likely to be regular but not active enough to be fully preventive. This level is common in retail businesses and office buildings.
- *Active:* Surveillance to prevent incidents. Personnel actively monitors the video and acts on alarms. Monitoring can also be for non-security purposes, such as predicting and tracking customer behavior. This level is common in retail businesses, banks, office buildings, and cities.

DOI: 10.4324/9781003412205-14

- *High security:* Surveillance for critical infrastructure and vital operations. Dedicated security operators monitor the video 24/7 and act immediately on incidents. This level is common in airports, hospitals, metro systems, and prisons.

This chapter provides an overview of the different types of video management systems, the platforms, and the most common functionalities available, in addition to some examples of integrated systems.

14.1 VIDEO MANAGEMENT ARCHITECTURES

There are three types of platforms for network video management systems:

- *Server-based video management:* A central server with software or an appliance manages the video from cameras and encoders. Server-based video management can be divided further, usually into these two categories:
  - *Server with video management software (VMS)*
  - *Network video recorder (NVR)* – a type of network appliance
- *Edge-based video management:* The cameras or video encoders manage the video individually. A client application presents the information in a single user interface, which makes the system look like a complete system of several devices.
- *Cloud-based video management:* A software application that runs on servers in the cloud manages the video from cameras and encoders. Cloud-based video management is also known as Video Surveillance as a Service (VSaaS) or hosted video.

Each platform has its pros and cons. In large operations, there might be hundreds of cameras at a central site, in which case it makes sense to manage them in a server-based system with IT servers and a VMS. In smaller remote locations, appliances such as NVRs might be easier to install and manage. In very small remote locations, where a few cameras store video on SD cards, edge-based video management might be the most efficient solution. Then one or two cameras in each location can act as backup to keep video available for management in a cloud-based application.

14.1.1 Server-based video management

In a server-based architecture, there is a central server, or appliance, that manages the configuration of the system and the video from each camera or video encoder. Server hard drives are usually the primary storage point, but other types of storage devices can also be used. NAS devices can function as main storage or backup storage. Some cameras and encoders have SD card slots for onboard storage. In a server-based system, you would typically use SD cards for backup video recording, so even if the network is down and you lose the connection to the network servers, you can still record video. When the network is up again, the video on the SD cards moves to the server. This is sometimes called video trickling, or failover recording. This use of SD cards or NAS should not be confused with edge-based video management, where the cameras and encoders actually manage video independently of any central servers.

Server-based video management architecture can be divided into two categories: PC-based systems with VMS software, typically used in larger scalable systems with hundreds of cameras, and NVR systems, which are suitable for up to 100 cameras. NVRs come with the software pre-installed on the hardware, which means the systems are easy to deploy but typically less scalable.

14.1.1.1 Server with VMS software

Server platform solutions are based on off-the-shelf hardware. Components such as multiple-processor systems and storage can be selected to obtain the maximum performance for the specific design of the system. A server platform solution can use standard components for increased or

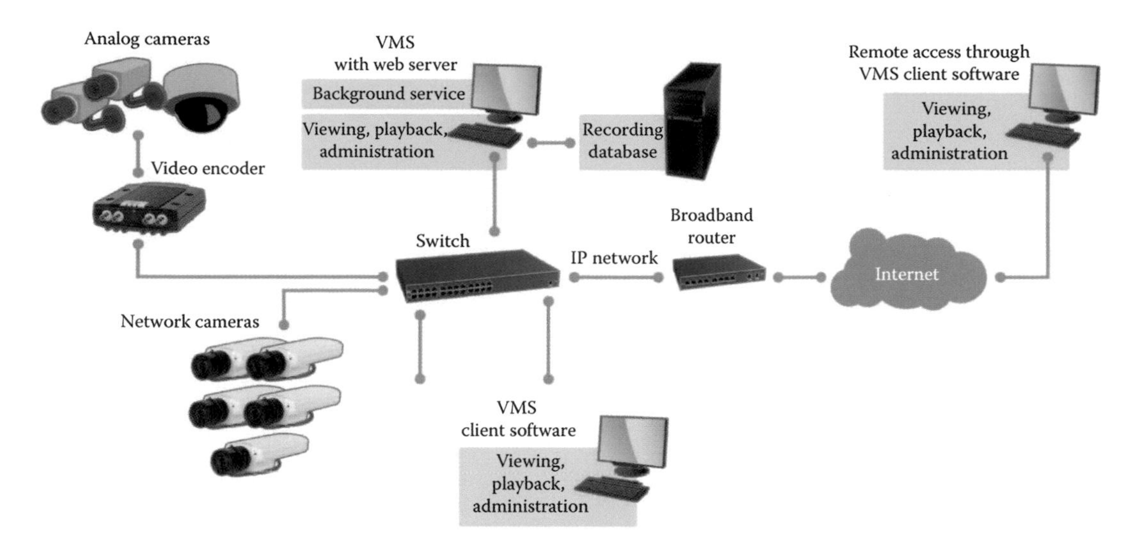

Figure 14.1 A network video system with server-based video management. The VMS manages the recordings and data and runs on a server. The VMS client, used for viewing live and recorded video, runs on regular computers.

external storage, for additional operator stations, and for running additional software, such as firewalls and virus protection, in parallel with the video application.

Usually, a Windows® platform or a UNIX®/Linux® platform serves as the basis for these systems, although there are also some that run on OS X®. Often, a system integrator is responsible for implementing the system and installing the software on the server, but users with the appropriate skills could install the software themselves.

Systems designed on a server platform are easily scalable (see Figure 14.1). The system hardware and software licenses can be expanded or upgraded to meet increased performance requirements. The server platform is suitable for scenarios with large numbers of cameras or when the IT department has standard requirements on the type of server hardware and software to allow them on the network. In demanding environments, the video management software can be installed on a ruggedized, industrial server.

Often, the viewing client software is installed on a computer that is separate from the recording server where the management software is installed. The server shares all the video management settings with the client, and the client's user interface looks the same and provides the same actions and options. In some cases, users of the client software can switch between different video management servers. This makes it possible to manage video at many remote sites or in a large system.

Video management software is available for all system sizes and monitoring activity levels. The choice of system also depends on camera compatibility and integration with other systems. Some software works best when the camera count stays within a certain range. Normally, a one-camera-one-license principle applies to the video management software, and licenses can be added one by one. System designers need to look carefully at the scalability of the system and factor in future camera counts, user licenses, and numbers of locations. They also need to consider the activity level and purpose. Will there always be someone monitoring the live video, or is the monitoring activity likely to be irregular and alarm-based? Is the ambition to prevent incidents or only to investigate these after they occur?

Software intended for large or high-security systems often provides full integration with other systems, such as physical access control systems or building management systems. More often than software intended for midsize systems, these are open systems that support more than one camera brand. In addition, they offer simultaneous access to multiple sites, with advanced functionality. This means that companies with multiple sites can streamline their video surveillance without sacrificing efficiency or security. There is great diversity in this group of software. Some are basic, and others are feature-rich; some are segment-specific, and others are complex enterprise solutions.

Most software intended for small and midsize systems has functionality that seems more basic compared to what users of large system are accustomed to. Some of these only support a single camera brand and usually have fewer possibilities for integration with other systems. As a result, they are mainly used for video surveillance. In return, they are generally quite rich in features supporting active, targeted, and frequent usage. Typically, when vendors release new products, they update the software to support the features of the new product. In other words, this type of software is perfect for homogenous systems because its features align so well with the product features.

The main type of storage used in server systems with management software is dedicated storage servers, although many support failover recording to edge storage devices, such as SD cards (also known as onboard storage) and network-attached storage (NAS). Today, users can expect support for multiple languages as well as remote internet access and viewing apps for smartphones for easy access to video from any location.

14.1.1.2 Network video recorder (NVR)

An NVR is a hardware box with pre-installed video management software. Usually, users view and manage video through client software, but many NVR and VMS solutions also offer web interfaces and apps for smartphones and tablets.

The most obvious difference between an NVR platform and a server-based solution is that an NVR is a hardware box with pre-installed video management functionality (Figure 14.2). In other words, the NVR is a self-contained system that includes the computer, software, storage, and sometimes a multi-port PoE switch in one unit. They are regarded as plug-and-play systems. NVRs are similar to DVRs (digital video recorder) but are used with IP cameras instead of analog cameras. So-called hybrid DVRs have network inputs as well as analog inputs for recording from both IP and analog cameras.

An NVR is dedicated to its specific tasks of recording, storing, analyzing, and playing back network video. NVRs do not allow any other applications to reside on them. The NVR hardware itself is locked to its application, and the unit can rarely be altered to accommodate anything outside its original specification.

Often, the NVR hardware is proprietary and specifically designed for video management. The operating system can be Windows®, UNIX®/Linux®, or proprietary. An NVR is designed to offer

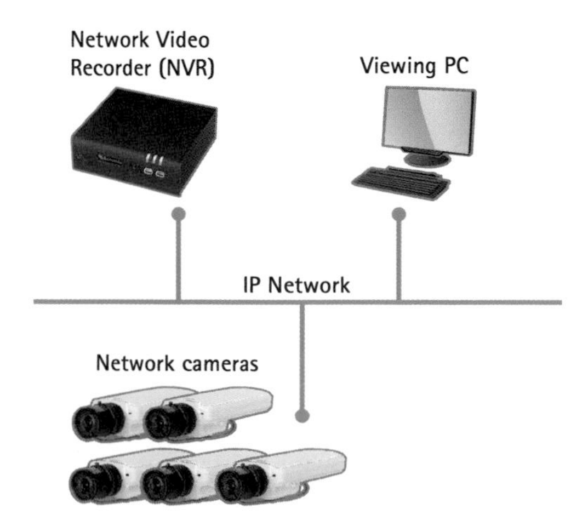

Figure 14.2 A network video system that uses an NVR.

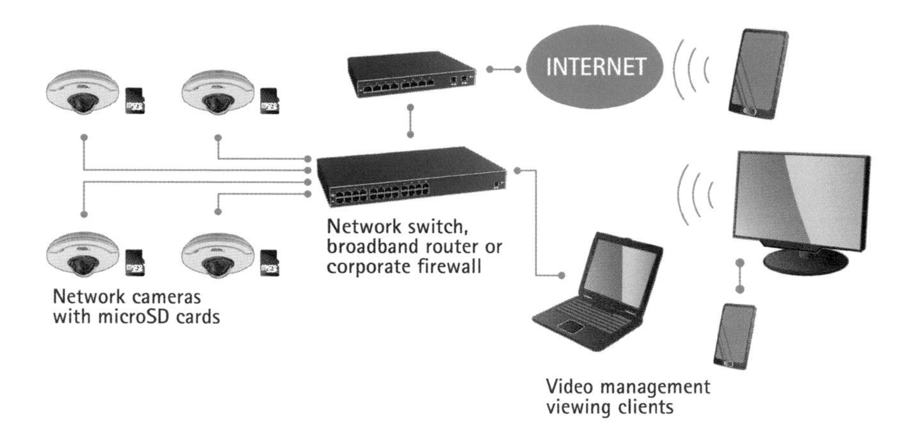

Figure 14.3 A small system with network cameras, SD cards, and multiple viewing devices.

optimal performance for a set number of cameras and is normally less scalable than a server platform system. This makes the unit more suitable for smaller system configurations, where the number of cameras is within the limits of an NVR's designed capacity. There are low-end and high-end NVR systems available. In comparison with a server platform, an NVR is normally easier to install, as the video management software comes pre-installed, and the NVR sometimes includes built-in PoE ports for quick connection of a few IP cameras.

14.1.2 Edge-based recording

Edge-based recording means that no centralized software is needed for the video recording. Instead, the camera or video encoder also acts as a recorder and saves the video to an SD card (see Figure 14.3). This type of system is cost-effective and scalable for smaller systems.
To make it as user-friendly as possible, the process of installing and configuring the system is optimized in a few steps. To view live or recorded video and to export video clips, users can use either the PC client or a smartphone app. This type of software is often free but limited to a single camera brand. Its main purpose is to provide easy and reliable video management, with the emphasis on recording video, for businesses with little time or money for video surveillance. This type of solution is best for small installations of one to ten cameras, and the sweet spot is somewhere in the middle of that range.

In edge-based video management, the video analytics are also based on the built-in intelligence of the camera. To view live and recorded video, the user signs into an application on a computer, smartphone, or tablet. There is no need for extra servers, as the camera itself acts as a server and keeps track of the recordings, schedules, and events.

Users can also choose if cameras should record video continuously, to a schedule, or when motion is detected. Through basic functionalities such as support for split-screen viewing, multiview streaming from multi-megapixel cameras, HDTV resolution, PTZ control, audio, multiple languages, and search filters, client–camera software meets most of the video management needs of its typical user.

14.1.3 Cloud-based video management

Cloud-based video management provides remote operation over the internet. Users can view and manage the system from any device with a browser and internet access. Most service providers of cloud video management include apps for smartphones in their offering. Cloud video management software is installed on a web server and is connected to one or more recording servers. The web server allows users access to the software and the video devices it manages from anywhere. All they need is an internet connection and a PC with a browser or an app on a mobile device (Figure 14.4).

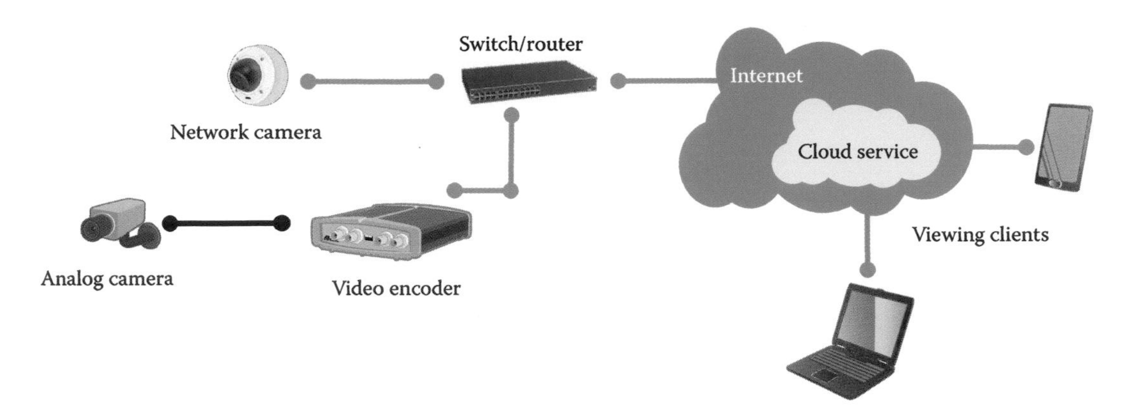

Figure 14.4 Cloud video management software enables users to view cameras and perform various operations with using any compatible device.

With cloud video management services, also known as hosted video or Video Surveillance as a Service (VSaaS), users do not have to worry about installation or maintenance of servers, clients, or storage devices. The service provider owns the storage servers and is responsible for maintenance, and they often have professional operators who can respond to alarms in case of an event. Sometimes, both video management and the cameras themselves are included in the monthly subscription fee. However, potential buyers of cloud video solutions need to make sure they have a good internet connection, as transferring video files can use a lot of bandwidth.

To compensate for limited bandwidth, you can use complementary technologies. SD cards and a local NAS can store high-resolution, high-frame rate recordings, while low-resolution video can be uploaded to the cloud. This provides for a very flexible solution and is often referred to as a hybrid solution, in which some functionality and recording is in the cloud and some is on-site.

Cloud video is a good introduction to video management at a relatively low cost and limited upfront investment. It is also a practical solution when a system consists of many locations (sites) with only a few cameras at each site.

One of the major benefits of cloud software is that users only need to pay for the services and the storage they use. Often, the service provider owns all the hardware and is responsible for its maintenance. A good interface should work and look good in most browsers, no matter the type of device and operating system. Often, the interface is less daunting than client-server software and is easy to use. With the proper safeguards, such as password protection and IP address filtering, cloud services allow for secure online management of video from anywhere in the world.

For more in-depth information about cloud video and its benefits and challenges, see Chapter 15.

14.2 OTHER ASPECTS OF VIDEO MANAGEMENT ARCHITECTURE

There are many different video management solutions, from simple onboard recording to fully-fledged systems for thousands of cameras integrated with other systems. When deciding upon the best solution, it is also important to consider other aspects of the video management architecture. Some of these aspects are the licensing of the system, if it supports multiple camera brands, which APIs are used for integrating the system, scalability, and support for viewing video on mobile platforms. These aspects are discussed next.

14.2.1 Open versus vendor-specific software

Most network camera vendors provide their own video management software. Some have only very basic recording and viewing functions with limited scalability and may even be given away free to boost sales of cameras. Others provide more fully-fledged systems. As mentioned before, many of these systems only support devices from a specific vendor, which limits choice and flexibility. However, one benefit may be that cameras and encoders can integrate with better optimization because all the built-in features can be used.

Nowadays, however, many systems support other camera brands with open application programming protocols such as ONVIF (Open Network Video Interface Forum) (see also section 18.2.7). Fully open systems that support multiple brands of network video devices also exist, but these are generally developed by independent companies. Hundreds of versions of open video management software systems are available in today's security market. Although the basic functions are similar, some software is more scalable and includes more advanced functionalities than others. Because the licensing and maintenance models can differ, it pays to factor in both current and future licenses in the investment cost.

Open systems that support network video devices from many different manufacturers often provide the highest flexibility for the user.

14.2.2 Protocols and APIs

To integrate different network video devices into a video management platform, a communications protocol or application programming interface (API) must be implemented in the video management platform. Some camera vendors use different proprietary protocols, but standardized protocols such as ONVIF have also been adopted by many vendors. While the ONVIF protocol has the benefit of being standardized, it does not always support all the advanced features of some cameras. Using the proprietary protocols of a network video manufacturer may provide for additional functionality and tighter integration.

14.2.3 Apps for smartphones and tablets

As mentioned, many video management software vendors complement their offerings with apps for smartphones and tablets.

The level of functionality varies, from the viewing of live video only, to some level of video management. Users now expect a higher level of intelligence and control, as well as more integration with other systems, such as the ability to greet visitors and open doors from a smartphone.

14.2.4 Scalability of video management software

The scalability of most video management software, in terms of the number of cameras, resolution, and frames per second that it supports, is in most cases limited by the hardware and bandwidth capacity rather than by the software itself. Storing video files puts strain on the storage hardware, as it may be required to operate on a continual basis, as opposed to only during normal business hours. In addition, video generates large amounts of data, which also makes demands on the storage solution.

The basic recording of video and audio is not particularly CPU-intensive. More processing power is needed when frames require decoding, as is the case with functionalities such as video motion detection, forensic search, and other server-based video analytics. Live viewing capacity is limited by the capacity of the video decoder and graphics card.

As a system grows and eventually exceeds the capacity of a single server, it is in many cases possible to add new servers to the system or to pay the cloud service provider for more capacity. Ideally, it should be possible to add storage seamlessly. For professional and scalable systems, factors such as redundancy, failover, and no single point-of-failure become critical. Redundancy in a storage system allows the saving of video or other data simultaneously to more than one location. This provides a backup for recovering video if part of the storage system becomes unreadable. There are a number of options for providing this added storage layer in a network video system, including

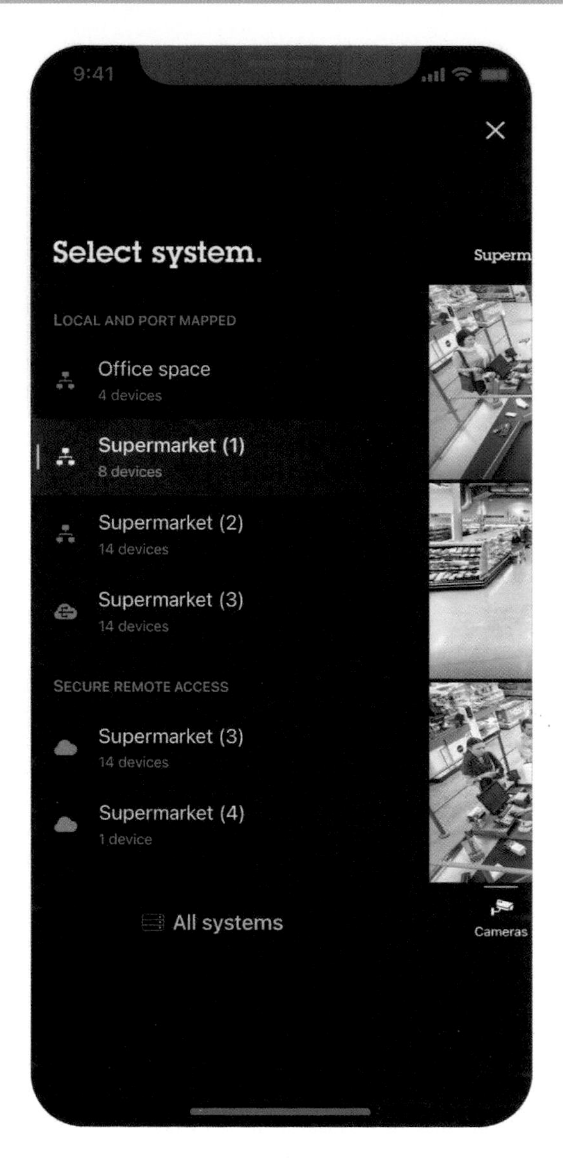

Figure 14.5 Viewing apps for smartphones provide mobile access to live and recorded video.

a redundant array of independent disks (RAID), data replication, server clustering, and multiple video recipients. See Chapter 13 for more on servers and storage.

14.2.5 Licensing of video management software

Most video management software applications are licensed products. Some licenses are sold as perpetual licenses, which means they can be used for as long as the user wishes, with no additional costs. More often, however, licenses are valid for a limited time, typically one to five years, after which, they need to be renewed. Licensing policies vary, but they mostly require one license per camera (or video source) in the system, as well as a base fee for the software itself. Some video management software only requires one license per encoder, regardless of whether it is a multi-channel or single-channel encoder, while others require a license for each channel. Sometimes, the license is a software key tied to the MAC address of a camera or a server CPU. Typically, there are also maintenance fees associated with technical support and the right to download and install new versions of the software. In the case of cloud video management, all services are included in the monthly fee to the service provider.

14.3 SYSTEM FEATURES

A video management system can include many different features:

- *Recording of video and audio*
- *Storing and searching recordings*
- *Exporting files*
- *Redaction*
- *Viewing of live video, recordings, and multi-streaming*
- *Mapping functionality*
- *Event management and responses*
- *Video analytics*
- *Inputs and outputs*
- *Log files*
- *Administration and management*
- *User access control and activity*
- *Health monitoring*

14.3.1 Recording

The primary function of a video management system is to record video and audio. Recording functionalities include setting up recording rules and intelligent ways of searching recordings and exporting them to other systems. The following subsections discuss the various recording aspects of a video management system.

14.3.1.1 Video recording

A typical video management system has four types of video recording:

- *Continuous recording*
- *Triggered recording* (by motion or other alarm)
- *Scheduled recording*
- *Manual recording*

Because they tend to cover longer periods, continuous recordings normally use more disk space than alarm-triggered recordings. An alarm-triggered recording can be activated by, for example, video motion detection or by an external input on a camera's input port. Scheduled recordings can combine timetables for both continuous and triggered recording instructions (Figure 14.6 and Figure 14.7).

After selecting the recording method, the quality of the recordings is determined by selecting the video format (for example, Motion JPEG, H.264, or H.265), resolution, frame rate, and degree of image compression. These parameters affect how much bandwidth and storage space is needed.

Each recording mode can have its own frame rate (frames per second). Because of previous limitations of analog recording technologies (DVRs), the specifications and requirements of some system designs are limited to lower resolutions, such as VGA or 720p, and frame rates of 7.5 frames per second (fps) or lower are common. With today's network cameras and video compression technologies, higher resolutions and frame rates are common. Video streams at 5 megapixels or 4K, with frame rates of 30 fps or even higher, are not unreasonable.

Storage can still be a significant expense in a video surveillance system, so making the right requirements on the system continues to be crucial. In a well-designed system, the camera or encoder requests the appropriate frame rate, which means that only the required frame rate is sent over the network.

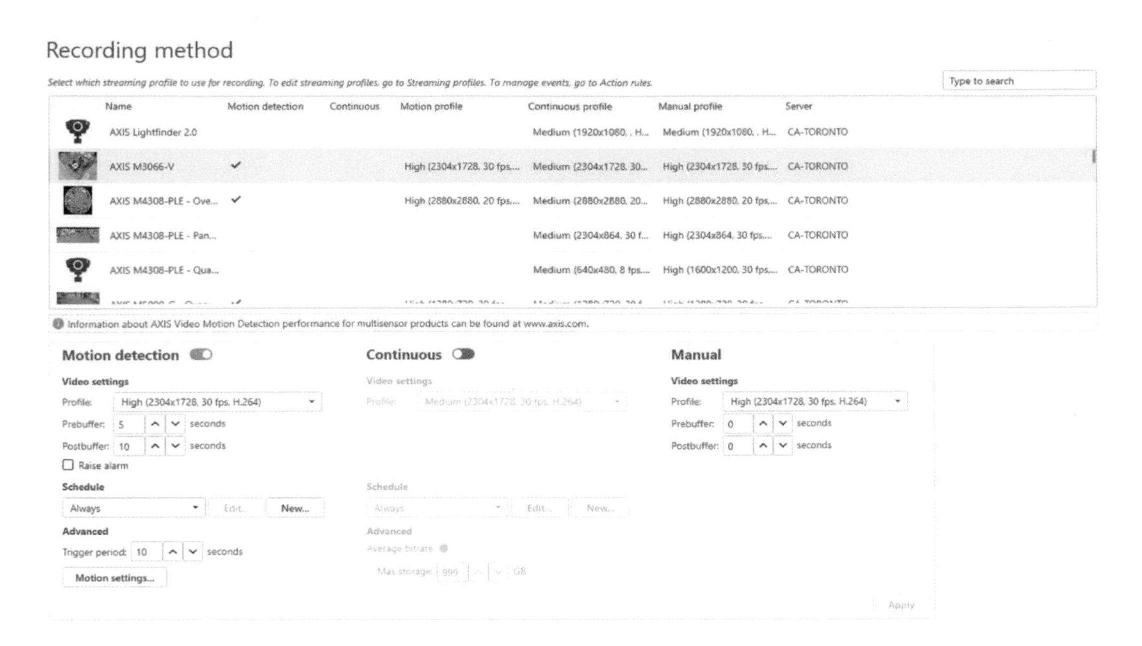

Figure 14.6 An example of an interface for configuring recording functions.

Figure 14.7 Scheduled recording settings.

14.3.1.2 Audio recording

Audio can be recorded using the built-in functionality in a network video device or through the video management software. If it is important that audio and video are synchronized, the audio and video packets must be time-stamped. Because a camera might not always support time-stamping of Motion JPEG video packets, using H.264/H.265 is recommended. These video streams are normally sent using Real-time Transport Protocol (RTP), which timestamps the video packets. For details on audio compression standards, see Chapter 7. Some areas have restrictions on audio recordings. For more information about legal considerations, see Chapter 18.

14.3.1.3 Recording and storage

Most video management software uses the standard Windows® file system for storage, so any system drive or network-attached drive can be used for storing video. Usually, a separate file or database keeps an index of the available video. The advantages of using a database for storing all settings and recording metadata and using a file system include the following:

- The ability to manage shared access and ensure data integrity
- The ability to efficiently search for recordings
- The ability to enable direct file access and record directly to disk

Video management software can support more than one level of storage. That is, a primary hard drive records video, but local disks, network-attached drives, or remote hard drives archive the files. Users may be able to specify for how long files should remain on the primary hard drive before they are automatically deleted or moved to the archive drive. Users also may be able to lock video files and protect them from automatic deletion.

14.3.1.4 Exporting files

Many video management software applications record video using proprietary file formats but still allow users to export recorded video to standard file formats, such as mp4, AVI (Audio Video Interleave), and ASF (Advanced Streaming Format). In many cases, the proprietary format can be exported along with a proprietary player. The advantages of using the native recording format are additional security, data integrity, and advanced playback features. The proprietary format can be encrypted and password-protected. Because the proprietary format is often more difficult to edit, it can be used to preserve the chain of evidence. A proprietary player may also include other features, such as multi-camera playback.

Of great importance in legal cases where video is presented as evidence is audit trail functionality, which records every change made to the files while the original file is left untouched. This helps secure the chain of custody and ensures reliable evidence that is admissible in a court of law.

Some VMS applications support this functionality. In other cases, you will need a separate application that does this for you. Examples of audit functions are digital watermarks and overlays with extra information to support the evidence, such as timestamps or case numbers.

14.3.2 Redaction

Also of interest when exporting video that will be used and seen by third parties, for example, in legal proceedings, is the ability to redact (conceal) faces and other identifying personal features or other sensitive information. This may be to protect the privacy of persons unconnected with the recorded incident or perhaps to provide protection for persons giving testimony anonymously. Redaction is often done by blurring out the relevant area, but other methods can also be used, such as superimposing boxes or shapes over these areas. Other situations that require redaction of personal information in recorded video are those that are used solely for analysis and where identification is not the goal, for example, when recording flows of people or traffic in city planning operations. In some VMS or other software applications, video redaction can be performed automatically, for example, for every human face that appears in the video.

14.3.3 Viewing and searching

Another key function of a video management system is the ability to view live and recorded video in an efficient and user-friendly way. Most video management software lets multiple users view several different cameras at the same time and allows recordings to occur simultaneously. Additional features include multi-monitor viewing and mapping, the latter meaning that camera icons can be overlaid on a map of the building or area, representing the true location of each camera. The following subsections discuss common viewing functionalities in video management systems.

Figure 14.8 An example of redaction, in this case, by blurring faces.

Figure 14.9 An example of live images in a split view.

14.3.3.1 Live viewing

Many video management software applications provide users with the option of viewing images in different ways: split views, single-camera popups, full screen, or camera sequences. In a split view, between 4 and 64 cameras are shown on the screen. A popup can also appear when a camera is triggered. Figure 14.9 shows an example of a split view. When using sequence mode, live views from different cameras will be displayed, one after the other, in a user-defined order. Users can choose to switch between cameras, view areas, and split views. They can also set a time for each view in the sequence, that is, the time to elapse before switching to the next view.

When using a PTZ camera, video management software may allow PTZ control through a number of input devices, including the following:

- *Joysticks*
- *Control boards, keyboards, and keypads*
- *Computer mouse*

Figure 14.10 A control board consisting of a keypad and a joystick.

Figure 14.11 Split views help get a better overview of when and how a certain incident evolved.

Many camera manufacturers supply joysticks and control boards especially developed for video surveillance (see Figure 14.10), but even a gaming controller could be used, as long as it is compatible with the VMS and the camera's PTZ protocol. Users can program control boards, keyboards, and keypads with shortcuts, or hotkeys, to move quickly between workspaces, camera views, and PTZ positions. They can also use a mouse to click on the image, move the camera, and use the scroll wheel to zoom in.

If a camera is equipped with audio capability, the VMS can also allow audio controls through the user interface.

14.3.3.2 Viewing recordings

Video management software usually includes multi-camera playback, which lets users view simultaneous recordings from different cameras (Figure 14.11). This makes it easier to get a comprehensive picture of events, which moves investigations forward and helps build cases in court.

14.3.3.3 Multi-streaming

Viewing or recording at full frame rate on all cameras at all times is more than what is required for most applications. In normal operation, frame rates can be set lower, say one to four frames per second, to limit bandwidth and storage requirements. The recording frame rate can be set to increase automatically when the camera detects particular types of events, such as motion in a monitored area or activation of an external sensor. It is also possible to send multiple video streams in parallel,

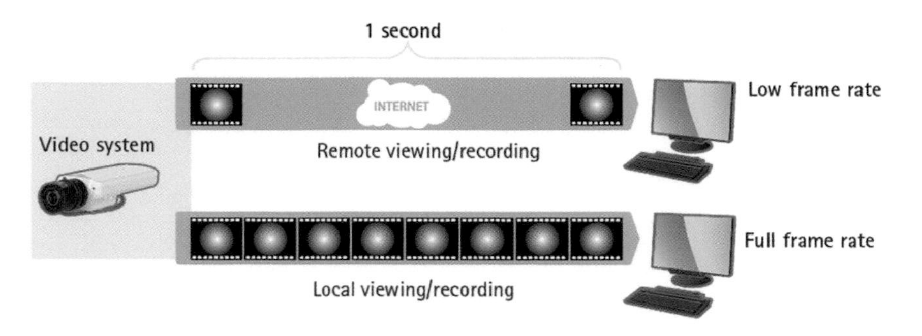

Figure 14.12 Video at different frame rates, resolutions, and compression can be sent to different recipients.

with different frame rates, codecs, compression, and resolution to different recipients (Figure 14.12), taking into consideration the available bandwidth and the performance of the devices used for recording and viewing. This ability to create, view, send, and record multiple streams with different sets of capabilities and values is called multi-streaming.

Multi-streaming is great when you want flexible recording, viewing, and storage options. For example, you can use a low-quality, low–frame rate stream for viewing in smartphones to save bandwidth when accessing video remotely or to offload a software client with poor performance. In an edge-based video management system, each camera can record parallel streams (one low-bandwidth and one high-quality stream) and keep track of its recordings. Only when you request a stream does the camera send it to your viewing device. Of course, streaming the low-bandwidth video saves data traffic over the mobile network, but you can switch to the high-quality stream whenever you want to. Say you are watching a low-bandwidth stream until you suddenly spot something of interest. By switching to the high-quality profile, you can watch the event in greater detail. You can also save video and images on your mobile device and share them with the police, your insurance company, or other investigators.

14.3.3.4 Search options

Searching through recorded video is an important feature in a video management system. There are different ways of finding the footage of interest, including the following:

- *Scrubbing through footage*
- *Selecting dates and times*
- *Activity-based or event-based searches*
- *Forensic searches based on metadata*

Scrubbing means that the user manually searches the recorded video to find certain activity in the scene. Because there are often many hours of footage recorded, scrubbing can be an ineffective method. If the user knows the date and time when an event occurred, they can enter that time and replay video sequences from one or more cameras. If the user only has evidence that something has happened but not when, they can do a motion- or object-based search (Figure 14.13). The video management system finds and replays video of movements or objects that occur only within a defined area of interest and ignores all events outside that area. This is the most advanced type of search and is also known as forensic search, advanced search, quick search, or smart search. Naturally, it is easy to find and replay video associated with a certain type of alarm or event trigger.

Forensic searches based on metadata (data about data) are typically much more efficient, as they do not require decoding of video, which is very CPU-intensive. The metadata is often generated by the camera and tags the video with the time and location, making it a simple matter to find and view the associated video. For more on forensic search and metadata, see Chapter 16.

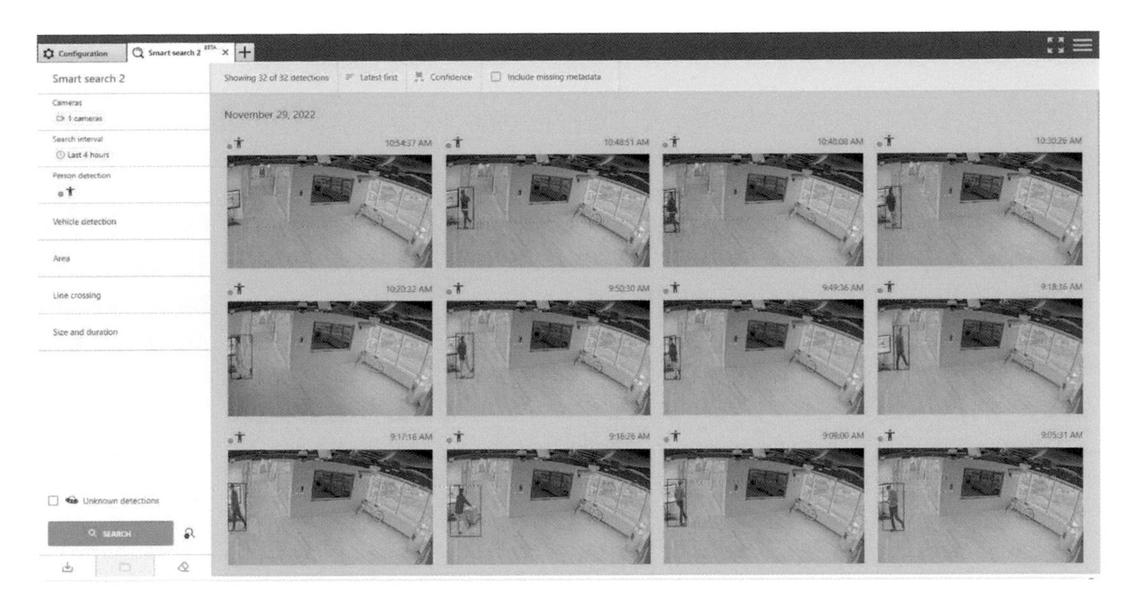

Figure 14.13 In video management systems, events in a defined area can easily be found in a forensic search.

Figure 14.14 The similarity search has been configured to search for a person wearing a red shirt, of which multiple instances have been captured on several cameras.

14.3.3.5 Similarity search

A similarity search is an AI-driven attempt to find all instances of the same object or all objects that display the same or similar characteristics, as defined in the search criteria. The search can be performed across multiple cameras, and often across multiple sites, to help reduce the time spent on investigating incidents. Figure 14.14 shows an example.

14.3.4 Mapping functionality

For easy identification and selection of cameras, users can place camera icons on a map of the monitored area. The map is imported into the video management software (Figure 14.15) and can be a photo or a drawing in a standard image format such as jpg, gif, or png.

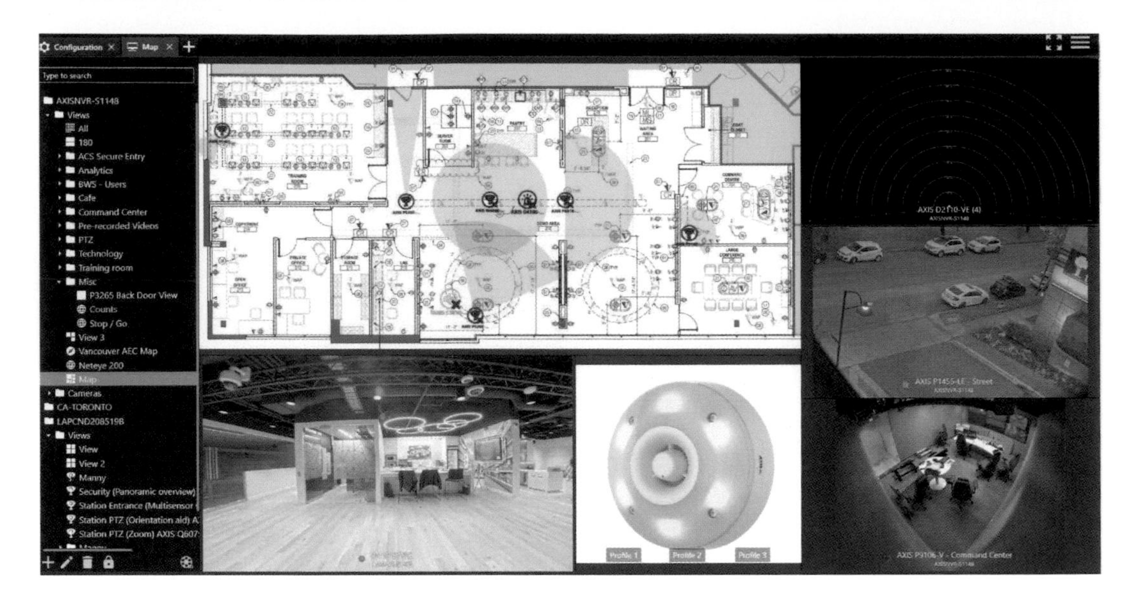

Figure 14.15 Using mapping functionality, finding the right camera becomes very intuitive.

Some video management systems have an icon library that makes it possible to drag and drop icons onto a map. By clicking on a camera icon, the user can view the live video stream from that camera. When an alarm occurs, a camera icon may change color to indicate that the alarm comes from that camera. More advanced systems also make it possible to show which area each camera covers.

14.3.5 Event management

Video management software and network video recorders can usually receive, process, and associate events from different sources, such as physical access control systems, point-of-sale terminals, video analytics software, and the network video products themselves. Once an event is triggered, the video management system can register the event, associate it with a video clip from a nearby camera, and alert an operator through a popup on the monitor or by sending a notification to a smartphone.

The following subsections provide more details about event and alarm management, video motion detection, input and output ports, and event log files.

14.3.5.1 Edge-based event handling

Video devices with built-in event handling have the benefit of enabling the efficient use of bandwidth and freeing up storage space. Unless an event takes place, there is no need for a camera or encoder to send continuous video streams for live monitoring or recording. When an event does take place, the device can send notifications, trigger other devices, and activate other responses automatically.

Event management, which includes alarm handling, involves defining an event that activates a device to perform certain actions, which can be scheduled or triggered. See Figure 14.16 for an example of a trigger setup.

There are many different types of triggers that can activate actions, including the following:

- *Inputs* from external devices can trigger actions. The input devices are connected to the input ports on a network camera or video encoder. Examples of input devices are motion sensors, door switches, and glass-break detectors. See Figure 14.17 for an I/O example.

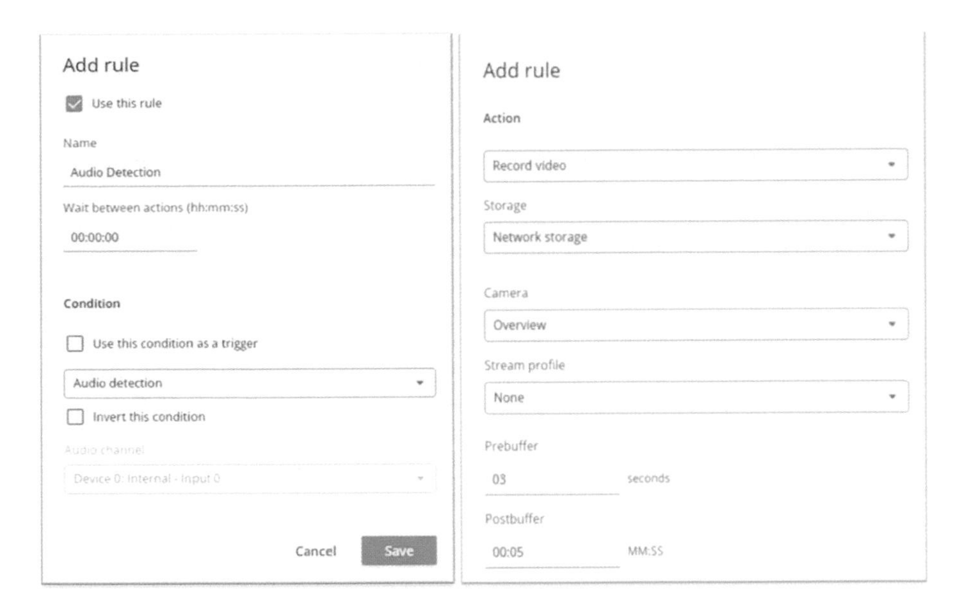

Figure 14.16 This sample action rule records video when the camera detects sound.

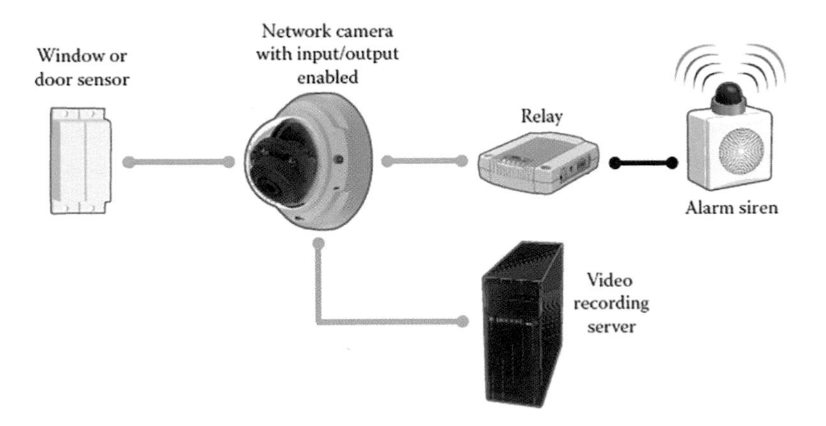

Figure 14.17 A window sensor can connect to the camera's input port and an alarm system/siren to the output port.

- *Manual triggers* are used by operators to initiate actions.
- *Analytics* can trigger actions when the camera detects activity or objects based on user-defined settings. One example is activity in motion detection windows.
- *Audio detection* can trigger actions when a camera with built-in audio detects sound below or above a set threshold value. For more on audio detection, see Chapter 7.
- *Temperatures* can trigger actions when the camera detects that the temperature rises or falls outside the operating range.
- *Camera tampering* can trigger actions when the camera detects that it has been covered, moved, or is no longer in focus. For more on camera tampering, see Chapter 16.
- *Shock detection* can trigger actions when the camera detects that someone has tilted it or hit it with an object.
- *Storage recording* can trigger actions when the camera starts or stops recording video to edge storage.

- *Storage disruption* can trigger actions when the camera detects that the storage device is unavailable, locked, full, removed, or if there are read/write errors.
- *Enter/exit detectors* can trigger actions when the camera detects passage from or to a user-defined area.
- *Fence detectors* can trigger actions when the camera detects that a subject or object crosses a virtual line.
- *Object removed* can trigger actions when the camera detects that an object disappears from a user-defined area.
- *PTZ errors* can trigger actions if the camera's PTZ functionality stops working correctly.
- *PIR sensors* can trigger an event in the camera if a PIR sensor, built-in or connected to the camera's I/O port, has detected activity.

14.3.5.2 Responses

Network video products can respond to events all the time or only at specific times. When a trigger is activated, some of the common responses and actions that can be configured include the following:

- *Recordings* can be used to save images, video, and audio in specified locations at specified frame rates and compression during the course of an event.
- *Outputs* can be used to activate external devices such as sirens, lights, and relays connected to the output ports on a camera or video encoder. (See Figure 14.17.)
- *Email notifications* can be sent to users, notifying them that an event has occurred. The email can include an image of the event.
- *HTTP or TCP notifications* can be used to send files and alerts to a video management system. In turn, this can trigger an action in the video management software, such as a recording or activation of another camera or auxiliary device.
- *SMS/texts* can send text or media messages with information about the event.
- *Edge storage* can be used to record video, and sometimes audio, to an SD card or NAS (network-attached storage).
- *PTZ presets* can be used to move a PTZ camera automatically to a specified position (such as a gate, fence, or entrance) when an event takes place.
- *On-screen popups* can be used to make a live view window pop up on the operator's monitor when an event occurs.
- *On-screen instructions* can be used to give the operator a procedure to follow when an event occurs.

Through pre-alarm and post-alarm buffers, recordings can include video not only of the event itself but also of a set amount of time before the trigger and after the event ends. This can give a more complete picture of the event.

14.3.6 Video analytics

Video analytics can act as detectors that answer yes-or-no questions and can trigger alerts to operators or automatic system responses. Video motion detection (VMD) is one example of analytics. Other common analytics include object detection, camera tampering, and crossline detection. For more examples, see Chapter 17.

Many network cameras have built-in analytics. For cameras without analytics, the video management software can provide this feature instead. Although the principle of analytics is much the same whether it is built into the camera or as part of the video management software, the two solutions affect the infrastructure and usage of the video management system in different ways. The following sections discuss some of these similarities and differences.

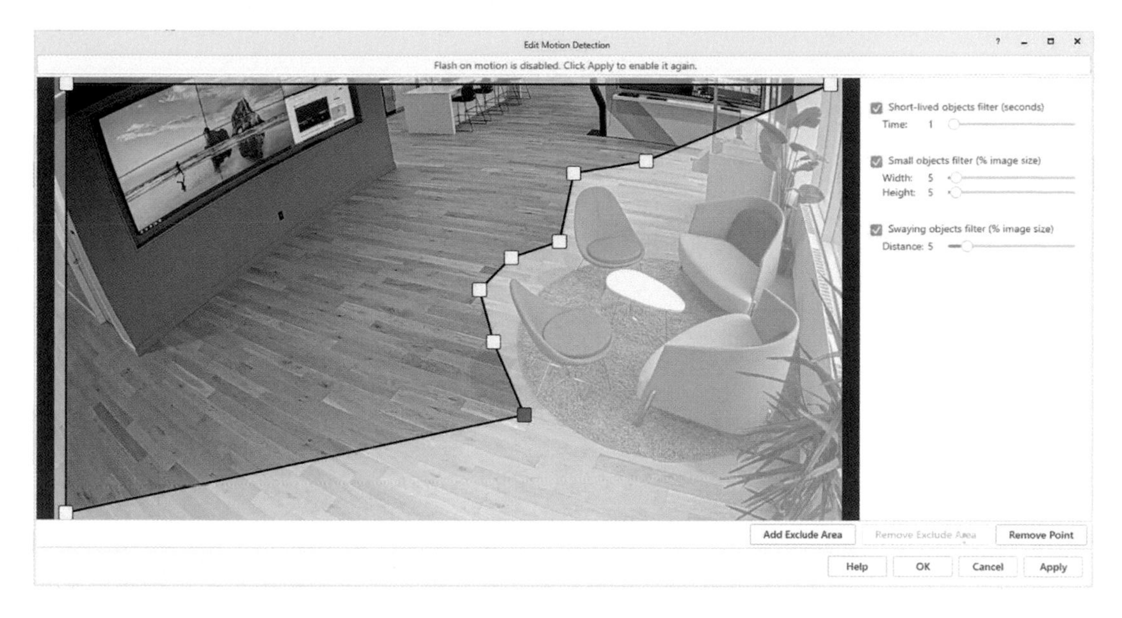

Figure 14.18 Configuring video motion detection in video management software.

14.3.6.1 Analytics in video management software

In the case of VMD, users usually define specific areas of interest in which the application should react to movement. They may also be able to set different sensitivity levels, adjusting for which object and movement size is typical for the monitored area (Figure 14.18). Upon motion detection, the application can trigger an external device to perform an action (such as open or close a door or turn a light on or off), start recording video from selected cameras, and send email. VMD can also be used to trigger actions if motion stops. This is useful in scenarios where a specific type of motion is expected, for example, in factories where robots and machines have very predictable movements.

Remember that running VMD in video management software is a CPU-intensive process and if used on many video channels it puts a lot of strain on the system.

14.3.6.2 Analytics at the edge

Using the basic built-in VMD or object detection in a camera or encoder offers substantial advantages over using the same applications in the video management software. Because the camera or encoder performs the motion and object detection, this frees up CPU power that can be used for other processes and recording devices in the system. Edge analytics helps make video surveillance more event-driven and lowers the demands on bandwidth and storage, as no video (or only low-frame rate video) is sent to the operator or recording devices while there is no activity in the scene.

The built-in VMD in network cameras (Figure 14.19) or video encoders is very similar to the VMD functionality found in video management software. Users can configure motion in certain areas while ignoring motion in others.

VMD data that provides information about, for example, the level of activity or the size of a moving object can be included in a video stream to simplify searches in the recorded material. Advanced object- and AI-based detection analytics are now widely available in network cameras and provide features that allow forensic search in the VMS. For more about analytics, see Chapter 16.

14.3.7 Input and output ports

Many cameras and video encoders have integrated input and output (I/O) ports. External devices can connect to these ports so that they can communicate with and be managed by the system. For

Figure 14.19 Configuring video motion detection in the interface of a network camera.

Table 14.1 Examples of devices that can connect to an input port

Device type	Description	Usage
Door contact	A magnetic switch that detects the opening of doors or windows	When the circuit opens (the door opens), the camera performs an action, for example, sending full-motion video and notifications.
Passive infrared detector (PIR)	A sensor that detects motion based on heat emission	When the sensor detects motion, it opens the circuit, and the camera performs an action.
Glass-break detector	An active sensor that measures air pressure in a room and detects sudden pressure drops	When the sensor detects a drop in air pressure, it opens the circuit, and the camera performs an action.

example, a camera that receives an input from an external alarm sensor can be configured to only send video when triggered by the sensor.

The range of devices that can connect to a network video product via its I/O ports is almost endless (Table 14.1 and Table 14.2). The basic rule is that any device that can toggle between an open and closed circuit can connect to a camera or a video encoder. The main function of the output port is to trigger external devices, either automatically or remotely by an operator or an application.

14.3.8 Log files

Video management software can provide event and alarm log files (Figure 14.20) that show various kinds of actions and events in the system. These can include camera and server events based on date, time, type, and source of the events. Users can sort or search for specific events with or without filters, for example, a list of I/O activities that occurred between two specific times or when motion was detected and where.

Table 14.2 Examples of devices that can connect to an output port

Device type	Description	Usage
Door relay	A relay (solenoid) that controls the opening and closing of door locks	A remote operator can lock and unlock a door.
Siren	Alarm siren configured to sound when alarm is detected	When motion is detected (through VMD or a digital input), the camera can activate the siren.
Alarm/intrusion system	A security system that continuously monitors a normally closed or open alarm circuit	The camera can act as an integrated part of the alarm system that serves as a sensor, enhancing the alarm system with event-triggered video transfers.

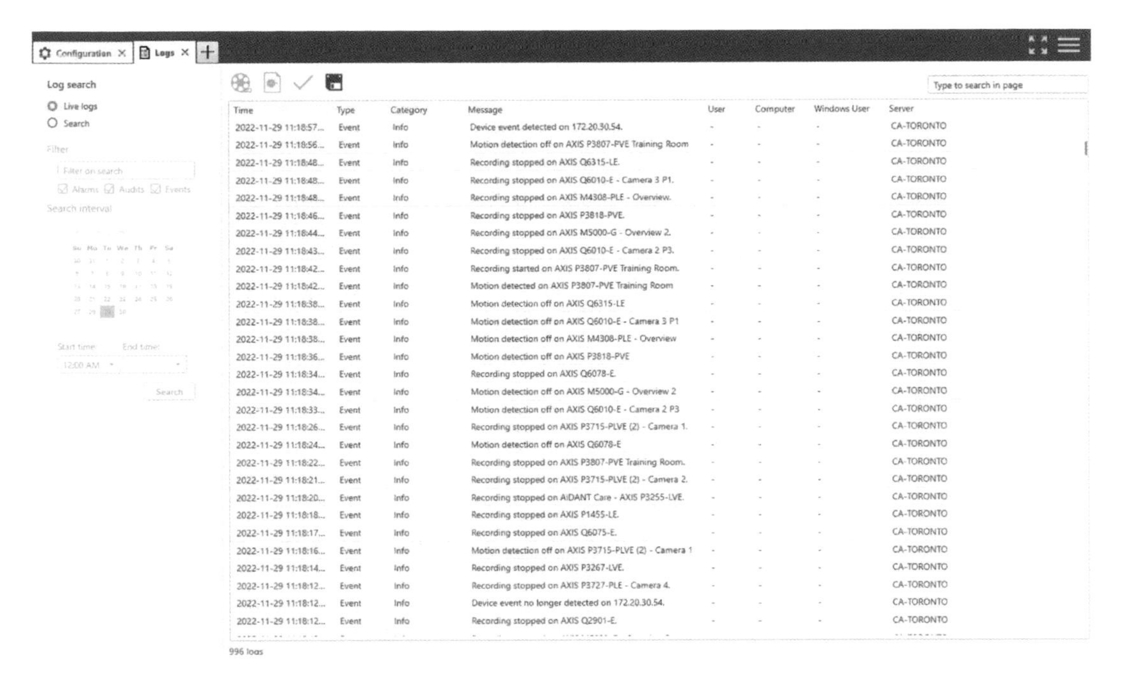

Figure 14.20 Example of an event log. The type of log to display is selected through filtering.

An audit log shows a list of user actions based on the user, time, type of activity, and video device. The audit log file is an essential function that provides proof of who used the system and when, what happened, and which actions the user took.

14.3.9 Administration and management features

Video management software should be able to handle the administration of cameras and encoders. This includes installation, firmware upgrades, security, audit logs, and parameter configurations. If the system includes devices such as cash registers, door controllers, and door stations, it simplifies things if the software can handle these too. The larger the surveillance system, the more important it is to be able to efficiently manage the networked devices.

14.3.9.1 Managing cameras

All video management software provides the ability to add and configure basic camera settings, such as frame rate, resolution, and compression format. Some also include more advanced functionalities, such as camera discovery and complete device management.

Software that helps simplify the management of networked cameras, encoders, and other devices in a system often provides the following functionalities:

- Locating and showing connection status of devices on the network
- Setting IP addresses
- Configuring single or multiple devices
- Managing firmware upgrades of multiple devices
- Managing user access rights

Video management software can provide a configuration sheet (Figure 14.21) that gives users an overview of all cameras and recording configurations.

14.3.9.2 Time synchronization

Conducting an investigation using multiple cameras or integrating different systems becomes easier if all networked devices have the same time. This is commonly achieved by using a network time protocol (NTP) server to synchronize all devices in a network. Most networked devices support NTP (Figure 14.22).

14.3.9.3 User access control

An important part of video management is security. Network video devices or video management software should have options for the following settings:

- Authorized users
- Passwords
- Multiple user-access levels
- User-differentiated access to specific devices

User-differentiated access means the system administrator can specify exactly which users and groups have access to each networked device. Access levels (Figure 14.23) define which information users can see and which changes they can make. The following user access levels are common in a network video context:

- *Administrators* – these have access to all functionalities
- *Operators* – access to live and recorded video
- *Viewers* – access to live video only

Many VMS applications can inherit a Windows® user database (local or LDAP/Domain). This feature eliminates the need to set up and maintain a separate database of users.

14.3.10 Health monitoring

Where supported, system health monitoring is used to verify the status of a VMS installation. The system automatically monitors the status of all the sites, devices, and storage. Notifications are sent if, for example, there is an issue with a connected device. Other features of health monitoring can include inventories of devices, including types, firmware versions, IP addresses, etc. It may also be possible to download system reports and notification logs, as well as overviews of used and available storage, retention times, etc.

14.4 INTEGRATED SYSTEMS

Video management systems that are based on a network video platform can be easily integrated into other IP-based systems, such as point-of-sale (POS), physical access control, building management, and industrial control systems. When video is integrated, information from other systems

Server configuration sheet for AXIS Camera Station – CA-TORONTO
General Cameras Schedules Recording storage Other devices
System information

Program version	5.49.449
Protocol version	5.0.56
Process	6 4b it
DirectX version	4 09.00 0904
NET CLR version	4 8 (release 528372, CLR 4 0.30319 42000)
Application culture	en-US
Install path	C:\Program Files\Axis Communications\AXIS Camera Station\Core\Server 5.49.449
Is AXIS NVR	Yes
NVR name	AXIS S1264 Rack 64 TB
NVR model	PURPLE1B
NVR service tag	F43QGR3
Machine name	AXISNVR-VKLC41F
OLE/DDE exchange	918332561189601627
Operating system	Microsoft Windows 10 IoT Enterprise LTSC
OS culture	en-US
OS version	10.0.19044 0
Is part of domain	No
Generated	2023-01-23 10:42:35' (UTC '2023-01-23 15:42:35')

Camera name	Manufacturer	Model	Product type	Firmware	Address	Port	MAC address	Security mode	IEEE 802.1X	HTTPS	Video view token	Camera ID	Device ID	Is enabled	Disconnects since server start
D2110-VE	Axis	AXIS D2110-VE	Security Radar	11.2.61	172.20.30.72	80	ACCC8EF336E4	HttpDigeet	Disabled	Disabled	1	137126	137068	Yes	0
M5000-G-Overview 1	Axis	AXIS M5000-G	PTZ Camera	11.2.61	172.20.30.129	80	B8A44F284AB2	HttpDigest	Enabled	Disabled	2	136176	136078	Yes	0
M5000-G-Overview 2	Axis	AXIS M5000-G	PTZ Camera	11.2.61	172.20.30.129	80	B8A44F284AB2	HttpDigeet	Enabled	Disabled	3	136171	136078	Yes	0
M5000-G-Overview 3	Axis	AXIS M5000-G	PTZ Camera	11.2.61	172.20.30.129	80	B8A44F284AB2	HttpDigest	Enabled	Disabled	4	136181	136078	Yes	0

Figure 14.21 Examples of configuration sheets.

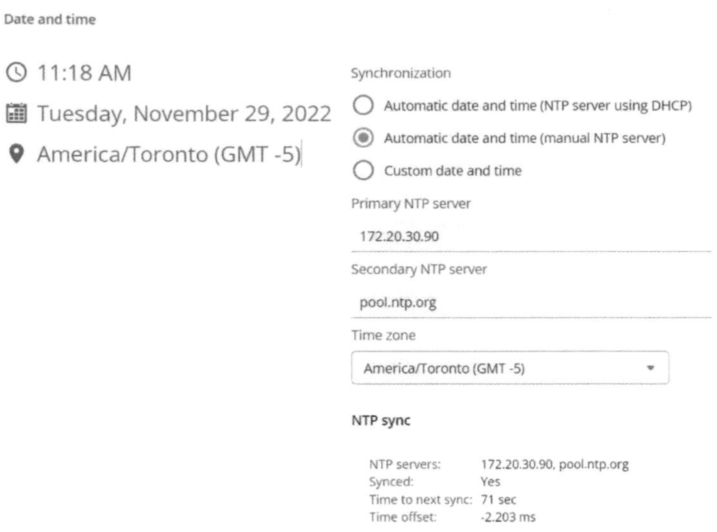

Date and time
11:18 AM
Tuesday, November 29, 2022
America/Toronto (GMT -5)
Synchronization
Automatic date and time (NTP server using DHCP)
Automatic date and time (manual NTP server)
Custom date and time
Primary NTP server
172.20.30.90
Secondary NTP server
pool.ntp.org
Time zone
America/Toronto (GMT -5)
NTP sync
NTP servers: 172.20.30.90, pool.ntp.org
Synced: Yes
Time to next sync: 71 sec
Time offset: -2.203 ms

Figure 14.22 Most network cameras and video encoders have support for NTP.

User permissions
Type to search
Name
Domain
Role
Details
Server
ACAP
CA-TORONTO
Administrator
ACS Operator
Operator
Toronto_100
Toronto_Pod_1
Remove
Add...
Role
Administrator
Operator
Viewer
Cameras
Views
I/O
System
Access
Video
Audio listen
Audio speak
Manual recording
Mechanical PTZ
PTZ priority
2N IP FORCE
2N IP STYLE
AV
Apply

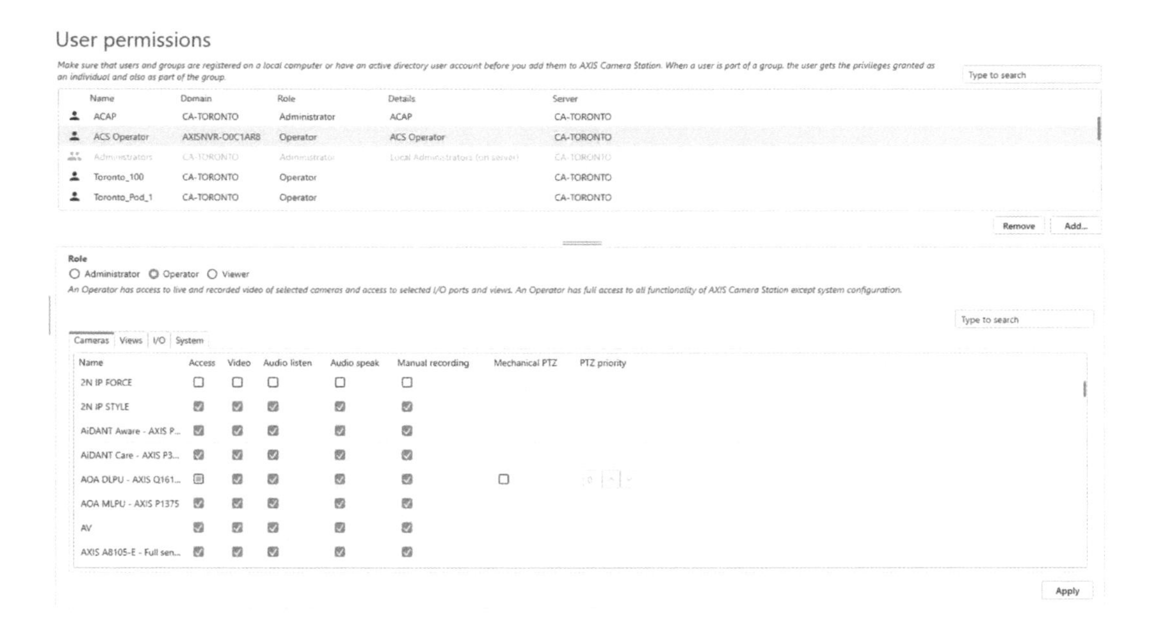

Figure 14.23 An example interface where user access can be set.

can be used to trigger functions, such as event-based recordings in the video system and vice versa. In addition, having a common interface for managing different systems makes it easier for users to do their jobs.

14.4.1 Application programming interface

An application programming interface (API) enables the development of customized applications. A video management system must have an API to integrate the video system into other systems. Once implemented, the subsystems can communicate with each other and perform actions, such as starting recordings, sending alarms, opening doors, activating microphones and speakers, and accessing live and recorded video.

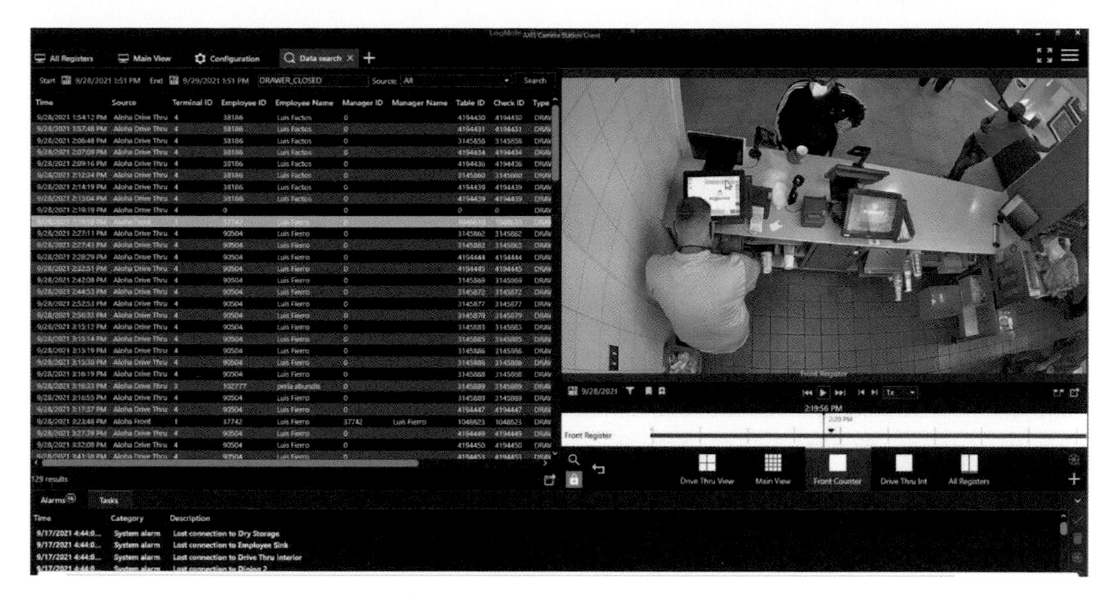

Figure 14.24 An example of a POS system integrated with video surveillance.

14.4.2 Point of sale (POS)

The introduction of network video in retail environments has made the integration of video with POS systems easier. The integration enables the linking of all cash register transactions to video footage of those transactions. This helps catch and prevent fraud and theft from employees and customers. It also makes it easier to search for and verify those suspicious activities. Through the captured video, managers can visually verify POS events, such as returns, manually entered values, line corrections, transaction cancellations, employee purchases, discounts, specially tagged items, exchanges, and refunds. It can resolve questions such as whether the right amount was entered for the products placed on the counter, whether all items on the counter were scanned, whether a return was handled properly, whether an employee discount was given to a friend, and show the face of a person who used a stolen credit card. High-quality video from network cameras can provide the necessary information to identify and verify details, such as the value of a bill or items handed to a cashier.

Some systems can store and show receipts together with video clips of the events. In such cases, searching and viewing of these events is also possible (Figure 14.24).

A POS event can be used to trigger a camera to record and tag the recording. For example, the opening of a cash register drawer can be used to trigger recordings. The scene prior to and after an event can be captured using pre- and post-event recording buffers. These event-driven recordings increase the quality of the recorded material, reduce storage requirements, and reduce the amount of time needed to search for incidents.

A POS system with integrated video should help achieve the following:

- Good insight into the various payment transactions
- Reduction of internal and external shrinkage (theft)
- Preventive negative behavior among coworkers
- Awareness of how and when mistakes are made

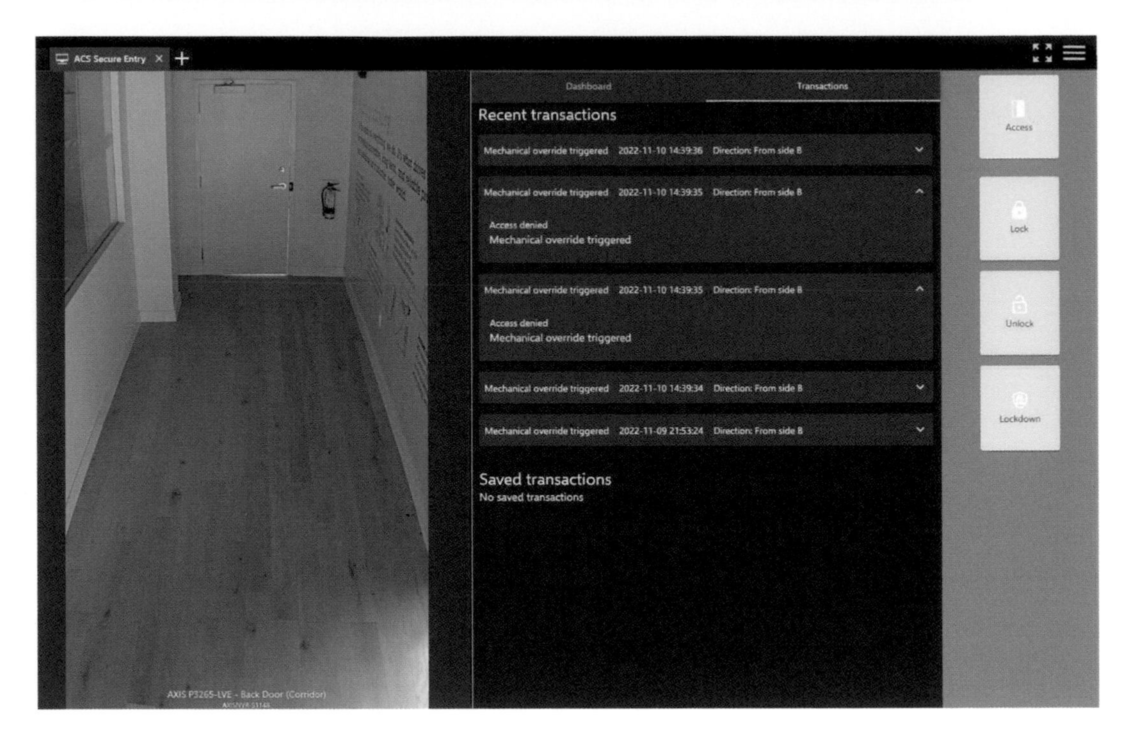

Figure 14.25 An example of integrated video surveillance and physical access control systems.

14.4.3 Physical access control

By integrating a video management system with a building's physical access control system, security managers can use video to log access to the building (Figure 14.25). For example, when someone enters or exits a door, the door controller can trigger a camera to record the event. This way, suspicious activity can be verified, and unwanted visitors can be identified.

Video also makes it easier to identify tailgating, which occurs when a person swiping their access card knowingly or unknowingly allows others who did not swipe a card to also enter via that door. To see if the system logged everyone who entered the area, an operator investigating an incident would look at the recorded video and compare it with the physical access control system. For rapid identification of employees or visitors, the operator can also have access to the images stored in the access card system.

14.4.4 Building management

Video can integrate into a building management system (BMS) that controls systems such as heating, ventilation, and air conditioning (HVAC), security, safety, energy, and fire alarm systems (Figure 14.26).

Examples of integration include the following:

- In addition to activating alarms in the BMS, an equipment failure alarm can trigger a camera to show video to an operator.
- A fire alarm system can trigger a camera to monitor exit doors and start recordings.
- Analytics can be used to detect a reverse flow of people into a building if a door is open or unsecured due to events such as evacuations.
- To save energy, video motion detection can be used to turn off lighting and heating systems when the room the camera is located in is vacated.

Figure 14.26 An example of a building management system that provides a single interface for monitoring a facility. The left-hand screen provides the operator with a quad view from different cameras and a floor plan showing camera locations. The right-hand screen shows information from ventilation and heating systems.

Source: Siemens

14.4.5 License plate recognition

License plate recognition (LPR), also referred to as automatic number plate recognition (ANPR), can easily be incorporated into video management systems. LPR is used in many different applications:

- *Parking management*
- *Toll road management*
- *Airport traffic management* – only authorized vehicles can access taxi/bus lanes
- *Access control* – gates are opened only for authorized vehicles
- *Vehicle alerts* – automatic alerts when a vehicle on a watch list is captured

LPR solutions often consist of a purpose-built camera with a pre-installed analytic, which runs on the camera, in the cloud, or on a server. The analytic automatically captures the license plate in real time and compares it or adds it to a list. The license plate is then used, for example, to open a gate, incur a charge, or generate an alert. For more on LPR, see Chapter 17.

14.4.6 Evidence Management Systems (EMS)

An EMS is designed for handling digital evidence recorded by, for example, body-worn cameras. Recordings are automatically imported along with the incident time and the location. Other digital evidence material, for example, from external sources such as public uploads can also be imported and associated with the same case or incident. An EMS can be integrated with a Record Management System (RMS) or Computer-Aided Dispatch system (CAD) for automatic case creation and the addition of evidence from other sources. An EMS also maintains the chain-of-custody by keeping a record of every change made in log files while preserving the original video files. An EMS can be hosted in-house by the user organization, or it can be provided by a third party as a cloud solution.

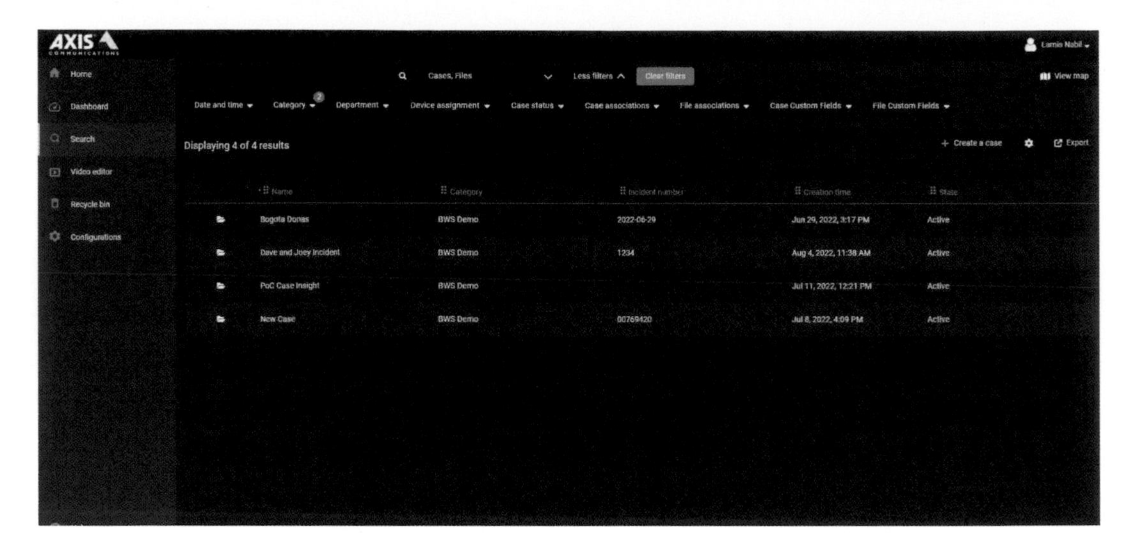

Figure 14.27 A screenshot showing an EMS application.

14.4.7 Industrial control systems

Remote visual verification is often required in complex industrial automation systems. By having access to a combined interface for network video and process monitoring, an operator does not have to leave the control panel to make visual inspections. In addition, when an operation malfunctions, the control system can trigger the camera to send images. In sensitive clean-room processes or in facilities with dangerous chemicals, video surveillance is the only safe or sustainable way to get visual access to a process. The same goes for electrical grid systems that have substations in very remote locations.

14.4.8 Radio-frequency identification (RFID)

Tracking systems that involve radio-frequency identification (RFID) or similar methods are used in many applications to keep track of items. An example is luggage handling at airports that keeps track of luggage and directs it to the correct destination. When RFID is integrated with video surveillance, there is visual evidence when luggage is lost or damaged, and search routines can be optimized.

14.5 BEST PRACTICES

The video management software is the gateway to the entire video surveillance system. Today, several hundred different systems are available from different vendors. It is a challenge to choose the right one, the one that has just the right combination of features, is scalable, works on a platform that suits the business, the IT solution, and the staff situation, and is easy to use. There is a lot to consider, including the following issues:

- *Size of the system:* How many cameras and sites are there today, and what size do you foresee in the years ahead?
- *System architecture platform:* Which platform best meets the requirements? Is it an NVR- or a Windows®-based hardware platform, an edge-based system, or a cloud solution? This choice is dependent on the collective requirements of the system.
- *Locations:* Is it a one-location or multiple-location business? How many devices are there in each location? How many operators will use the interface?

- *Storage and maintenance:* What are the recording and storage needs? Is there staff to maintain a hardware-based system? What is the cost of additional storage?
- *Scalability:* Some systems have limited scalability but are easy to install and operate, whereas others scale to thousands of cameras but may be complex to use in a small system. What is the current and estimated future device count? How many simultaneous cameras views (split view) are needed? What are the licensing options? How many operators will use the system at the same time?
- *Complexity and functionality:* Is a basic system that enables recording and viewing of a few cameras good enough? Or do you need an advanced system with, for example, event handling, mapping functionality, and support for integration with other systems?
- *Integration:* Do you need integration with a point-of-sale system, a physical access control system, a building management system, or an industrial control system? Examine the possibilities for merging systems or support for adding devices other than video products. Does the software have an open API?
- *Analytics and intelligence:* Does the system need specific built-in analytics to support surveillance operations? In retail and transportation installations, segment-specific analytics such as people counting or license plate recognition might be required.
- *Open or vendor-specific:* Most camera and encoder vendors supply their own video management systems that are normally limited to a single product brand. An open system from an independent company might provide better flexibility.
- *Activity level:* How active will the monitoring be? Is video checked on a casual basis or only when there is an alarm? Perhaps there will be dedicated staff watching the video at all hours of the day. Different activity levels require different features, such as split screens, smart search, and effective alarm handling.
- *User level:* Who are the users, and what are their levels of knowledge? Is it more important that the interface is simple and straightforward or that it is complex enough to meet every possible need? Ease-of-use is even more important if many operators are going to use it and their skills vary. Trained and skilled operators may have more detailed requirements on functionality and prefer a complex system. What are the operators' working patterns? Is a desktop client sufficient, or do they need to be able to use mobile devices and browsers as well?

Chapter 15

Cloud-based video solutions

As discussed in Chapter 14, there are many ways to manage video. Earlier solutions stored video on a recorder, typically a DVR or NVR, at the same location as the cameras. When IP cameras and video encoders are connected to an IP network that is connected to the internet, an alternative way of managing video becomes possible, one that has become increasingly popular over the years. Rather than doing everything on-site, video systems can be managed and stored – partly or completely – off-site in a data center. This type of video management is referred to variously as cloud video, cloud-managed video services, or cloud video surveillance.

So cloud video can be summarized as running or accessing some or all functions of a system on off-site hardware, usually at a data center run by a third party.

A video surveillance system consists of several components, including the camera that captures and streams IP video, a rules engine, video storage, and a user interface to manage the system, view live video, and access recorded video. Obviously, the camera (or other sensor type) must be located on the premises being monitored, while the remaining parts can all potentially be in the cloud.

However, direct cloud solutions where all functionality, apart from the sensors, is in the cloud are rare. Most solutions offload only some functions to the cloud and perhaps not all the time. Many solutions advertised as "cloud" actually perform many functions on-site. Inversely, many traditional VMS solutions use cloud-based components or integrations, even if this is not immediately apparent to the user. Solutions that combine on- and off-site functionality are often referred to as hybrid solutions.

15.1 MARKET DRIVERS, SIZE, AND TRENDS

Though many use the term cloud for any type of cloud video surveillance service, there are different types of services:

- *Managed video:* On-site video recording, storage, and remote management by the cloud service provider.
- *Hosted video:* Off-site recording, with data transfer to the cloud provider's site for storage and management.

Systems often have both video streaming to the provider's site and storage at the camera site, for example, in onboard (edge) storage or NAS devices. Nevertheless, these systems can still be referred to as cloud video systems. And whatever the type of service, the factors driving the trend towards cloud-based systems are the same.

DOI: 10.4324/9781003412205-15

15.1.1 Simplicity and scalability

The more distributed an enterprise's video surveillance system is, the more complicated it often becomes to implement and run. Cloud-based services can take the effort out of these processes, and the user can manage more sites with fewer resources. In the simplest scenarios, all they need do is provide suitable connection points for the on-site cameras. On-demand access to the system from any location at any time, and across multiple platforms and devices, is also part of the simplified user experience.

15.1.2 Video Surveillance as a Service (VSaaS)

VSaaS refers to video surveillance provided as a web-based service, which better describes the business model than the technology involved. VSaaS is characterized by the customer paying a recurring fee for services over time. Payment for hardware components such as cameras can either be through upfront payments or simply rolled into the recurring service fees. Cloud solutions lend themselves to VSaaS business models more easily than on-premises systems.

The user's costs for the installation and maintenance of cloud video can be lower than for an NVR- or server-based system for several reasons. There is no need to buy, install, or maintain a local recording and monitoring station because the service vendor can store the recorded data and maintain the system. Some vendors even take care of the setup and network configuration. The system can grow one camera at a time as needs arise. Another reason for low maintenance costs is that problems can often be solved remotely. For example, there is no need to dispatch technicians to adjust camera focus when this can easily be done centrally and remotely.

15.1.3 Responsibilities and purchasing decisions

Installing and managing a video surveillance system can be daunting for enterprises that need such a system but which don't have the skills or interest to manage all the various associated technical issues. Offloading responsibility for the system's function to a specialized vendor can be extremely attractive for a small company with limited resources. But even larger organizations familiar with IT matters can see the advantages of cloud, as these services include tools for deploying and managing the requisite network devices, and the vendor will also be responsible for maintaining the cybersecurity of the system.

As regards the investments required to purchase a video system, the wide range of cloud video services available today gives the customer multiple options for financing a system. Earlier, traditional systems usually required a significant initial outlay of capital, in which all components were owned by the customer. These days, prospective buyers can choose among offerings in which they will own some, all, or none of the required hardware, irrespective of where this hardware is physically located. Video surveillance is moving from being an upfront investment to being a recurring operating expense.

15.2 WHY USE CLOUD VIDEO SERVICES?

Cloud-managed video is a service that is hosted and sold on-demand, is accessed through the internet, and is elastic in the sense that its subscribers can use as much (or as little) of the service as they want. Subscribers pay for a worry-free monthly service rather than procuring and maintaining a system of their own, which also means organizations can limit their upfront investment.

In very broad terms, cloud-managed services can now bring scalable video surveillance solutions to organizations of any size. Early cloud video installations were limited by bandwidth and edge device compression and were mainly confined to small businesses with just a few cameras per site. These limitations have now been overcome, and camera counts per system are increasing steadily.

The following are some of the functions requested by users:

- *Viewing and operating:* View sites and receive alarm and notifications from anywhere, investigate incidents remotely, open doors, or trigger I/Os from anywhere.
- *Forensic review and verification:* Searching for video associated with intrusion, safety, or environmental alarms.
- *Multi-streaming:* Automatic video quality selection dependent on user location to provide an optimized experience, for example, lower resolution video for remote viewing and high-resolution video when accessing the system on-site.
- *Manage and maintain:* Manage users remotely, update sites, manage devices, and limit on-site infrastructure and support.
- *Storage:* Varying needs for storage of video and metadata, cost-efficient scaling of storage capacity, retrieve information at any given time, more uptime.
- *Analytics:* Tools to analyze activity and manage incidents, including perimeter violations, crossline breaches, or motion detection.
- *Business intelligence:* Integration of business-intelligence-focused analytics, such as people counters, queue monitoring, and heat mapping.

Apart from these functions specifically requested by users, a cloud system has several other advantages:

- Customers get a user interface that has its backend in the cloud. They do not need to run or maintain it themselves.
- Users are less limited by local hardware.
- Responsibility for the maintenance and cybersecurity of the system is offloaded to the platform provider.
- Ease of installation, with fewer hardware devices and better "plug-and-play" functionality.
- A wide developer ecosystem makes a variety of solutions available to the user.
- Continuing the trend developed in traditional on-premises VMS, cloud-managed systems continue to enable the video system to be integrated with other systems, such as access control, point-of-sale, etc.

Cloud-managed video offers similar functionality to a traditional surveillance system but with varying levels of upfront investment: Different vendors offer various packages in which the required hardware is either paid for upfront or is rolled into the contracted recurring payments. Cloud also brings flexible and low-maintenance storage options. Subscribers can combine cloud and local storage with less trouble and less cost than in traditional surveillance systems. In most designs, the costs and maintenance associated with DVRs, NVRs, and VMS servers are no longer necessary. As part of the service, video can be recorded in the cloud, so the system is not dependent on local recorders that need to be maintained and which could get broken, stolen, or confiscated.

The interface is one of the areas in which cloud video management stands out against many traditional video management applications. This is usually web-based or specially developed for mobile devices, and it focuses on basic functionality and ease-of-use (see Figure 15.1). Users need only open a browser or smartphone app and enter their username and password. Maintenance is also problem-free, as that all happens behind the scenes. The cloud service provider takes care of the system and handles all upgrades.

15.2.1 Different needs, different services

The first step in finding the right video surveillance solution should be to identify the user's needs and the system's purpose. Cloud video can be used for many different levels of video surveillance, and it is up to the service provider to package solutions that meet those requirements. A good

Figure 15.1 One of the characteristics of most cloud video applications is a clean, functional, and modern-looking interface that is easy to use. The focus is on system overviews, such as split screens and thumbnail sliders, timelines, event searches and alarm management, pop-ups, and video export.

Source: Genetec Inc., Montreal, Canada

provider should also offer plenty of flexibility so that users can add services as their needs change and their business grows. Figure 15.2 illustrates a ladder of video surveillance needs.

15.2.1.1 Basic video surveillance

A user's main objective might be to protect property and prevent loss. They probably also want to feel safe and confident that the system has them covered. They find comfort in that the cameras make staff, managers, and business owners feel safe and that they can get an overview of the premises and can easily share video from any location. They know they can easily follow up on incidents and perform basic forensic investigations. Storing video is easy: It is available in the cloud and can be stored locally by plugging in a NAS or inserting an SD card. If needed for forensic purposes, the cloud storage provides theft-proof storage of critical data.

15.2.1.2 Integration with central alarm stations

The alarm monitoring industry has been around for many years and has traditionally been based around central alarm stations that monitor remote alarms, which are often simple contacts or passive infrared (PIR) sensors that detect motion. The main goal of central stations nowadays is the video verification of alarms, that is, to manually confirm that an incident warrants an active response, all according to its type and severity.

Also, as the use of standard phone lines (POTS) is in decline in favor of IP-based communications, central stations look for systems such as cloud video surveillance solutions that can use the same IP network.

So at the second level of the video surveillance needs illustrated in Figure 15.2, some users want to know that there are trained operators who can verify activity through video, making visitor and contractor entries more efficient and secure. When an alarm occurs, security personnel will act on it. Users can even watch the whole scenario remotely if they want to. Security operators monitor the system and keep an eye on the premises, deliveries, cash registers, and deposits. The system helps users and security centers make better decisions on manpower distribution, and it allows them to

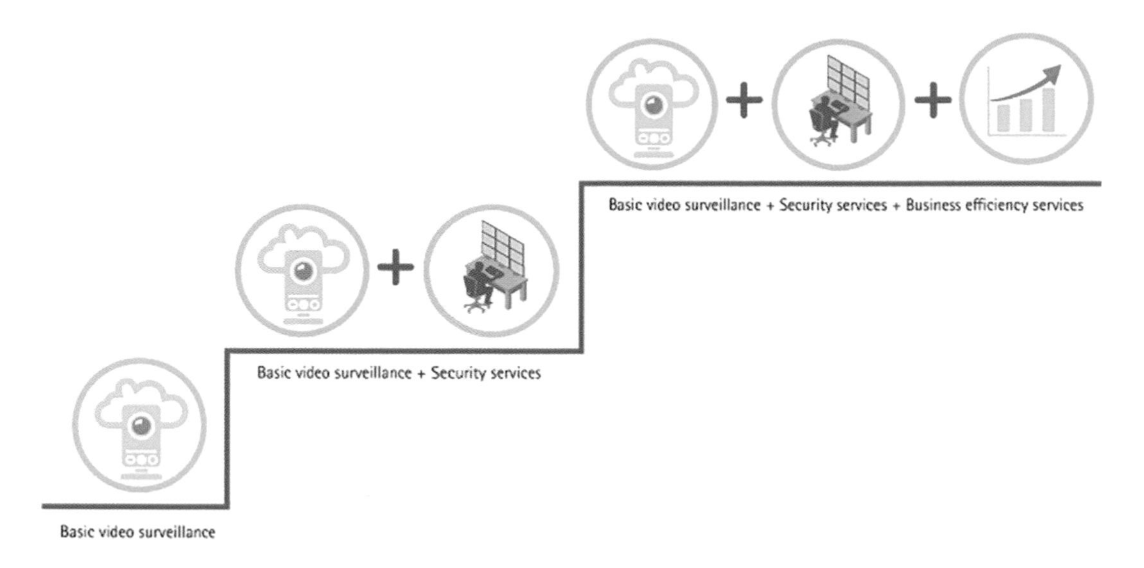

Figure 15.2 A ladder of video surveillance needs, starting with basic surveillance, then adding services such as alarm monitoring, and services that help make the subscriber's business more efficient.

Figure 15.3 Integrated central station software and a cloud video solution can provide a very effective solution for the operator to decide if the alarm requires a police response or not.

centralize the guard force, even if they need to cover several locations and buildings. A shopping mall operator might want to deploy a video system and provide it as a service to the tenants.

For many of these scenarios, cloud video verification can be a valuable addition. To make the solution more efficient, it may be possible to integrate the alarm station software and the cloud video

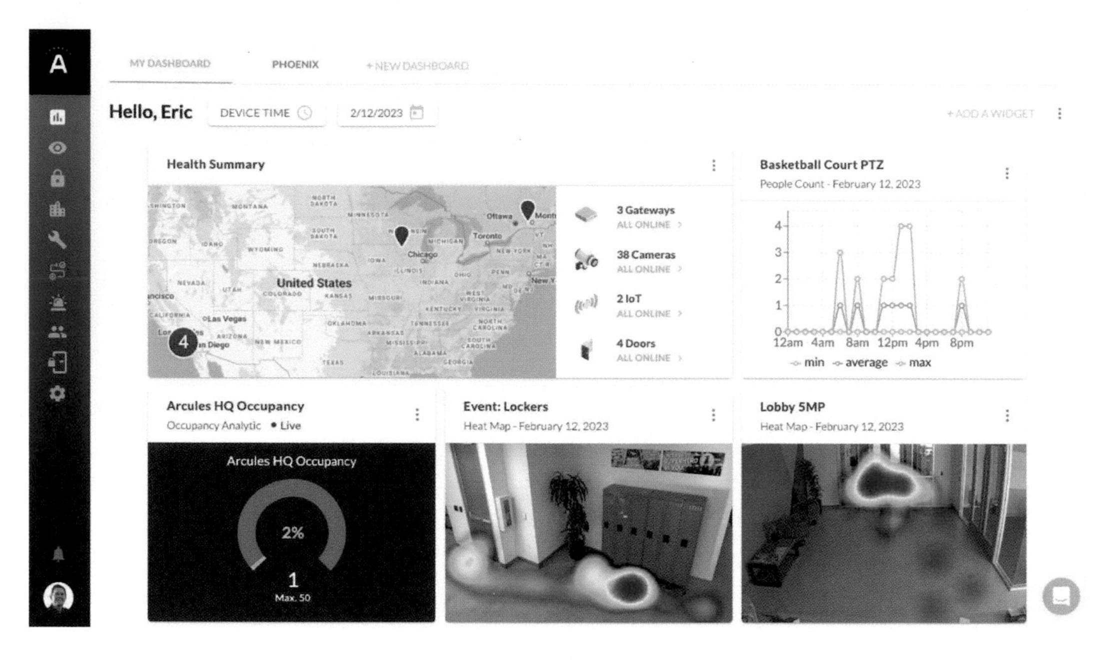

Figure 15.4 An example of a cloud-based solution for people counting and occupancy.
Source: Arcules, Irvine, CA

system, although the feasibility of such an integration should always be examined closely before making any purchasing decisions.

15.2.1.3 Business intelligence

Many business owners have understood that because it is easy to scale and deploy, cloud video can help them build a more successful business. Critical data can be stored safely in the cloud and easily shared with the police and judicial system, should this be necessary. Businesses can use video to monitor sales campaigns, find ways to make merchandizing stand out more, as well as see the status of different areas. They can use analytic tools that help them resolve incidents faster, predict behaviors to prevent inventory loss, follow movements through heat mapping, customer traffic and dwell time monitoring, identify patterns based on parameters such as gender, and use people counting as scheduling input and as a tool for evaluating merchandise positioning. (See Figure 15.4.) To read more about analytics for business intelligence, see Chapter 17.

15.2.2 Edge analytics

Running analytics within the camera (at the edge) allows for faster alerts and responses. This also reduces the need to stream video to the cloud, which in turn means that less bandwidth and storage is consumed, making the analytics more economical than when performed on cloud-based servers. When analytics do need to be run on cloud-based servers, in some cases, this is possible to do using the streams of video metadata produced by the camera, that is, the video itself does not need to be sent to the server.

Edge-based analytics also scale up very well, as each new instance (that is, each new device running analytics) takes care of its own load in the overall system.

Edge analytics are also advantageous due to the fact that, on some devices, these can be run on uncompressed video, which is always preferable to running them on compressed streams.

Figure 15.5 An example of using video analytics in a cloud-based system.

Source: YourSix

15.3 CLOUD CONSIDERATIONS

Although cloud video is an excellent choice in many situations, there will be some cases in which the user might decide against using it. For example, if there is uncertainty that stored data will remain in approved geographical locations due to company policy or legal requirements. Some data and cybersecurity governance policies may mean that the network cannot be connected to the internet, which obviously means that cloud services are out of the question. Some services might have requirements on using vendor-specific hardware that are incompatible with customer wishes.

Lastly, there is cost to consider. Although the price of storage and broadband has generally dropped over the years, transferring and storing large of amounts of video material on cloud servers can still incur considerable costs, which may or may not be acceptable for a particular application. Aside from storage, also consider the fact that video operator functionality, for example, searching and fast-forwarding, usually requires data either to be transferred to the client or to be done in the cloud. As downloading data is significantly more expensive than uploading it, this makes redundancy storage in the cloud the use case of most interest, as little or no data is ever downloaded.

15.4 COMPONENTS OF CLOUD VIDEO

Cloud video solutions have brought modern network video technology to small and midsize installations. These can now benefit from resolutions up to 4K, edge analytics, and the scalability of network camera technology that previously was attainable only by large organizations with in-house IT departments that were involved in the design and maintenance of these advanced systems. Earlier small systems would typically use an NVR-and-IP-camera solution, which needed active management. With cloud video, these systems were able to grow one camera at a time while keeping investment costs to a minimum.

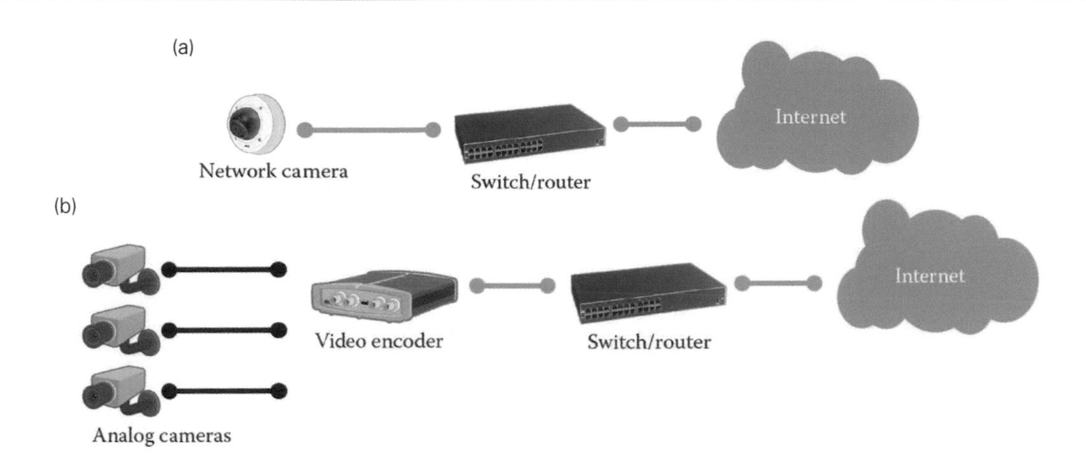

Figure 15.6 (a) In its simplest form, the only equipment needed on-site is a network camera and an internet connection. (b) Existing analog cameras can easily be integrated using a video encoder.

Figure 15.7 One of the benefits of cloud video is the ability to easily monitor video from any device and to get alerts via text messages or push notifications.

Source: YourSix

One of the keys to any digital security deployment is ease of installation, as well as efficient system operation and maintenance. Cloud video solutions are available for a relatively small investment, so users can limit their video surveillance costs. They only need a camera and an internet connection (see Figure 15.6), and they can let the system expand, add functionality, and add new services to meet the challenges of a growing business.

Monitoring video is easy and secure. Web clients and smartphone apps (see Figure 15.7) allow easy access to video from anywhere, and since viewing is over the internet, the user does not have to invest in video management software. Live viewing, recordings, and notifications are available directly in the browser.

The number and variety of cloud-based services is still growing, and much effort is going into establishing best practices for securing private, organizational, and governmental data. A benefit of storing video in the cloud is that data centers are experts at handling sensitive data. They continuously service the servers, keep them updated, and replace old hardware. Through video audit trails, records of access, and processes, they keep track of all the data and what happens to it. Many of these regulations and practices apply to cloud video as well, so users can trust that their video data is safe. Users with particular security requirements may need to have in-depth discussions with potential cloud providers to ascertain how various functions have been implemented. For more information about data security in the cloud, see Chapter 12, "Cybersecurity".

Cloud video can often be combined with other systems to provide broader security and protection. For example, a cloud video solution can be integrated with a central station alarm platform. For more information, see section 15.7.

15.5 STAKEHOLDERS OF CLOUD VIDEO

The main stakeholders in cloud video systems:

- Cloud video service providers
- Installers and/or integrators
- Users

15.5.1 The cloud service provider

The cloud service provider is responsible for running the system that connects and manages all the cameras and stored video. Some providers focus exclusively on these services, while others, for example, traditional security operators, may offer cloud-based video systems as a complement to their other operations. Some cloud providers run their own data centers, but many rent space in large data centers run by companies, such as Amazon (AWS), Microsoft (Azure), and Google. Some larger enterprises may set up their own cloud solution to support many different facilities. This is called a private cloud solution, as opposed to the public cloud solution that is usually implied when talking about cloud video surveillance.

In general, the cloud video vendor provides some or all of the hardware, as well as a management interface, and sometimes integrates the video with other systems. For examples of integration with other systems, see section 15.7.

The cloud service provider usually charges a monthly fee for the video surveillance service. This fee varies depending on the number of cameras per site and the amount of storage and bandwidth needed.

Another fundamental responsibility of the cloud provider is to guarantee data backup, recovery, performance, system availability, and security. Cloud providers must demonstrate that the servers they use – their own or as provided by a third party – have the necessary security to protect private and confidential data.

15.5.2 The installer/integrator

Most cloud service providers do not provide the required services for physical installation and integration, so in many cases, they still rely on traditional security integrators/installers to physically install the cameras and other on-site equipment. This obviously requires training and knowledge of cloud-based systems and design versus traditional on-site solutions. Most cloud providers have a set of recommended or preferred integrators.

A cloud-based video system can be easier to install than other types of surveillance systems, partly because they need very little equipment on-site and also because some systems have a very easy on-site installation – which often involves simply pressing a button on the camera to make it automatically find and connect to the cloud service. The installation is even easier if the cameras are staged, that is, preconfigured and connected to the service provider's system before the installer gets them. The installer then only needs to install the cameras, connect them to the network, and point them in the right direction. Many cameras have remote focus, which means the user can adjust focus and field of view without leaving the office.

15.5.3 The user

Cloud video services can be a good alternative for a small or midsize business with a single store to monitor remotely, or for a franchise with multiple sites that need simultaneous monitoring, or for a large corporation looking for secure off-site archiving to satisfy internal policies or compliance regulations.

Cloud video surveillance systems allow users to leverage their system's scalability and cost-effectiveness. Because the need for on-site storage and maintenance is greatly reduced or eliminated, users can reduce both the capital investment in physical security technology and the total cost of ownership. It is up to the cloud provider to always make sure that users have the hardware and software resources that best serve their purpose.

15.5.4 Other uses for cloud video

The flexibility of cloud video makes it ideal for surveillance that is temporary in nature. Public events such as carnivals, marathons, demonstrations, and outdoor concerts have very different needs and altogether different types of infrastructure than a parking garage or government building. Video surveillance for public events must be deployed quickly, be easily accessible, and provide the ability to manage video in a scalable and secure way for a limited time. Just as quickly, they need to be removed from the site again. Some organizations need surveillance that can be moved frequently to meet changing site needs.

There are also enterprises that operate large on-site surveillance systems that need to move critical video data off-site to a secure cloud-based location because they cannot afford to lose that data. Cash rooms, server rooms, and pharmacies are examples of areas that have critical video content that could be problematic for the business to lose. Enterprises can isolate these cameras and simply host them in the cloud, so the video is secure and no longer at risk of loss due to video manipulation, local recorder failure, or theft.

15.5.4.1 Camera streaming services

A camera streaming app allows you to pull a camera feed and broadcast it via live streaming services, such as YouTube, Facebook, Vimeo, and others, without requiring additional servers or software. In many cases, all that is required is an internet connection and a network camera with the streaming application installed directly on it. This allows an organization to broadcast, for example, live sporting events, training sessions, meetings, religious services, and so on. Some streaming services also provide features for overlaying text, images, and live data over the video stream, for example, live scores or weather updates.

15.6 CLOUD VIDEO ARCHITECTURES

Exactly what constitutes a cloud video surveillance system is a matter of debate. Each definition will be a combination of the included functions combined with where these functions and their supporting hardware are located. As any video surveillance system must have cameras physically installed at the site being monitored, we can just as well disregard the cameras (or other installed sensors) for the purposes of our discussion.

- *Direct cloud:* Solutions that claim to be "direct cloud", sometimes also called "cam-to-cloud" or "pure cloud", usually have three of the major functions in the cloud – viewing, management, and video storage. Some may also run analytics as cloud applications, although it is often more cost-efficient to have these at the edge. These solutions do not generally require an on-site bridge or gateway but stream instead directly to the cloud backend.
- *Hybrid cloud:* This term is often used to describe a system in which some functions are in the cloud, and some are local. One example is an *on-demand* cloud, where most functions can be performed in the cloud, but in practice, some or all are performed at the edge to reduce latency

Figure 15.8 An example of a camera streaming app showing live images, with graphics and text overlays providing other relevant information. The setup shown here uses a PTZ camera to cycle between preset positions showing various points of interest.

Source: Camstreamer, Prague, Czech Republic

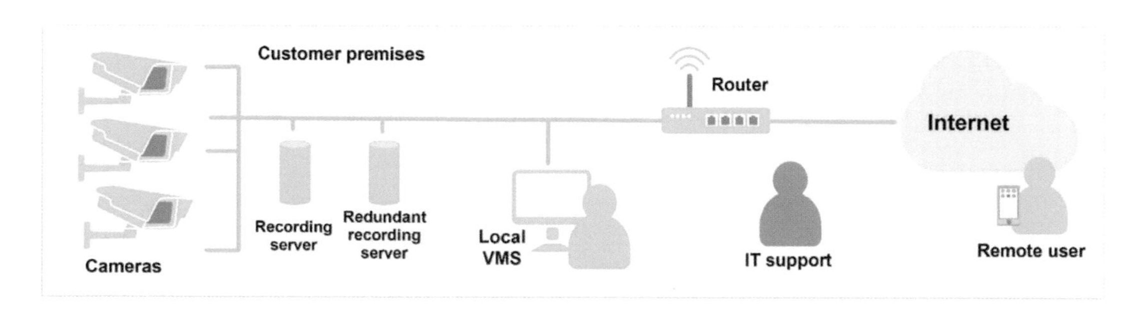

Figure 15.9 An overview of a traditional on-premises video surveillance setup, in which all hardware and the VMS is located on-site, along with support functions. Remote user access is possible via custom setups or third-party apps.

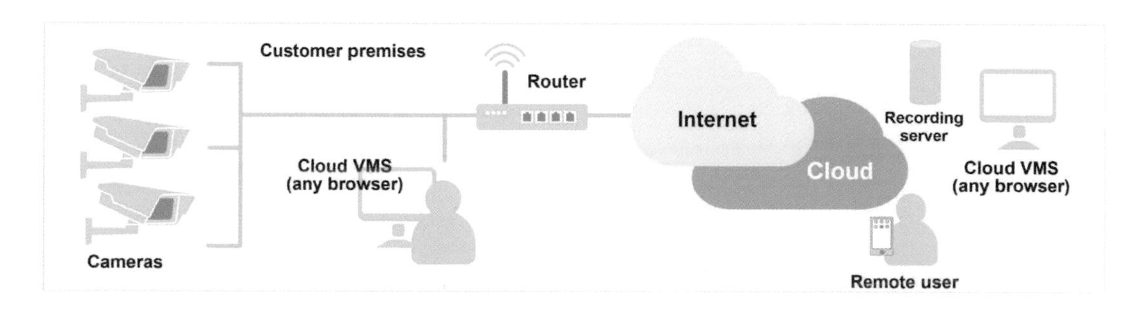

Figure 15.10 A cloud-based video surveillance solution, in which video storage is moved to the cloud, along with the VMS and all support functions. Remote user access is integrated into the solution.

or costs. Another example of a hybrid system is a *bridge* or *appliance-based cloud*. This setup uses a local NVR-like appliance as a bridge between the cloud and the local system, and some of the four main functions will be performed on that appliance. There is a significant overlap with the on-demand cloud model.

- *Relay cloud:* A solution that only provides cloud viewing and sometimes light cloud management functionality. All other functions are local or available through other solutions.

15.7 INTEGRATION WITH OTHER SYSTEMS

While cloud video is most commonly used as a stand-alone application, it can become even more powerful when integrated with other systems. Cloud solutions are very dynamic and make new functionality and services easy to approach. For example, the subscriber starts out by signing up for video services and could then later add access control. In the next step, they might add video analytics, such as heat mapping or people counting to certain cameras, or perhaps something more innovative, such as biometric analytics. Many services can be added remotely without even having to go to the site.

Two common systems to integrate with are central station automation platforms used in alarm monitoring and hosted physical access control systems.

15.7.1 Integration with physical access control

Cloud access control has been around for some time now and has become an integral part of the security industry. One of the reasons it was adopted so fast was because of the system's remote management capabilities. The benefits of an outsourced model for access control are the same as with video, and time has shown that the data security measures are sufficient also for physical security applications.

Video can add another dimension to the access control system and can be a new business opportunity for the provider of the hosted access control system. The benefit for users is that they can get visual verification for all access control transactions, along with general video surveillance (see Figure 15.11).

Figure 15.11 Integrating cloud access control with video gives a single system that covers several physical security needs.

Source: Genetec Inc., Montreal, Canada

15.8 BEST PRACTICES

Cloud video services have some unique characteristics that call for careful consideration before deciding if it is the right solution:

- *Distributed locations:* Typically, cloud video solutions have the most benefits when there are many different locations with few cameras.
- *Full cloud services:* Some vendors refer to remote access to a DVR or NVR as a cloud-based service. While the internet is certainly used for accessing them, the recordings are not actually stored in the cloud. With this solution, you miss out on many of the benefits of a true cloud video setup.
- *Off-site redundancy:* The need for a redundant recording solution with video recorded off-site is often a deciding factor when opting for a cloud solution.
- *On-site redundancy:* Different local storage solutions can be used to limit the need for bandwidth and to create redundancy, prevent data loss in the case of a network disruption or unreliable bandwidth, and add recognition or identification capabilities to a cloud video verification system.
- *Multi-site redundancy:* Large amounts of data (that is, high-quality video or long recordings) affect storage and archiving requirements. Some users need a redundant system with both remote and on-site storage.
- *Bandwidth estimates:* Bandwidth is an important factor for a cloud video solution. For the video results to be useful, you first need to determine the requirements on image quality. Detection requires less bandwidth than recognition or identification. High-resolution images need more upstream bandwidth than low-resolution images.
- *Bandwidth handling:* Some cloud providers offer hardware devices and software that help manage streams, recordings, and storage while consuming less bandwidth.
- *Security:* Any serious cloud provider should be able to prove that their data storage solutions are secure, whether these are their own or as operated by a third party. They should use approved authentication processes and comply with the latest security standards so that data is always streamed, recorded, and stored without risk of eavesdropping.
- *Alarm monitoring:* Consider alarm-monitoring services for increased comfort and help with verifying and acting on alarms, centralizing guard forces, and making the right decision at the right time.
- *Integration:* Through integration with, for example, physical access control or central station automation software, the video system can become even more powerful. Control doors, map movements, and handle alarms more efficiently.
- *Terms of contract:* Business models in cloud video differ, as they usually involve monthly service fees rather than upfront capital expenditures. The contract normally defines service levels and for how long the contract is binding. The cloud service provider needs to lock users into agreements so that, over time, they can earn back their own investments.

Chapter 16

Video analytics

Many of today's surveillance systems record massive amounts of video, but the sheer volume of recordings and lack of time means that much of the material is never watched or reviewed. As a result, events and incidents are overlooked, and suspicious behaviors are not detected in time to prevent crimes from being committed.

Furthermore, although video compression has come a long way, a typical camera at full frame rate still generates around 5 GB of data per day, which is similar to streaming a movie online. Considering that there are hundreds of millions of cameras installed around the world and that their retention times are often weeks or months, there is still a need to reduce the amount of data.

Video analytics, also known as analytics, video content analysis (VCA), intelligent video, intelligent video analysis, and sometimes even artificial intelligence (AI), can bridge these gaps. Analytics is a broader term that includes audio and non-video sensors, such as passive infrared (PIR) sensors and radar. Analytics applications convert video, audio, and other types of input to data and analyze it to find events of interest. For example, some applications recognize car license plates, while others focus on protecting critical infrastructure through virtual lines that trigger alarms in the event of an intrusion. The development of new video analytics systems is ever-growing, matching many types of security and efficiency needs.

Analytics-based systems are never idle. They scan video and audio in real time, around the clock, looking for information, events, or threats and responding immediately by starting recordings or alerting security staff. Analytics can significantly reduce demands on network bandwidth and storage space, they can use the cloud for storing data, and they can free up staff for tasks other than the constant monitoring of numerous cameras. Analytics can also enable smart search to quickly pinpoint specific events.

Analytics-based systems can also extract data from surveillance video and integrate the information into other applications, such as retail management or access control systems, creating new benefits and opening up a wide array of business opportunities.

This chapter describes the basics and history of analytics, as well as the applicable architectures and standards. Chapter 17 discusses some common analytics applications.

16.1 WHAT ARE VIDEO ANALYTICS?

Video analytics means the process of analyzing video data with the goal of transforming it into actionable information. Analytics-based systems use complex algorithms to analyze video and convert it into data. Typically, they extract moving objects or other recognizable forms while filtering out irrelevant movements (Figure 16.1).

DOI: 10.4324/9781003412205-16

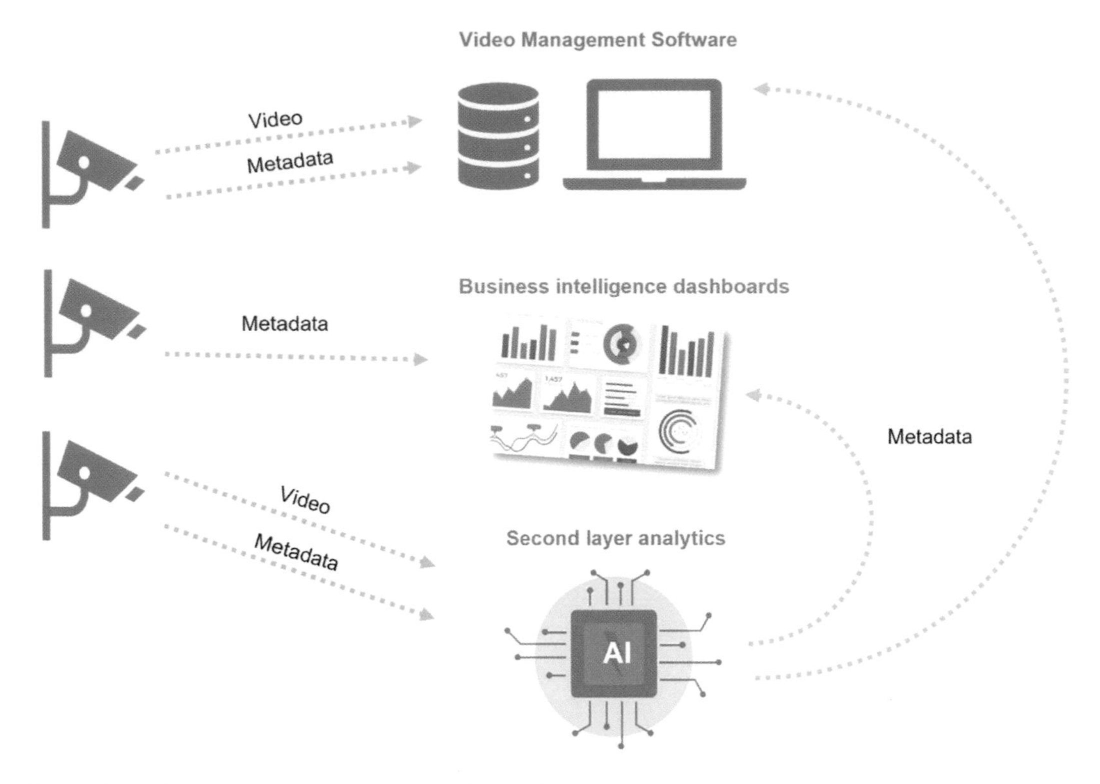

Figure 16.1 Video analytics extract actionable data from video images or streams.

The resulting data is stored in databases that can be searched with rules applied, for example, an object passing a virtual line in the video or more than ten cars waiting in a drive-through line. Rules can be programmed to help determine if the events observed in the video are normal or if they should be flagged as alerts to security staff.

Video analytics are a vital component of critical security installations, supporting timely decision-making in critical situations. Applications such as people or traffic counting also open up new effective ways to manage businesses.

16.2 THE GENESIS OF VIDEO ANALYTICS

In 1997, the Defense Advanced Research Projects Agency (DARPA) Information Systems Office in the United States began a three-year program to develop Video Surveillance and Monitoring (VSAM) technology. The objective of the VSAM project was to develop automated video and enable it to understand and evaluate the information received for use in battlefield surveillance applications. The technologies developed under this project enable a single human operator to monitor activities over a large area using video analytics, which were designed to be autonomous and to only notify the operator if security threats occurred.

Many researchers at universities such as Carnegie Mellon University (CMU) and Massachusetts Institute of Technology (MIT) were chosen to develop a wide range of advanced surveillance techniques. They include the following:

- *Real-time moving object detection and tracking from stationary and moving camera platforms*
- *Recognition of generic object classes (for example, human, sedan, truck) and specific object types (such as a police car or a courier van)*
- *Object pose estimation with respect to a geospatial site model*

- *Active camera control and multi-camera cooperative tracking*
- *Human gait analysis*
- *Recognition of simple multi-agent activities*
- *Real-time data dissemination*
- *Data logging*
- *Dynamic scene visualization*

Many of the early video analytics companies and technologies were spin-offs from the VSAM project. Today, most manufacturers of VMS applications and cameras offer some type of video analytics, which are often well-integrated into their offerings. Some also provide the platform necessary for supporting the use of third-party applications.

Although the early days of video analytics brought many new developments and applications, there were also many unreasonable expectations placed on the technology, which often failed to deliver. More recent developments – generally within artificial intelligence and specifically within deep learning – have now given us much more robust solutions.

16.3 WHY VIDEO ANALYTICS?

The security industry is a growing and evolving industry. As video surveillance installations expand both in number and size, there is a demand for smarter software systems that enable management and security staff to tackle their surveillance challenges successfully. In addition, as the security market shifted from proprietary, analog closed-circuit television (CCTV) systems towards open and fully digital IP-based network video, new possibilities for harvesting non-security-related information from surveillance systems emerged and provided new user benefits. As the network video surveillance market matures, the expectations and demands evolve as well, resulting in a huge variety of analytics applications that are more sophisticated, more efficient, more specialized, and more cost-efficient than ever before. Network video and analytics greatly simplify the process of integrating video with other IT and IS systems.

The three main market drivers for video analytics can be summarized as follows:

1. *Streamlining video surveillance operations*
2. *Managing stored video efficiently*
3. *Improving business operations*

The following sections examine each of these drivers.

16.3.1 Streamlining video surveillance operations

The efficiency of the video surveillance in a typical security installation is limited because of a single major issue: It is almost impossible to watch all the video all the time, even in large installations with multiple operators. Furthermore, studies show that even after a short period of continuous video monitoring, operators often miss a large portion of the activity shown on the screen. After only about 20 minutes of uninterrupted viewing, test subjects overlooked almost all activity. If nobody is actively watching and acting upon the events shown in the video, there is obviously an increased security risk to persons and facilities.

Analytics present a solution to this problem. Analytics applications analyze and filter the massive amount of information in multiple continuous video streams so that only relevant alerts are presented to security staff. With analytics, fewer operators need to spend their time monitoring video. Security managers can reduce personnel or invest the freed-up hours to have guards patrol the premises or perform other preventive measures. Even very large video surveillance installations benefit from analytics because staff will not be expected to attentively watch dozens or even hundreds of monitors for hours to detect undesired activity or suspicious persons. Instead, an analytics-based system can do the job of alerting operators, for example, when people move into restricted areas, when cars drive the wrong way, if crowds gather, or if someone tampers with a surveillance camera.

16.3.2 Managing stored video efficiently

Finding incidents in stored video would be extremely time-consuming if the operator actually had to sit and watch the recorded video to find the incidents. Even if the operator was experienced and could watch video at four times normal speed, large archives of video would still take a long time to search manually. Searching also implies that you know that an incident has occurred. The odds of finding an incident of which there is no physical proof are low. With smart search, operators can find the right video clip much easier (see Figure 16.2). For example, in a camera view, they can draw a shape around the area of interest and specify a timeframe during which they suspect or know that there has been activity. The VMS finds and displays the relevant video clips. There is also scope for more advanced searches based on objects and their attributes, for example, the direction of travel.

By using metadata, forensic searches can be even more powerful. Metadata is data about data (data about video in our case), and it describes the content of the video in terms other than images. For example, metadata can list the objects in the video, such as vehicles and people, as well as their associated attributes: colors of vehicles and clothing or the direction of travel. As the metadata also tags the video with the time and location, it is then a simple matter to find and view the associated video. For more on metadata, see section 16.5.

Many studies have shown that only a very small proportion of recorded video is ever watched. Another possibly significant issue is the cost of streaming and storing video. Continuous high-resolution recordings require a lot of bandwidth and server space. Most of the time, there are no incidents, so saving hours and hours of continuous recordings can be a waste of time, storage, and money. Given these factors, video files are often simply stored and deleted after 7 to 30 days. Although today's advanced video compression algorithms now compress video very efficiently and storage is now less of an issue, as systems also get larger in terms of camera count, higher resolution, and longer retention times, storage can still be a major outlay for many systems. Analytics applications such as motion detection ensure that only relevant video footage is recorded and stored, thus reducing the amount of relevant video data that needs to be stored or searched. In addition, some

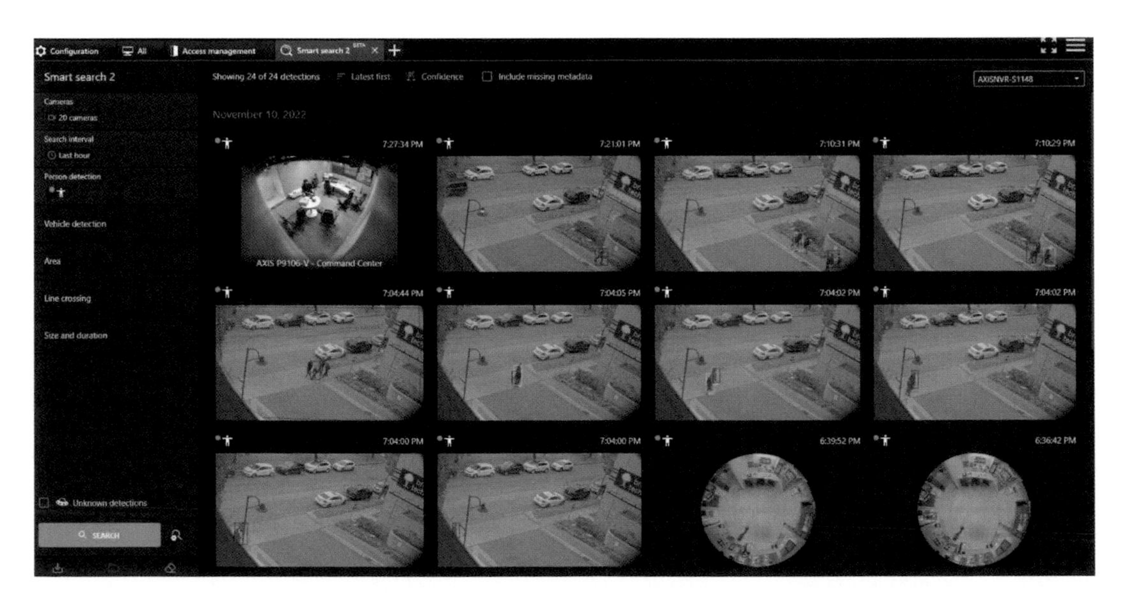

Figure 16.2 A smart search returns results based on parameters such as area, object attributes, or time, as set by the operator.

analytics-based systems can automatically search through days or even months of stored video to find incidents in a matter of seconds.

16.3.3 Improving business operations

Analytics also make it possible to use video for other things than security. Retail is one of the segments that is pushing the frontier of analytics. Retail uses video intelligence for business intelligence purposes, such as consumer behavior analysis. For example, through people counting, dwell-time tracking, and heat mapping, users can see how many people stopped at a particular shelf, which register is the most popular, what time of day the customer traffic peaks, if cash is handled the right way, or which route customers take through the store. In this way, analytics makes it possible to extract greater benefits from a video surveillance installation, enabling a higher return on investment.

Big data is one of the major trends in today's technology and business environment. Modern businesses are bombarded with huge amounts of data from various sources, and big data is a great opportunity for businesses to improve their operations if they can find effective ways to manage and analyze all this data. Video content analysis, combined with point of sales data, loyalty card data, and the like, gives retailers a great opportunity to evaluate behaviors and improve operations. City and traffic planners can gain great insights through the analysis of traffic patterns and flows of pedestrians.

16.4 ARTIFICIAL INTELLIGENCE

Artificial intelligence (AI) is a broad concept applied to machines that can solve complex tasks while demonstrating seemingly intelligent traits. The advantage of AI over traditional programming is the ability to process much more data than humans can, resulting in a more accurate application. However, the more complicated the application, the harder it is for the machine to produce the required result.

Machine learning and deep learning are subsets of AI.

16.4.1 Machine learning

Machine learning uses statistical learning algorithms to build systems that can learn and improve during training without being explicitly programmed, for example, to determine what is happening in images or video streams (computer vision).

Traditional programmed computer vision is based on calculations of an image's features, which need to be manually predefined by a human. Machine learning algorithms automatically build a mathematical model using training data to gain the ability to make decisions without being programmed to do so. The features are still predefined, but the algorithm learns how to combine these features through exposure to large amounts of annotated training data collected and annotated by humans. The data is fed into the system until the program has learned enough to detect what is needed. When finished, the program will not learn anything new.

16.4.2 Deep learning

Deep learning is data-driven learning of features and how to combine them. Much like the human brain, the algorithm can learn very deep structures of chained feature combinations, which are simulated in artificial neural networks, the most common type of algorithm in deep learning.

As deep learning can learn from greater amounts of more varied data, it can build intricate visual detectors and automatically train them to reliably detect very complex objects. In most cases, these will significantly outperform hand-crafted computer vision algorithms. This makes deep learning suitable for complex problems where the combination of features cannot easily be formed by human experts, as in image classification and object detection.

A deep learning algorithm typically uses much more feature combinations than a classic machine learning algorithm, making it more flexible and able to learn more complex tasks. For surveillance analytics, however, a dedicated and optimized classic machine learning algorithm can be sufficient. If well-specified, it can provide similar results to a deep learning algorithm but at lower cost and effort.

Most machine learning today is supervised, but unsupervised learning is also possible, where algorithms analyze and group unlabeled datasets. This is not yet common practice in the surveillance industry because the model requires a lot of calibration and testing and results can still be unpredictable.

The datasets must be relevant for the analytics application but do not have to be clearly labeled or marked, thus eliminating the manual annotation. However, the number of images or videos needed for the training is far greater. The model being trained identifies common features in the datasets, which when deployed, allows it to group data according to patterns but also to detect anomalies that do not fit into any of the learned groups.

16.4.3 Neural networks

Neural networks are algorithms used to recognize relationships in datasets through a process similar to how the human brain works. A neural network consists of a hierarchy of multiple layers of interconnected so-called nodes or neurons, and information is passed along the network connections, from the input layer all the way to the output layer.

The assumption for neural networks to work is that an input data sample can be reduced to a finite set of features, creating a good representation of the input data. These features can then be combined and will help classify the input data, for example, describing the contents of an image.

Figure 16.3 illustrates how a neural network is used to identify which class an input image belongs to. Each pixel in the image is represented by one input node. All input nodes are coupled to the nodes in the first layer. These produce output values that are passed along as input values to the second layer and so on. In each layer, weighting functions, bias values, and activation functions are also involved in the process.

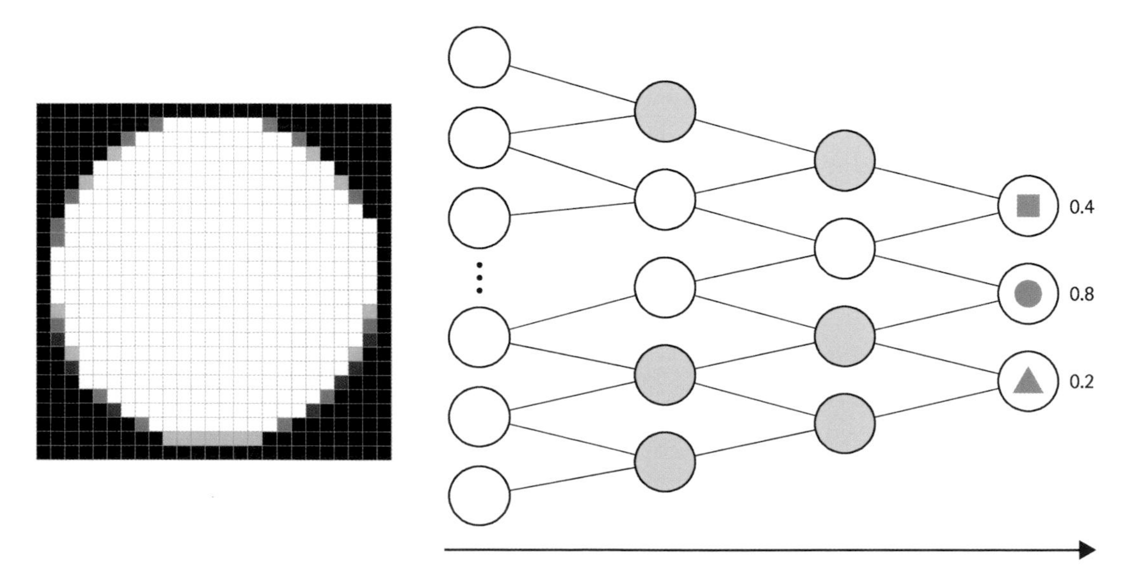

Figure 16.3 Example of an input image (left) and a neural network (right). When the output layer is reached, the network has concluded the probabilities for each possible category (square, circle, or triangle). The category with the highest probability value is the most likely shape of the input image.

This process is called forward propagation. If the result of the forward propagation is mismatched, the network parameters are slightly modified through back propagation. During this iterative training process, the performance of the network is gradually improved.

After deployment, a neural network generally has no memory from previous forward passes. This means that it does not improve over time, and it can only detect the types of objects or solve the types of tasks it was trained for.

16.4.4 Processor platforms for AI applications

The deployment of an analytics application is also called inference or prediction, which is the process of executing a trained machine learning model. The algorithm uses what it learned during the training phase to produce the desired output. In the surveillance analytics context, the inference phase is the application running on a surveillance system monitoring real-life scenes.

To achieve real-time performance when executing a machine learning–based algorithm on audio or video input data, specific hardware acceleration is generally required, especially when power is limited. High-performance video analytics used to be server-based because they required more power and cooling than a camera could offer. Algorithm development and the increasing processing power of network video devices have, however, now made it possible to run advanced AI-based video analytics on the edge (in the device). Edge hardware accelerators can be complemented by server and cloud resources when suitable.

- *Graphics processing unit:* GPUs were mainly developed for graphics processing applications but are also used for accelerating analytics on server and cloud platforms. GPUs are not generally optimal from a power efficiency standpoint in edge analytics applications.
- *Machine learning processing unit:* MLPU is an umbrella term for integrated chipsets designed for machine learning algorithms. These processors are accelerators integrated on an application-specific integrated circuit (ASIC) or on a system-on-chip (SoC), with data memory and architecture dedicated to machine learning, in applications designed for real-time object detection of a limited number of simultaneous object types, for example, humans and vehicles.
- *Deep learning processing unit:* DLPU describes chipsets – on ASICs or SoCs – dedicated to accelerating deep learning inference, which allows for more granular object classification, more different circumstances, and more object classes. For example, different vehicle types, crawling or crouching persons, and fewer limitations, such as detecting an object even though only part of it is visible.

16.5 METADATA

Metadata means "data about data", and it provides a solution to the challenge of sifting through volumes of recorded video to find, filter, and retrieve the right information efficiently. Metadata allows the tagging of images with information about those images. The metadata tags are what enables automatic analysis of video streams and users to quickly find exactly what they are looking for in a recording.

For example, an analytics application that counts people passing through an area can tag an image with a number as part of the process of streaming it to a central server for further analysis. The analytics application can also be designed to send only the tag information – the number of people – and to not stream images at all.

For video surveillance of a bridge or a highway, an analytics application could tag vehicles appearing in an image according to certain criteria and store the tags together with the video. This would make it possible for a user to search based on these criteria and instantly have access to the correct video sequences. For example, the operator could search for all blue cars heading north in the past 24 hours, instead of manually searching or watching 24 hours of recorded video. Figure 16.4 shows an example of images tagged with metadata.

Figure 16.4 A video stream tagged with metadata, for example, the object, color, and location of a car in a parking lot.

Metadata can efficiently collect, organize, and store content of interest, as well as identify patterns and trends – which can be used to improve operations and optimize businesses.

Metadata can describe many details about objects of interest in video:

- *Location*
- *Time*
- *Colors*
- *Sizes*
- *Shapes*
- *Coordinates*
- *Speed*
- *Duration in the scene*
- *Direction of travel*

16.6 APIS AND STANDARDS

Analytics applications work on digital streams of video from network video devices. For analytics to work as part of a video surveillance system, there must be standardized formats for digital video streams. A number of video compression standards exist, some of which are more relevant for analytics than others. These include Motion JPEG, MPEG-4, MPEG-4 AVC/H.264, and HEVC/H.265. For more information about compression, see Chapter 6.

In addition to the requirement for video compression standards, there is also a need for open standards applicable to metadata. When the metadata functionality is based on open and standardized ways of describing information, it can easily integrate and scale with other open systems. Early examples of standards for attaching metadata to data streams include Extensible Markup Language (XML) and Scalable Vector Graphics (SVG). Many other standards have been developed for specific areas, such as libraries, geospatial use, and database management.

One of the most important metadata standards for network cameras is the ONVIF Profile M, which allows the streaming of metadata and events from edge-based analytics applications to be standardized. Conformance to ONVIF Profile M enables easier integration of metadata and events with conformant clients, such as video management software. Conformant clients will be able to query, filter, and receive metadata to trigger automatic responses and effectively store and retrieve video content.

16.7 VIDEO ANALYTICS ARCHITECTURES

Analytics can reside in different parts of a video surveillance system (see Figure 16.5). There are several different architectures for implementing analytics-based systems:

- *Server-based architecture:* Video and other information is collected by cameras and sensors and sent to centralized on-premises servers for analysis.
- *Cloud-based architecture:* This is a variation of server-based systems in that the servers are usually owned and maintained by some other organization and are mostly accessed over the internet.
- *Distributed architecture:* Also known as edge analytics, in this case, it is the network video devices that process some or all of the video and extract the relevant information from it.
- *Hybrid architecture:* This type of system uses some combination of edge-based analysis with server- or cloud-based analytics.

Through rational design and load distribution, analytics-based systems can improve performance and lower overall costs substantially. The following subsections look at the various systems and how different components of a video surveillance system can be used to implement analytics.

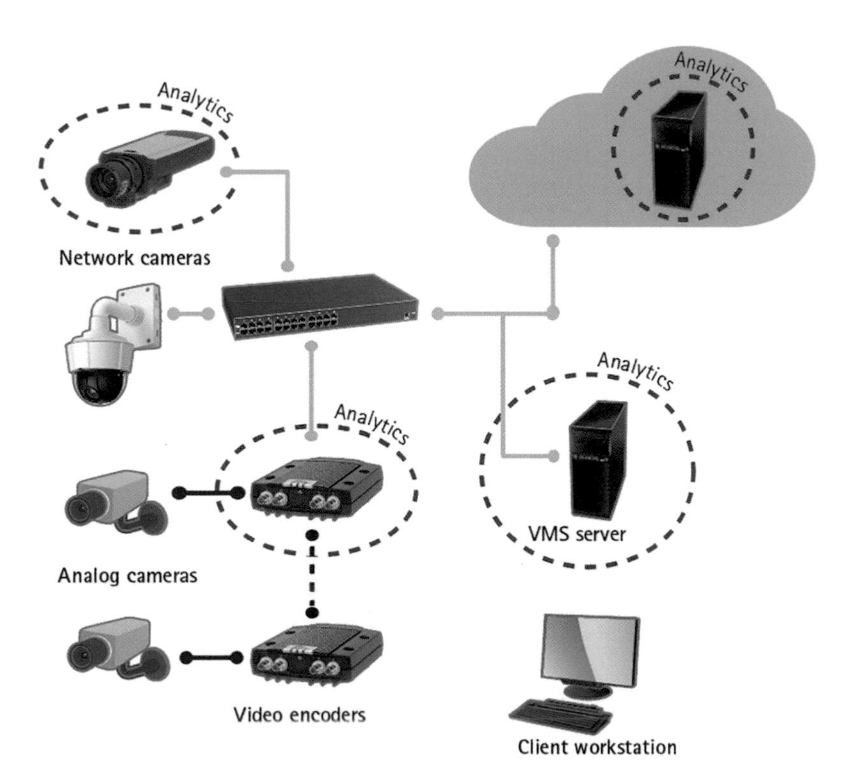

Figure 16.5 The analytics can be in different parts of a video surveillance system, creating centralized or distributed architecture.

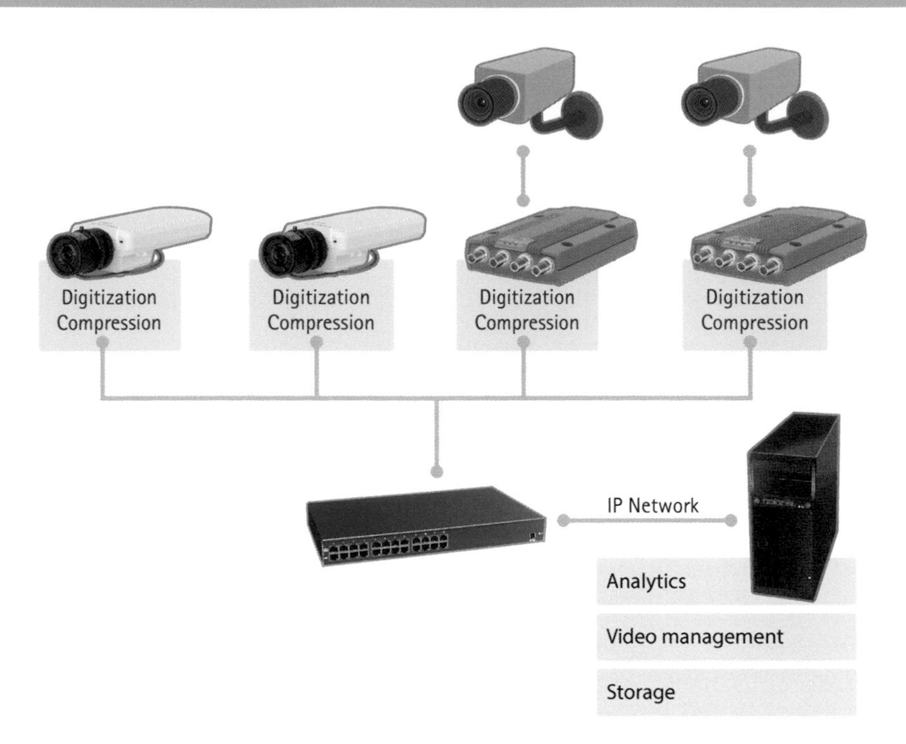

Figure 16.6 In a server-based installation, the analytics functionality is in the server, along with all other functions, such as video management and storage.

16.7.1 Server-based architecture

Demanding analytic processes often require video to be analyzed on dedicated servers. Video from the cameras is digitized, compressed, and sent over the network for centralized analysis on standard IT servers. Analog cameras can be used with video encoders to digitize the video before transmitting it to the servers, as shown in Figure 16.6.

As these servers perform many processor-intensive tasks (transcoding video, managing storage, and more complex video analysis applications), they often require considerable processing power. However, these servers can be equipped with multiple processors and, in some cases, can support hundreds of cameras. One potential drawback of server-based analysis is that the video has already been compressed on the camera and is of a lower quality. The video also needs to be decompressed before any analysis can be performed.

16.7.2 Cloud-based architecture

Analyzing video in the cloud involves connecting cameras directly to the cloud servers, and depending on the number of cameras and the available bandwidth, this might be limited to low-resolution video. A solution based purely on the cloud is not always optimal or even possible. Cloud analysis of any sort requires a robust and reliable internet connection, which might not always be available.

Many organizations run operations that involve sensitive information, for example, banks, power plants, hospitals, governments, etc. Such organizations may be required to retain all data on-premises – due either to their own governance or to compliance with regulations – and are unlikely to consider a commercially available cloud solution, although they may consider a private cloud setup.

The main benefits of cloud-based analysis:

- Sharing data across services and solutions is easier.
- Service upgrades become the supplier's concern.

- Greater processing power allows more complex analytics to be used.
- Demands for extra bandwidth can be met immediately.
- Processing multiple video streams from many sources become easier.

The potential drawbacks:

- High demands on internet bandwidth may limit the number of video streams that can be analyzed.
- Cloud processing can be prohibitively expensive when analyzing large numbers of streams.
- Higher energy consumption: Up to 90% of the energy consumption in cloud data centers is spent on cooling the server racks. Analytics at the edge typically make the overall solution more sustainable.
- Live monitoring becomes more complex as the video must be streamed securely to the cloud and back again for monitoring.

16.7.3 Edge analytics

The most scalable, cost-effective, and flexible architecture is based on "intelligence at the edge", which means that the network video devices do all or most of the video analysis themselves. This fact can be exploited by, for example, only sending video when certain objects or motion is detected in defined areas of a scene (see Figure 16.7). If no objects or motion are detected, no video is sent. The load on the infrastructure drops dramatically, and there is less risk of overloading a central point, such as a server.

Another advantage of performing video processing at the edge is that it significantly reduces the cost of the servers needed to run analytics applications. Also, when analysis is performed at the edge, the servers can handle many more video streams. In some applications, where only the data is

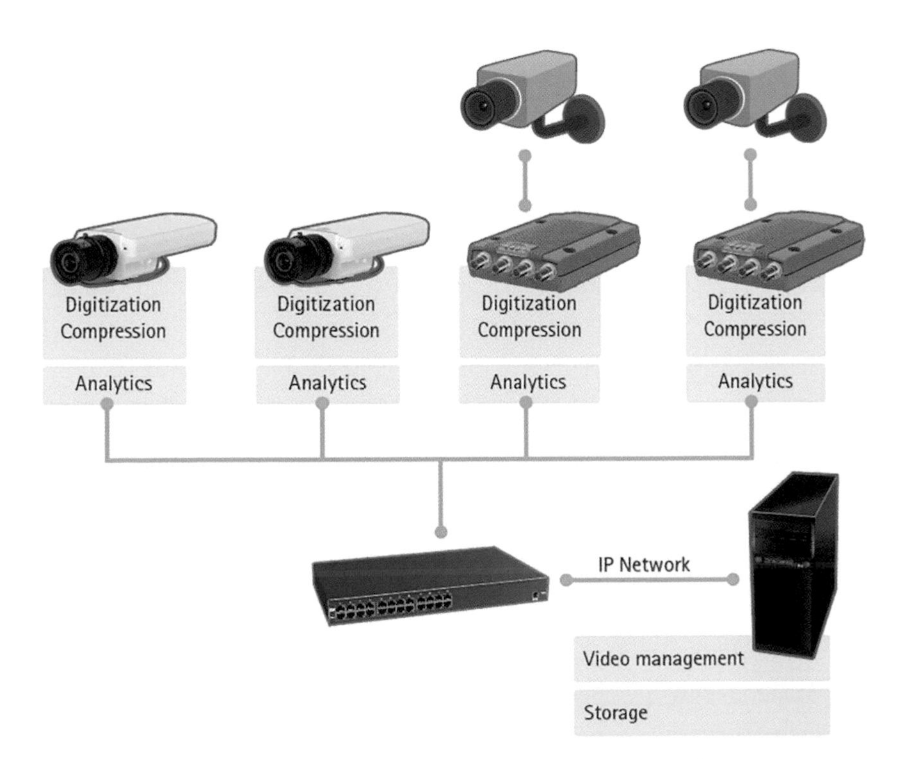

Figure 16.7 A truly distributed system with intelligence at the edge (in the cameras and encoders) is the most scalable and cost-effective.

needed and not the video (for example, people counting or license plate recognition), the resulting data can be sent directly to a database, which further reduces the load on servers.

Algorithms that process video are more effective when processing raw video rather than compressed video. However, streaming raw video to centralized servers for analysis would require excessive amounts of bandwidth, which is why cameras with edge intelligence process the raw video directly in the camera. Furthermore, in centralized architectures, when compressed video arrives on the servers, it needs to be decompressed or transcoded, requiring more servers for each given number of cameras.

The benefits of architecture based on intelligent edge devices can be summarized as follows:

- *Fewer servers needed for video processing:* This means lower power consumption and maintenance costs. In environments that lack a server room, installing large numbers of servers is simply impractical.
- *Lower demands on bandwidth:* Streaming only essential information means lower data rates, and lower-cost network components can be used.
- *Greater accuracy.* In edge analytics, the video available to analyze is uncompressed and, hence, includes more details than compressed video, giving more accurate analysis.
- *Faster response time due to low latency:* An edge solution reduces the risk of delay versus a server- or cloud-based system.
- *Easy to scale up:* The system is scalable, as each new load (video stream) added to the system is processed at the edge and does not necessarily increase the load on a central server.
- *Improved privacy:* Running video analytics at the edge gives the option to only send anonymized data and/or alerts over the network, which can enable compliance with strict privacy requirements

Analytics at the edge also give the opportunity to create a self-contained recording system, where the video management software is hosted on a PC, NAS, or on the camera itself. Users control and manage the video through browsers or apps on mobile devices. These systems are completely edge-based and are perfect in environments that have limited network connectivity and no on-site server resources.

Edge analytics are also essential in cloud-based systems and are especially cost-effective in applications where the metadata is more important than the video. Cameras and encoders with edge analytics process the video and send only the metadata and a low-bandwidth stream to the cloud. Edge storage (NAS or onboard storage) can be used to store a stream with higher resolution and a higher frame rate. Read more about cloud solutions in Chapter 15. The implementation and acceptance of edge analytics is expected to accelerate as more and more IP cameras support deep learning applications.

16.7.4 Hybrid architecture

A hybrid solution involves using both edge-based and server- or cloud-based analytics, with pre-processing performed at the edge and further processing, where required, on the server. This allows each component to be used according to its individual strengths, for example, where cameras perform the first step – detection of people, vehicles, or objects – and a server or the cloud performs more complex analysis, for instance, the granular classification of objects and attributes. This architecture is common with proprietary systems, as they are often complex in nature.

Possible drawbacks include little control of the costs associated with cloud processing and greater system complexity.

16.8 INTEGRATING VIDEO ANALYTICS APPLICATIONS

Thanks to developments in deep learning, increased accuracy and greater ease-of-use has driven the increased implementation of video analytics, along with a desire to make the most of new and existing video surveillance systems.

Many manufacturers of video surveillance equipment and video management software include object-based analytics with their products. These analytics, for example, video motion detection, are suitable for most installations. Other analytics for specific applications may also be provided, such as license plate recognition and people-counting applications.

Building robust and commercially viable analytics applications with a high level of functionality requires a great deal of expertise in specific sciences. For example, image analysis and biometrics use advanced mathematical algorithms. In addition, applications often require specialized knowledge in a certain field, such as retail, public transportation, or customs control. For this reason, some software companies focus their skills on developing and supplying analytics applications that solve specific needs in specific markets, allowing for tailored video surveillance solutions.

Although this progress brings greater freedom of choice for the user, it also makes it necessary for manufacturers to ensure compatibility and ease of integration between cameras, video management software, and analytics applications. To be commercially attractive, cameras, software, and analytics must be easy to use. They must also build on open, published platforms and application programming interfaces (APIs). This enables easy installation of plug-in camera software (see Figure 16.8) and communication with video management systems.

From the analytics suppliers' point of view, using open platforms and standards provides the greatest flexibility and makes the software easier to sell. It is this openness that allows vendors to install analytics applications from different manufacturers in their cameras and systems.

When analytics apply metadata tags to objects and activities in a video stream, this helps applications trigger automated actions or search through vast amounts of video. This potentially allows operators to perform searches such as "Find all video in area A with red cars, 8:00–10:00 a.m. on Wednesday, 25th March".

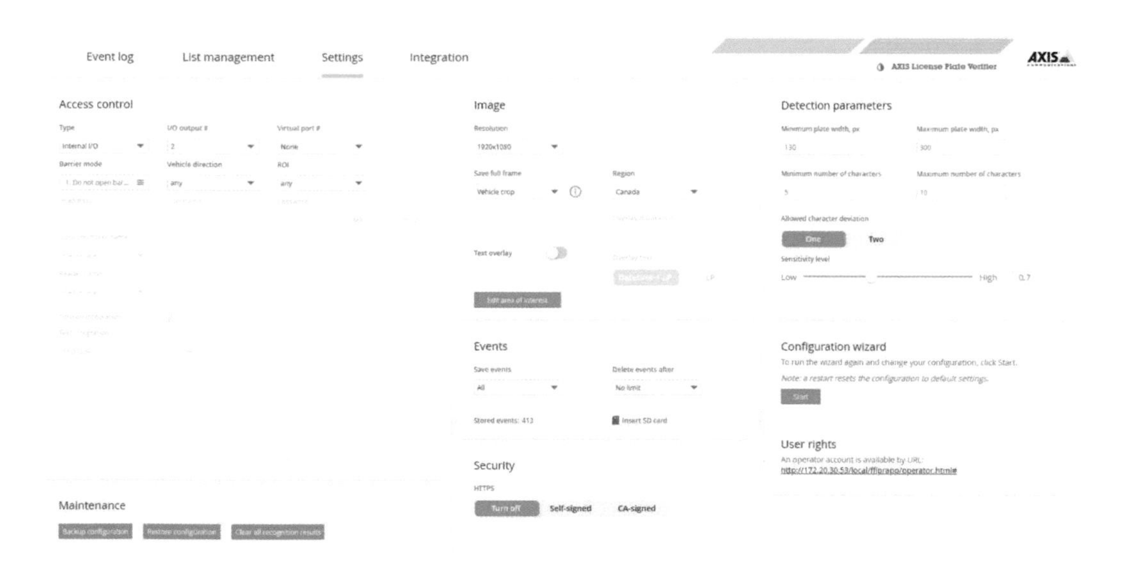

Figure 16.8 An example of plug-in software for license plate recognition.

Figure 16.9 Pixel-based analytics sometimes raise both true alarms (left) and false alarms (right).

Figure 16.10 (Left) A false negative and (right) a true negative.

16.9 ALARMS: TRUE OR FALSE?

Before AI technology, there were typically only two terms used to discuss alarms: *true alarms* (being useful to act upon) and *false alarms* (being not useful at all, especially when these alarms are sent to an operator in real time).

In Figure 16.9, earlier pixel-based analytics would react to movement in both scenes and would raise an alarm for each case. However, although the alarm is technically **positive** (an alarm was triggered), the detected person would usually be considered a true alarm, while the rabbit would be seen as a false alarm. Today's AI-based analytics can considerably reduce the number of false alarms, but there is still a potential for missed alarms.

So in fact, we have four types of alarm (or rather alarm result), of which only two are useful.

- *True positive:* An object that should trigger an alarm does so. The person in Figure 16.9 is an example of a true positive.
- *False positive:* An object that should not trigger an alarm does so anyway, meaning that resources are wasted since alarms normally call for action. The rabbit in Figure 16.9 can be considered a false positive.
- *False negative:* An incident or object should have triggered an alarm but did not. This is serious, as important incidents may be missed entirely. The person walking undetected behind the fence in Figure 16.10 is an example.
- *True negative:* Nothing of interest happens in the scene, and no alarm is triggered.

16.10 ETHICAL CONSIDERATIONS

Privacy considerations and legislation are well-known issues for the surveillance industry. For example, the placement of cameras may be restricted, signage is often mandatory in areas where cameras are permitted, privacy needs to be protected, and there are the aspects of data management and cybersecurity.

When developing AI-based applications, manufacturers also need to be mindful that the collection and management of the training data must also conform to the law and ethical standards. Data – especially personal data – must be collected, handled, and protected according to applicable legislation, for example, according to the General Data Protection Regulation (GDPR) in the European Union. Furthermore, the data needs to be relevant, it should address the use case it aims to solve, and it should be diverse enough for that purpose. This is important in order to remove or minimize any biases in detection and actions. A good guiding principle is to not simply allow automated processes to make all the decisions but to always keep people involved in the decision-making. This improves the relevance of the data based on automated alerts as provided to operators and should help prevent unwanted results.

16.11 BEST PRACTICES

After seeing the first applications for video analytics 20 years ago, the market has evolved relatively slowly and often not meeting expectations in large scalable systems. With the advent of deep learning processing at the edge, the market is expected to develop much more rapidly. There is a wide variety of offerings from many companies. Although the architecture of the solution is often overlooked, it is important to understand it fully, or you might end up with the wrong solution. When implementing analytics, the solution should be based on open standards and be able to scale as an installation grows.

Things to consider include the following:

- *Reliability of the system:* Does the architecture minimize the risk of system failure and the downtime associated with it?
- *Scalability and flexibility:* Can the system scale effortlessly from a few to many cameras? Can it intelligently distribute processing on different components of the network?
- *Interoperability:* Are the analytics applications and video management applications tied to specific hardware? Is it possible to combine system components from different vendors?
- *Data format used by the analytics application:* Is the metadata based on an openly published standard or API so that it can be easily incorporated into other systems?
- *Accuracy:* No video analytics system is 100% accurate, although as regards impact, the difference between 90% and 95% accuracy is large. An operator can only deal with a limited number of false positives a day.
- *Integration:* Is the system stand-alone, operating without or alongside a VMS? Is the system integrated and tied to a central recording platform? Is the system centralized or distributed? Depending on the system, different opportunities and choices arise.
- *Processor platform:* A camera or server using processors capable of running deep learning applications provide much more capable and accurate analytics than when using regular processors.
- *Built-in or plug-in analytics:* Are the analytics built into the camera, or is there an open camera platform? Open platforms allow users to choose from among the best plug-in applications but at the expense of added complexity.

CHAPTER 17

Video analytics applications

Today, there is a wide range of video analytics applications to choose from. Video analytics are also known as video intelligence, video content analysis (VCA), analytics, and sometimes artificial intelligence (AI). As the technology becomes ever better, smarter and smarter applications are being developed. Some analytics such as video motion detection and camera tampering alarms are relevant to most video surveillance installations. Others are specially developed to address the needs of specific industries, such as retail, transportation, and critical infrastructure. This chapter provides an overview of the most common types of analytics applications.

Video analytics got off to a shaky start in the security industry in the early 2000s. Demo installations looked good but often failed to deliver the same level of accuracy when implemented on a larger scale. Since then, the reputation of analytics has been restored, especially with the advent of deep learning applications running at the edge, and they are now used frequently and by many, and their benefits are well recognized. It is still important to have realistic expectations and to understand exactly what the technology can and cannot deliver. This is discussed at the end of the chapter, along with some best practices.

Developments in artificial intelligence (AI) are also helping to improve video analytics. Although today's AI solutions cannot replace the human operator's experience and skills, they do mean that we can increase human efficiency. For an introduction to AI, see section 16.4.

17.1 CATEGORIZING VIDEO ANALYTICS

The following sections discuss the different types of analytics technologies and their uses, but first, an introduction about different ways to categorize analytics.

17.1.1 Categorizing video analytics by technology

One approach to categorizing applications is to describe them based on the technologies they use. Video analytics are typically divided into the following technology categories:

- *Pixel-based analytics:* The first video analytic functions available on network cameras were very simple and purely pixel-based, with the earliest, most basic video motion detection being one example. Because changes in the scene were based solely on changes in pixels, it was often hard to know if the change was caused by actual motion or merely due to changes in the lighting.
- *Motion-based analytics:* Sometimes known as object-based analytics, these applications use chunks of pixels to detect movement, and they add algorithms to differentiate background from foreground. This enables a very basic object recognition, making it possible to use these objects as trigger conditions. Although this was a big step in the field of video analytics, the accuracy

DOI: 10.4324/9781003412205-17

provided by some of these analytics did not always meet expectations. People counting and advanced video motion detection are examples of motion-based analytics.

- *AI-based analytics:* These applications offer real classification of objects based on training, meaning the object is classified based on its appearance and irrespective of how it moves. AI-based analytics typically offer significantly more accurate results, with fewer false alarms, for example, in busy scenes, in poor lighting, or when coping with partly hidden objects. Although AI-based analytics are less "trigger-happy" than pixel- or motion-based analytics and generally mean fewer false alarms, there is also the risk that objects are not detected unless they match very closely the objects the AI model was trained on. AI-based analytics are based on either machine learning or deep learning.

17.1.1.1 Pixels, blobs, and objects

At a basic level, video analytics analyze every pixel in every frame of video, characterize those pixels, and then make decisions based on those characteristics. In, for example, pixel-based motion detection, when a certain number of pixels have changed, based on criteria such as size, color, and brightness, this can be used to raise an alert or to trigger other actions.

Blob recognition involves a level of intelligence beyond the detection of pixel changes. Blobs are essentially collections of contiguous pixels that share particular characteristics, and they have boundaries that delineate them from other parts of a video frame. Blobs can be analyzed and characterized as particular objects. For example, a blob can be identified as a person or a car by analyzing its shape, size, speed, or other parameters, see Figure 17.1. Applications based on object classification and tracking require sophisticated software algorithms.

Deep learning is a refined version of machine learning in which an algorithm learns both feature extraction and how to combine these features in a data-driven manner. The algorithm can automatically define what features to look for in the training data. It can also learn very deep structures of chained combinations of features. This makes it possible to build intricate visual detectors and automatically train them to detect very complex objects that are resilient to scale, rotation, and other variations. For more on machine learning and deep learning, see Chapter 16.

17.1.2 Categorizing video analytics by use

Many video analytics are designed to extract different kinds of information from video, to process information in different ways, and to apply different rules for making decisions. As described earlier, analytics can be categorized based on the technologies behind them, but what really matters in the end is how they can be used to make life easier and more secure for organizations and businesses.

The analytics applications described in this chapter are categorized based on how they are typically used:

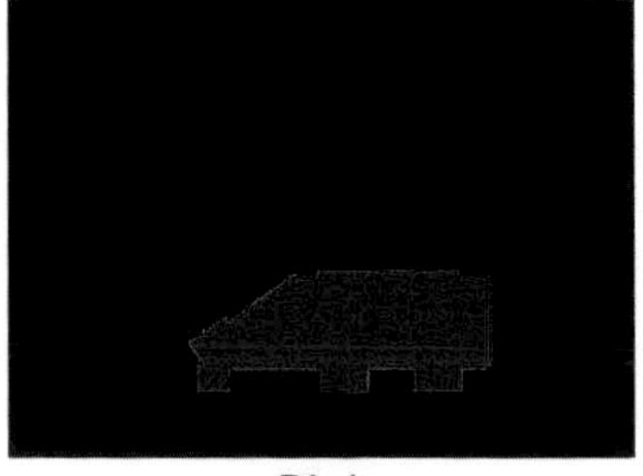

Figure 17.1 Examples of detection and tracking of pixels (left), blobs (center), and objects (right).

- *Analytics for security and safety:* Primarily used in security management applications and often in real-time, event-based analytics use video data to improve surveillance system efficiency and to facilitate investigations. The camera reacts to real-time events and triggers alarms accordingly. An operator analyzes the alarms and decides on any actions required.
- *Analytics for operational efficiency:* Primarily used for management and evaluation purposes and not for immediate action. For example, customer flows in a store, traffic flows in cities, object tracking in logistics centers.
- *Analytics for hybrid operations:* Used by both real-time security operations and business intelligence. For example, license plate recognition (LPR) can be used for real-time purposes as well as reporting purposes. LPR applications with approved license plates can be used to open gates when approved vehicles approach (an event-based operation) but also to create parking invoices (a reporting operation).

17.2 ANALYTICS FOR SECURITY AND SAFETY

For users, these analytics act as detectors that answer yes-or-no questions: Is something there? Did something move? Did something new happen? Did someone tamper with the camera? Whenever the system detects an activity, it can trigger alerts to operators or automatic system responses.

Examples of analytics for security:

- *Video motion detection (VMD)*
- *Camera tampering detection*
- *Object detection and tracking*

Examples of analytics for safety:

- *Fire and smoke detection*
- *Missing safety equipment (e.g., hard hats)*
- *Pose-based analytics (e.g., fall detection in care facilities)*

17.2.1 Video motion detection

Video motion detection (VMD) is the original and the most basic and widespread analytics application in the surveillance industry. It is used primarily to reduce the amount of video stored. By only storing video in which changes occur, video can be stored for a longer period at a given storage capacity. However, as today's advanced video compression algorithms compress video with no motion so efficiently, this makes it feasible to run continuous recording in parallel with VMD so that nothing is ever missed.

So how does the VMD application know which video to record and which to ignore? The application flags video that includes changes (motion) and ignores video that has no changes. VMD can also flag events to operators for immediate action, for example, when persons enter restricted areas.

VMD is the foundation for a large number of more advanced analytics, such as people counting, digital fences, and object tracking. Worth noting is that VMD is very different from detecting motion with a passive infrared (PIR) sensor. A PIR sensor detects heat by measuring the infrared light emitted from objects and people in its field of view. For more information about infrared technologies, see Chapters 4 and 5.

17.2.1.1 The evolution of VMD

Algorithms continually compare images from a video stream to detect changes in an image. The first generation of applications that recognized motion in the camera view did so simply by detecting pixel changes from one video frame to the next. Although this certainly helped reduce the amount of storage required (by not storing static video), it was not very useful for real-time applications because it gave too many false positives, that is, it raised too many alarms based on pixel changes caused by events of no interest, such as minor light changes, slight camera motion, or trees swaying in the wind.

More advanced VMD systems can exclude pixel-based changes from known sources, such as natural changes in lighting conditions based on the time of day or other known and repetitive changes in the camera's field of view. This drastically decreases the number of false alarms.

Later generations of VMD applications are more sophisticated and deploy more advanced algorithms. They not only detect individual changes in pixels but also group pixels together or detect motion on an object level. This way, the system can understand that many pixels together actually constitute larger objects, such as people or cars. This decreases the number of false alarms even further.

17.2.1.2 Tuning VMD parameters

VMD applications usually have controls for tuning the detection parameters. Examples of parameters:

- *Ignore filters for different object types*
- *Object size thresholds*
- *Lingering times*

The balance between different settings can directly affect the number of false alarms and whether all alarms in the scene are detected. For best results, you should fine-tune the VMD after installing the cameras. To ensure a robust implementation, keep a log for a while, and observe how the settings affect the VMD events in each camera.

Advanced network cameras often allow more exact placement of detection areas (see Figure 17.2). It is also possible to create multiple detection areas, where each area can use different parameters.

Figure 17.2 Video motion detection applications are especially suitable for low-traffic areas, such as office corridors, parking lots, and unattended shop areas. The user can draw polygons around the areas in which to detect and also around areas that should be ignored.

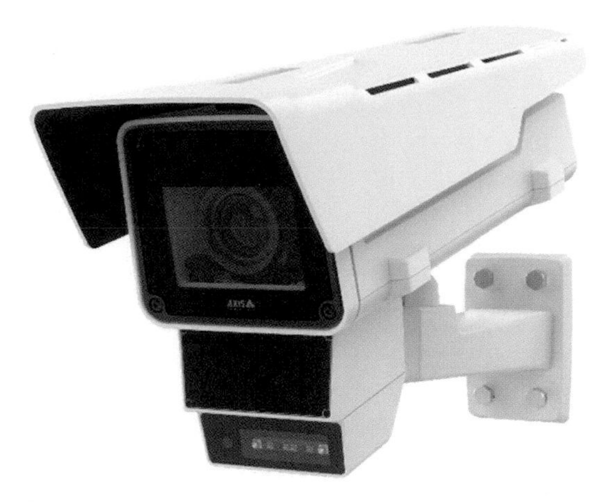

Figure 17.3 An example of a camera with a built-in radar detector.

Figure 17.4 The view from a camera with built-in radar. (a) Shows the regular view as seen by the optical unit. (b) This image shows the radar data superimposed over the camera view.

Whenever there is an incident, operators and investigators usually find it helpful to see what happened before and after the main event. Many network cameras and encoders keep pre-alarm buffers and post-alarm buffers that, for example, start 30 seconds before the event and continue 15 seconds after the event. This ensures that when the VMD detects motion, the system always has video of what happened in the seconds before the event, during the event, and the seconds after the event.

17.2.1.3 VMD and radar

Specialized cameras that feature an integrated radar detector offer even greater possibilities for motion detection, at distances that are generally unfeasible for regular pixel- or object-based VMD. The optical and radar modules are precisely aligned and calibrated, and the video stream incorporates radar-detection data, such as object class, true speed, distance, direction, and position.

17.2.2 Object analytics

Edge-based object analytics in general can detect and classify moving objects, for example, humans or vehicles. Exactly what can be detected and classified depends on the camera and its analytics capabilities. A camera running machine learning applications should be capable of classifying humans and vehicles.

Analytics on a camera that supports deep learning applications should offer more granular object classification and be able to distinguish between different types of vehicles, including cars, trucks, buses, and motorcycles or bicycles. It should also be better at detecting and classifying humans in unusual positions, for example, when crouching, as well as objects that are only partially visible.

Busy scenes and more demanding surveillance requirements require object analytics running on deep learning cameras.

Figure 17.5 Some analytics can detect the color of clothing worn by persons in the scene, and some can even differentiate between upper and lower clothing.

Source: BriefCam

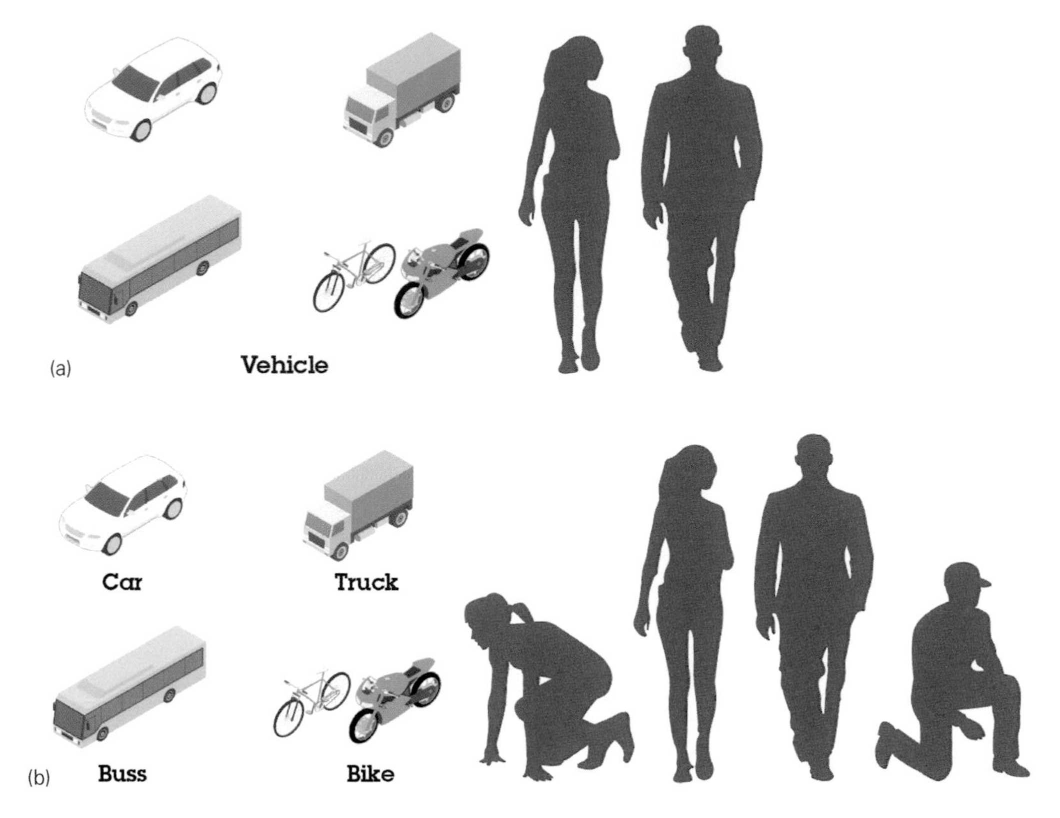

Figure 17.6 (a) Examples of machine learning analytics. (b) Deep learning object analytics offer more nuanced classifications than machine learning. For example, vehicles can be specified by type, and humans can be detected in several different poses.

Figure 17.7 Examples of object classification in the video stream from a deep learning analytics application.

Object analytics allow you to select various conditions for detection, such as objects that move or stay longer than a set time within a predefined area or which cross a virtual line. Upon detection, the camera can perform different actions, such as recording video, playing an audio message, or alerting security staff.

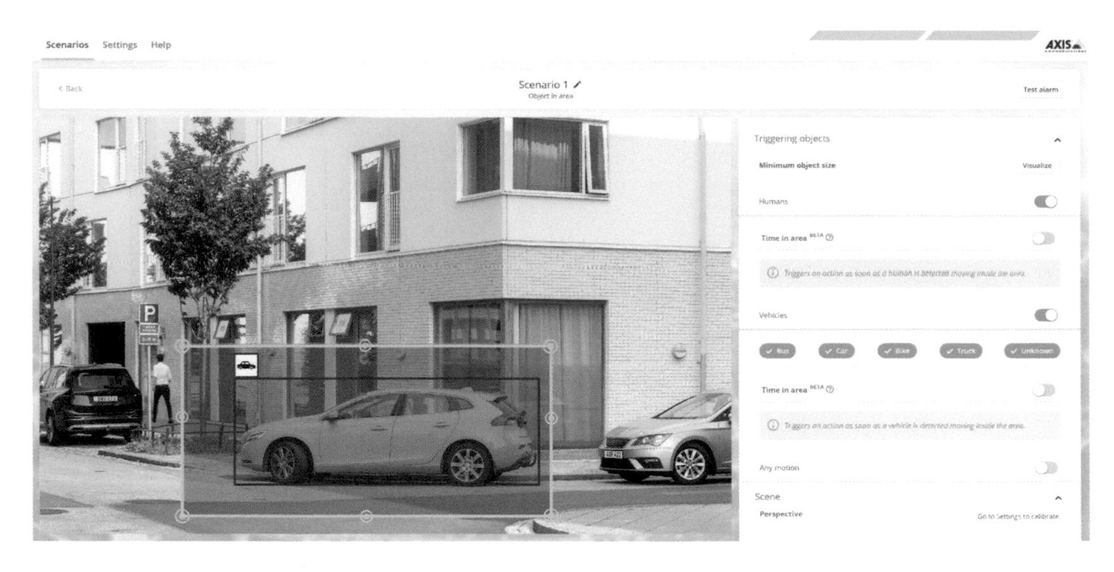

Figure 17.8 An example of a detected object triggering an alarm after staying in a defined area longer than the configured time.

Figure 17.9 Camera tampering quickly detects when a camera is repositioned, loses focus, or is covered.

17.2.3 Camera tampering detection

Imagine a video surveillance installation in a subway with thousands of cameras. One of the cameras has been mistakenly redirected. For weeks, it has been pointing in the wrong direction. Then one day, there is an incident and investigators need to review video from that location. However, the video from the camera in question is useless because no one has noticed that the view has changed.

Moving or covering cameras can be discovered by camera tampering detection. Because it is the obstruction or view change that the application detects, it does not matter to the system whether the interference was intentional or not. For example, the lens of a camera could be deliberately or accidentally covered by paint, powder, moisture, a sticker, or other material (see Figure 17.9).

Cameras could be deliberately or accidentally redirected, perhaps even removed or unfocused. Without camera tampering detection, these threats would go undetected, and the recorded video would be useless.

Camera tampering detection is applicable in any installation but is used predominantly in environments that are potentially exposed to vandalism, such as schools, prisons, or public transportation.

A tampering detection application must be capable of telling the difference between predictable changes in a camera view versus unexpected changes caused by tampering. Otherwise, false alarms would counteract the benefits.

As with many analytics applications, there are different ways of implementing tampering detection. It can be implemented directly in the camera or centrally in software or on a server. By installing tampering detection algorithms in each camera, the system scales more easily than when run through a central server.

17.2.4 Object tracking

In most security operations, it is necessary to keep track of objects and people, inside a facility, or around its perimeter. Analytics applications based on object tracking process images in a specific way. First, they detect a particular object in a camera view and then they track that object as it moves around in the view or from one camera to the next.

Video motion detection used to be pixel-based, but most modern applications are object-based instead. Object tracking is better than pixel-based detection at eliminating false positives in outdoor environments, such as flags, bushes, and other swaying objects, birds and animals, headlights, reflections, and other short-lived objects.

Note that in a deep learning object tracking application, there is a risk of missing objects that do not fall into the specified categories being used, for example, humans and vehicles.

17.2.4.1 Crossline detection

Crossline detection applications, also known as tripwire, perimeter guard, fence guard, digital fence, or virtual fence are used to alert security personnel to possible perimeter breaches. Setting up such alerts usually involves designating a line or area and then configuring the system to generate

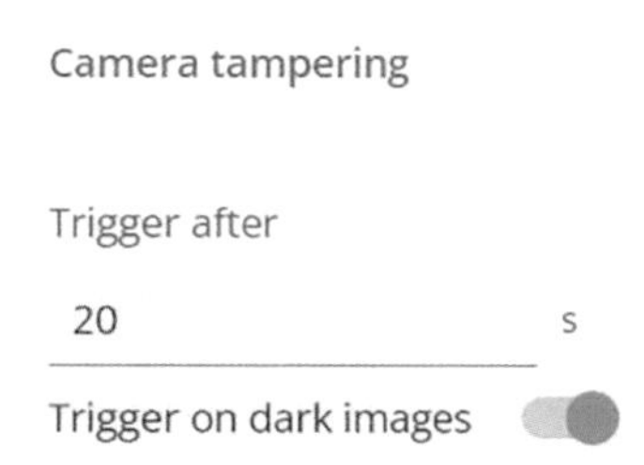

Figure 17.10 Some camera tampering detection applications will learn the scene automatically, making setup very easy. Simply set how long to wait before camera tampering detection should be activated and if the application should trigger on dark images or not.

Figure 17.11 Crossline detection can be used to detect and send an alert if someone crosses the boundary between an open and a gated area.

an alert if objects pass that line in a particular direction or if objects enter or leave a certain area (Figure 17.11).

Another example might be someone trying to enter through an exit, in which you could set up a system that allows staff to *leave* the building but raises an alarm if someone tries to *enter*.

17.2.4.2 Intrusion detection

Intrusion detection is fundamental to the protection of critical infrastructure, such as transportation systems, oil and gas plants, power and water treatment plants, police and military buildings, and hospitals. The protection object is typically a fenced area, and it needs protection against unauthorized persons entering or moving in the forbidden area. Advanced intrusion detectors adapt to the perspective in the video image and use trajectory algorithms. This means that they can detect which direction people are moving in and where they came from, even if they pass behind things like trees and pillars.

It is possible to set up multiple detection zones with different rules for allowing movement, zone-crossing, sending alarms, and recording video. For example, anyone is allowed to move within the warning zone, but no one is allowed to cross from the warning zone to the intrusion zone. Persons may also move within the intrusion zones. Because they originate from the intrusion zone, the system identifies them as "friends", but if they originally came from a warning zone, then the system identifies them as "foes". If a foe tries to enter an intrusion zone, the system sends an alarm to the operator, see Figure 17.12. Typically, the system starts recording whenever it detects movement in any of its detection zones.

17.2.4.3 Object-left-behind

Object-left-behind is often a critical application for the security of common areas. The application targets threats from explosives hidden in bags or packages. It watches an area and keeps track of all objects in it. When an object that was previously moving becomes stationary and stays that way for a certain time, the system alerts the operator (Figure 17.13). Note that applications of this type are usually motion-based (not AI) and cannot be taught what to look for in advance. This means that the scene needs to have a low level of activity, as the application will not know what it is tracking. A lot of traffic in the scene will likely also mean a lot of false alarms.

17.2.4.4 Loitering detection

Loitering functionality tracks the number of people and the time they spend in a certain area. For example, in a parking lot or in front of an ATM, loitering persons might indicate malicious intent. Dwell time software, which is used for business intelligence, analyzes similar patterns (see section 17.3.2.6).

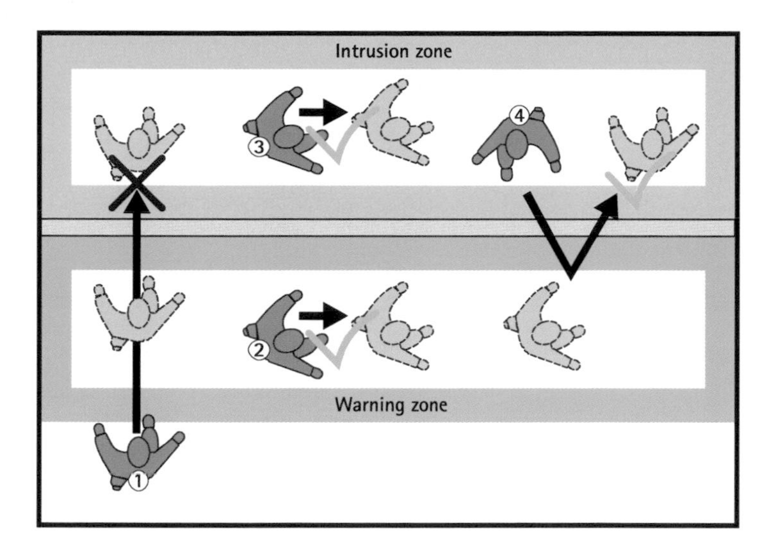

Figure 17.12 Foes may not cross to an intrusion zone (1), but they may move within a warning zone (2). Friends may move within the intrusion zone (3) and also move from an intrusion zone to a warning zone and back again (4).

Figure 17.13 An example of a system for left object detection.

Source: intuVision Inc, USA

Figure 17.14 An example of a system for loitering detection.
Source: AiDANT Intelligent Technology, Canada

17.2.5 Fire and smoke detection

Traditional smoke detectors trigger when there is a certain amount of smoke in a room. Fire and smoke detection as part of analytics applications process video streams and look for visual cues to fire or smoke, that is, combinations of color, light, and movement typical for fire or smoke. This gives the potential advantage of being able to react faster than a smoke detector, as soon as flames are visible in a room, rather than later when the smoke has reached the ceiling. Such an application can be a good complement to traditional types of smoke detectors. It can also send alarms together with relevant video images to an alarm monitoring central.

17.2.6 Hard hat detection

Analytics can also establish and maintain safety on-site, using cameras and smart analytics to confirm the presence (or absence) of required personal protective equipment (PPE). Integrating PPE detection with network speakers allows workplaces to play audio messages to persons not wearing the right gear. Linking this detection to an access control system can even prevent site access without the proper PPE.

17.2.7 Blocked exit detection

The blocking of emergency exits is a common problem, one that often results in heavy fines when fire code regulations are violated. Video analytics for blocked exit detection can alleviate this problem by triggering events when an object is left in an area of interest for a configurable period. Examples of areas of interest are emergency exits, access aisles, control panels, and sprinkler valve areas. When an object is seen to be blocking the area, the application will trigger the camera's event engine, for example, to play a pre-recorded audio message via a network speaker to alert personnel to the problem.

Figure 17.15 Fire and smoke analytics help protect lives and property. They are essential in high-risk environments containing flammable materials.

Source: Araani NV, Kortrijk, Belgium

Figure 17.16 A PPE detection application can help identify and warn persons not wearing the required safety gear.

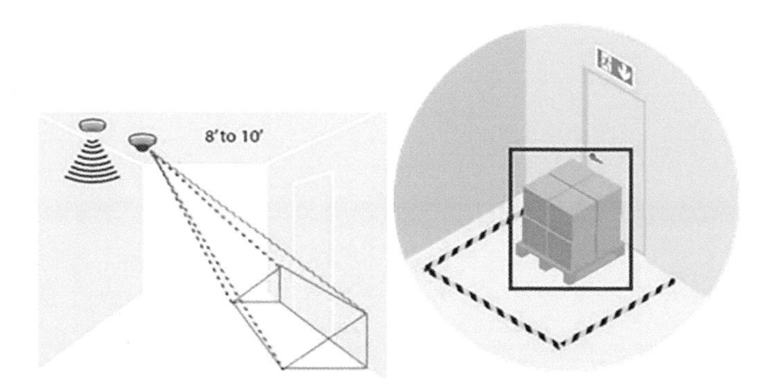

Figure 17.17 An example of an analytic designed to detect objects blocking an exit.

17.3 ANALYTICS FOR OPERATIONAL EFFICIENCY

Video analytics for operational efficiency are becoming a popular way of extracting further value from video surveillance systems, for example, in transportation, in urban planning, and in retail environments. The software produces reports with information about, for example, people's movements in buildings or on streets or if they are displaying unusual or unwanted behavior.

These analytics deliver many types of information, help analyze business strategies, and make studying customer behavior easier.

Analytics for operational efficiency include the following:

- *Object counting*
- *People counting*
- *Dwell time and heat mapping*
- *Traffic management*

17.3.1 Object classification

Analytics software generally perform the following (Figure 17.18):

- *Analysis:* Analysis of the pixels in the video frame. Traditional algorithms compare pixels to a reference frame to figure out which objects are moving. Deep learning algorithms always analyze all the pixels irrespective of movement. Note that it is also possible to, for example, ignore stationary objects if only interested in moving objects.
- *Detection and classification:* Identifies the objects of interest and assigns them a classification, a timestamp, and a position in the image as defined by a bounding box. Many other descriptive attributes can also be applied, such as color, size, and direction.

The ability of video analytics-based systems to recognize object types and isolate the object of interest greatly enhances their accuracy and usefulness. Does the application count all objects in a scene, or can it single out and only count particular objects, such as people or vehicles? (See Figure 17.19.) Can the program tell apart two people holding hands, or would they be interpreted as one object?

A specific challenge for some analytics applications is that objects can appear in other configurations than expected. For example, the system may be able to distinguish a human being from a

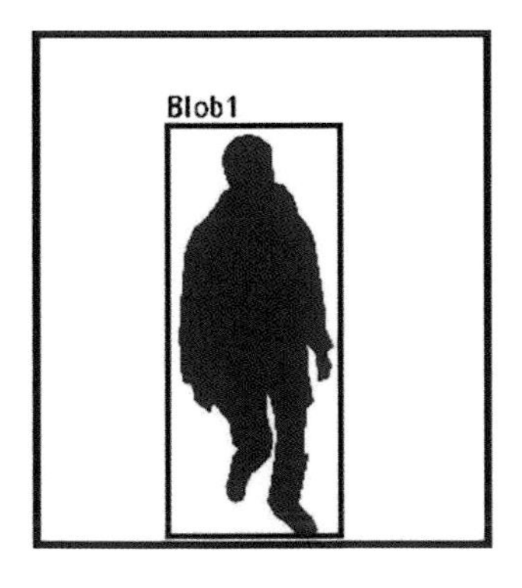

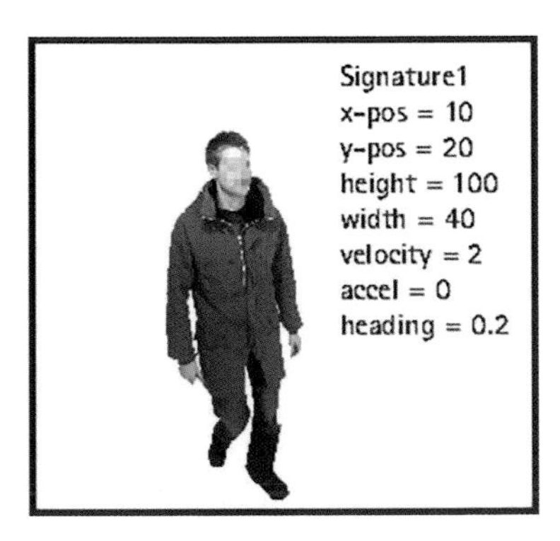

Figure 17.18 Object-based analytics software detects a moving object (left), classifies it (center), and generates metadata that describes it (right).

Figure 17.19 Video analytics need to distinguish between object types so that they only call attention to objects that matter to the user.

dog. This is based partially on the knowledge that human beings have a particular aspect ratio, that is, they are substantially taller than they are wide. However, if a person is crawling along the ground, their proportions differ from the norm. To be reliable, a video intelligence system needs to compensate for these types of deviations and be able to recognize a human being regardless of whether the person is crawling, crouching, standing, or running. This ability is often called aspect ratio compensation. However, this compensation does not apply to deep learning detection algorithms, as the ability to handle different poses is based solely on the training data used during development.

The classifications and characteristics of the objects in an image are called metadata. Once metadata is collected, it can be compared to a set of criteria for action, such as a person walking the wrong way, a bag left behind, or a car entering a restricted area. If the criteria are fulfilled, the system can raise an alert in real-time. For more information about metadata, see section 16.5.

17.3.2 Object and people counting

After detecting and classifying an object, the analytics system can use the information for different purposes. One purpose is to count the number of objects that behave in a certain way, such as a person walking in the wrong direction. Another application can be to determine the total number of objects of a particular type in an area.

When using people-counting software (see Figure 17.20), the camera is placed above the area to monitor, at an angle of 90° in relation to the ceiling. A person passing under the camera usually needs to be larger than the software's specified percentage of the camera's total horizontal field of view, and the camera must produce images with sufficient quality to distinguish one person from another. Although other technologies (such as infrared) can also count people, video-based people counting can, in many cases, provide better accuracy.

17.3.2.1 People counting in 3D

Some locations for people counting are more demanding than others, for example, where the lighting gives a person a large shadow, which could be misinterpreted as being another person. This situation can be alleviated by using a 3D people counter, which is often supplied as a complete product, that is,

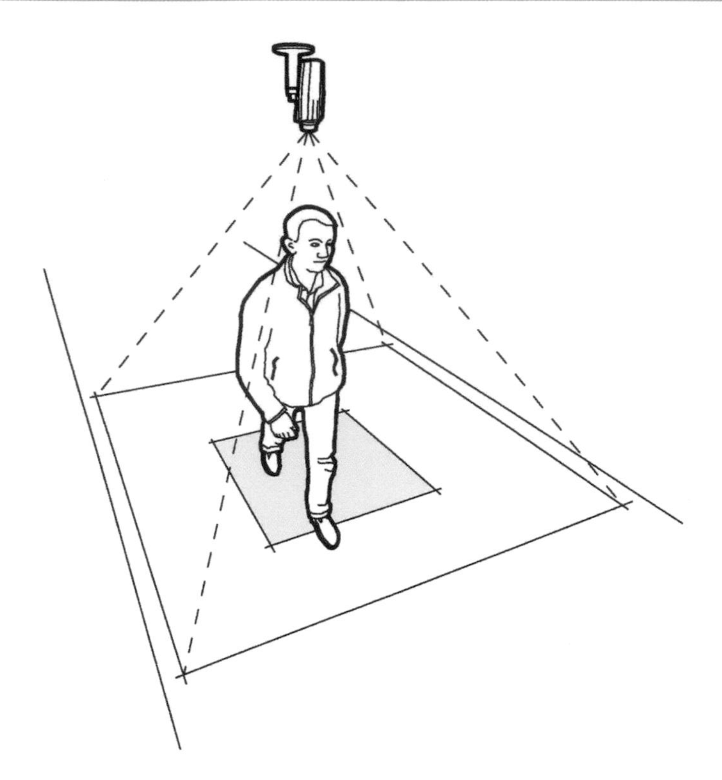

Figure 17.20 The placement of the camera is important to provide the appropriate accuracy.

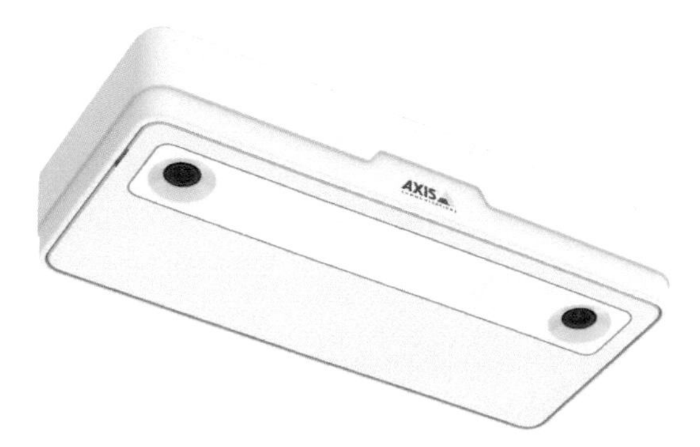

Figure 17.21 An example of a 3D people-counting camera.

a stereoscopic camera with the counting application pre-installed and configured (see Figure 17.21). A 3D counter calculates the depth within the field of view, meaning that the height of the person can also be calculated. It is this depth that also recognizes shadows for what they are.

People counting is useful in retail stores or other environments where it is important to know the number of people entering or exiting an area. This data can be used to understand customer behavior, to better plan product placement, promotions, and advertisements, and ultimately, to increase the return on investment.

The ability of video intelligence systems to accurately count the number of people is at the core of a range of applications, including the following:

- *Customer traffic* monitoring
- *Queue management*
- *Tailgating*

17.3.2.2 Customer traffic monitoring

A retail store can use people counting to track the number of people who enter and exit the store, who walk through certain aisles, or who stop by a particular merchandising display (Figure 17.22). By comparing the traffic count with point-of-sale data, store managers can calculate conversion rates. A franchise or chain can use the data to compare performance between different stores.

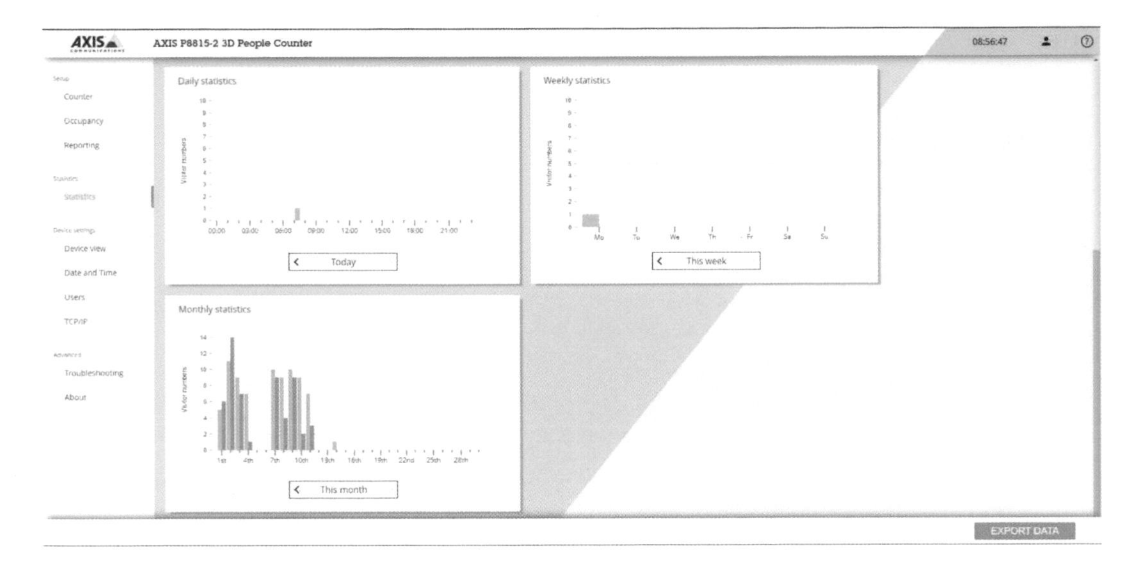

Figure 17.22 Example of people-counting software for statistics over customer traffic.

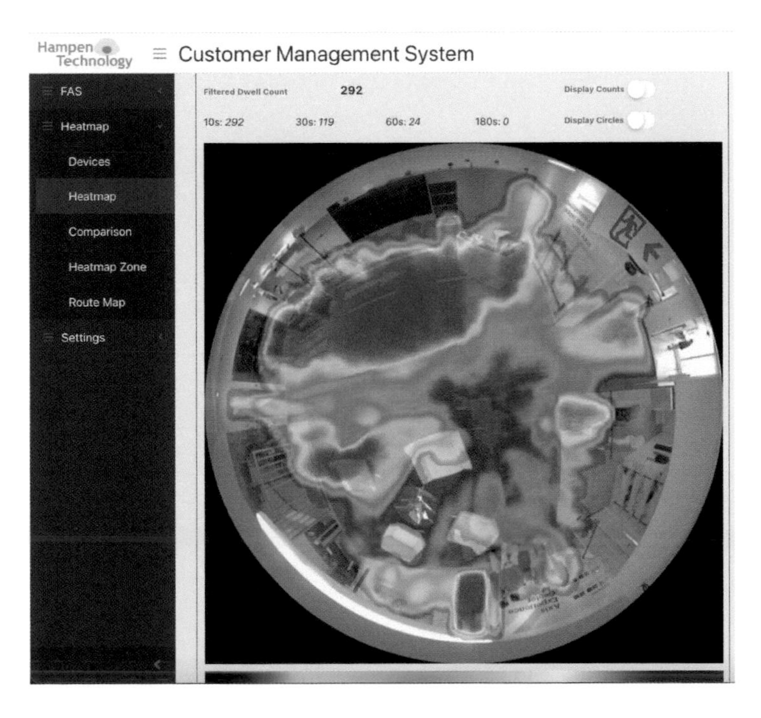

Figure 17.23 Example of customer traffic monitoring. The application shows which routes visitors take after entering a store.

Source: Hampen Technology

17.3.2.3 Queue management

Also known as staff optimization, queue management software counts the number of people standing in various queues, for example, retail checkout counters, customer service desks, airport check-in counters, passport, and security controls (Figure 17.24). This information is often used to improve customer service. For example, when a queue passes a specific threshold, the software can prompt the staff to open more checkouts. The software also keeps statistics over time so that managers can improve their understanding of customer traffic across different seasons, get indicators of how good the staff response is, and make smarter decisions when planning work shifts.

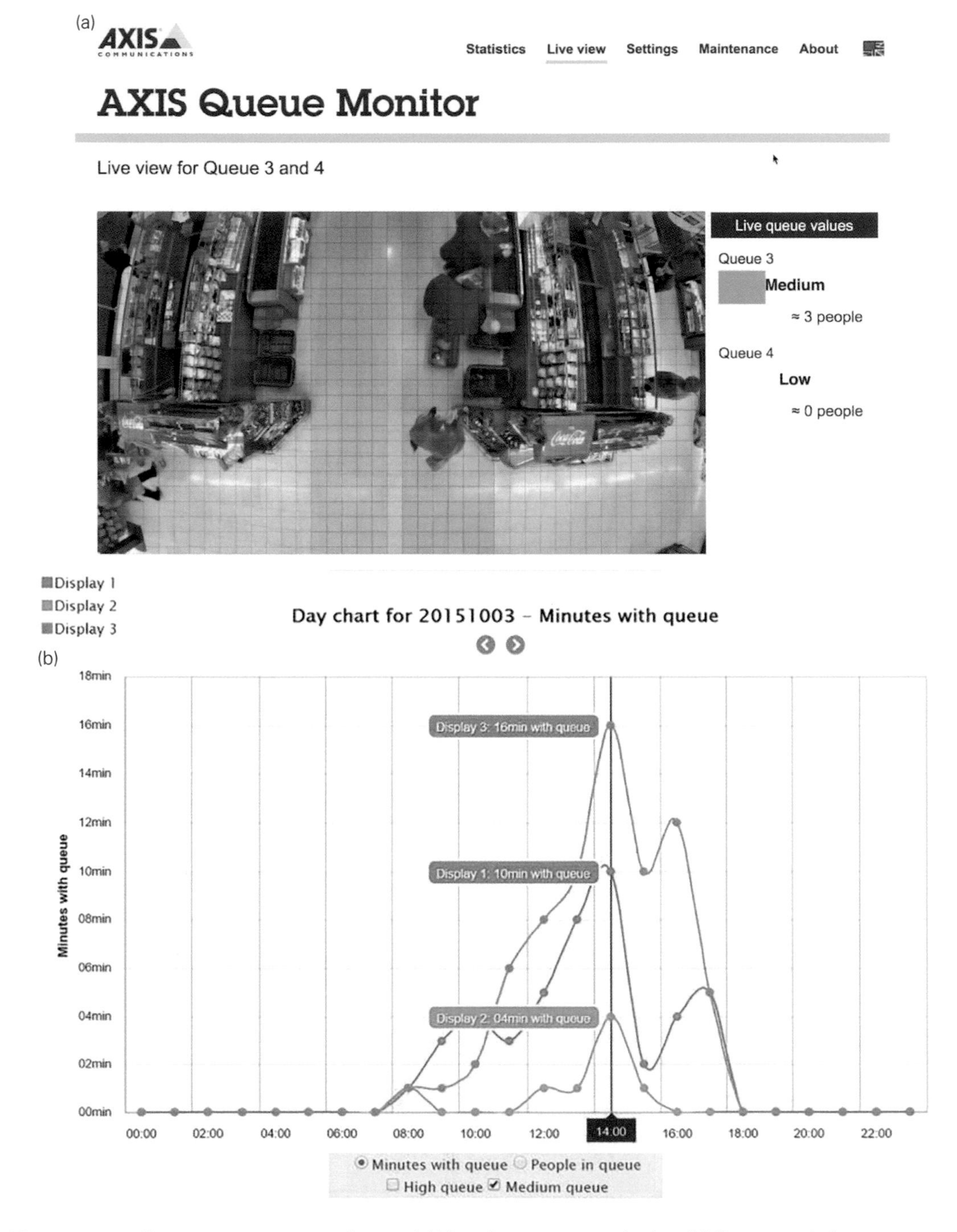

Figure 17.24 Queue management software. (a) Live view queue monitoring. (b) Queue statistics.

17.3.2.4 Tailgating

People counting can be used with access control systems to make them more secure. The main principle of access control is that only people who have authorized access to a building should be able to enter its doors. People-counting software can send an alert when several people enter a facility, but only one person has swiped their badge.

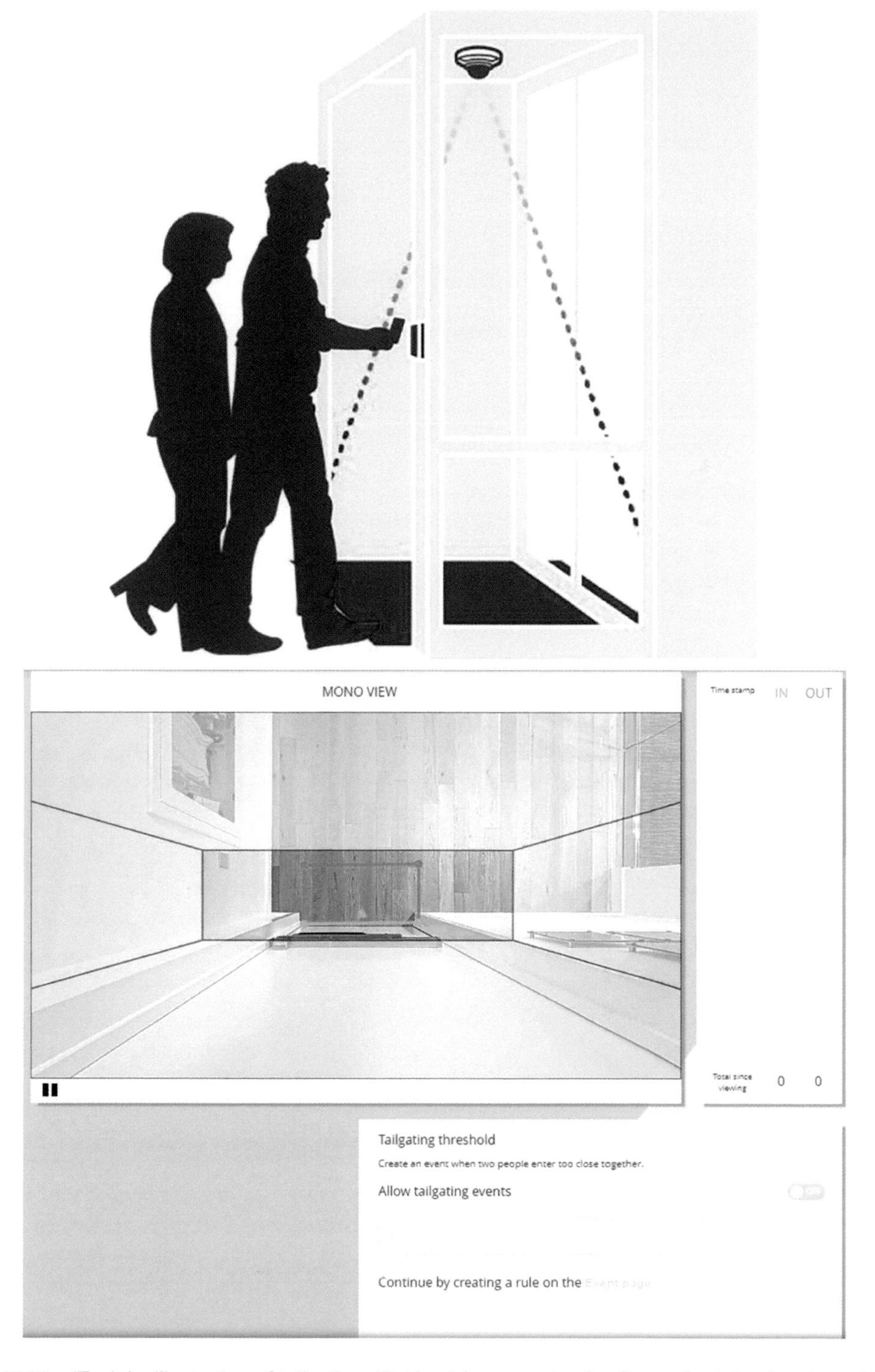

Figure 17.25 (Top) An illustration of tailgating. (Bottom) An example of software for tailgating detection.

Figure 17.26 An example of using dwell time software to create a heat map, which supports merchandising and advertising decisions.

Source: BriefCam

17.3.2.5 Dwell time and heat mapping

Dwell time software keeps track of how long people stay in an area and shows the typical traffic paths. However, rather than preventing unwanted behavior as in loitering detection, dwell time focuses instead on how customers connect with products and promotions, how many persons pass a specific display or area, their level of engagement, and how many of them actually purchase the exhibited product.

Heat maps, also known as shopper activity maps, show customer traffic patterns in a very visual way (see Figure 17.26). They put colorful overlays over the video stream, showing which zones are hot and which are cold. Their color palettes look very similar to those of thermal imaging, where blue is cold and red is hot. But of course, the analyzing algorithm is not based on infrared radiation but instead on object classification and people counting.

Dwell times and heat maps provide data that help when optimizing the layout of a retail store. For example, the software can give visual proof if a product display is not attracting the expected amount of interest. The manager can then compare it to areas that have the opposite effect and change the displays accordingly. Just as you want to steer the traffic to specific displays, you want to avoid bottlenecks that frustrate customers and deter sales. It's all about creating the best flow and getting the most out of the space, customers, and staff.

17.3.2.6 Other object counting applications

Some analytics are intended for counting applications that make it possible to keep track of movements and/or the current occupancy of a defined area. Examples include crossline counting applications and area-counting applications. See Figure 17.27. If the application is to provide useful results, it must be able to accurately follow an object in the scene, that is, it must not confuse it with other objects or lose track of it. The ID applied to the object must be retained for the whole time the object is within the application's view.

17.3.3 Traffic management

Traffic management software is used to monitor and improve the overall traffic flow of urban areas. It efficiently replaces induction loops and other types of sensors that detect and count vehicles and

Figure 17.27 Counting the number of vehicles in a parking lot. The total occupancy is updated live, as vehicles come and go.

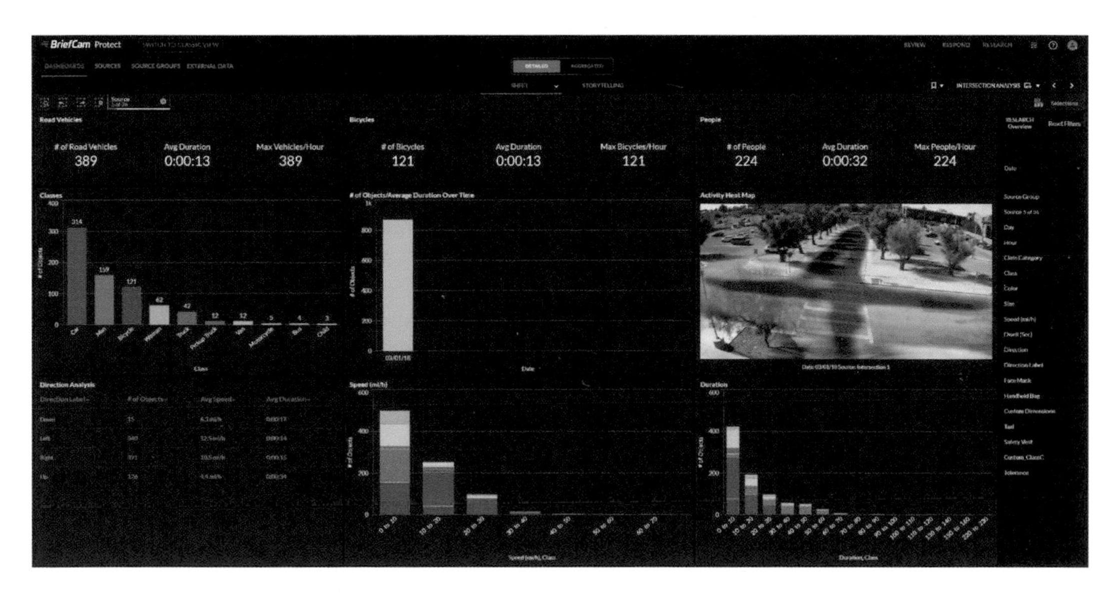

Figure 17.28 An example of a traffic analytics dashboard, illustrating traffic flow, vehicle types, speeds, etc.

Source: BriefCam

pedestrians. Beyond the obvious benefit of not having to dig up the road to improve the system, traffic management software gives both statistics and video that show the status of traffic in real time. How many cars are passing through each lane? What are the average speeds? In which direction are they moving? See Figure 17.28 for an example. Traffic management analytics can make a distinction between cars, heavy trucks, and motorbikes, and some can also detect pedestrians and smaller vehicles such as bicycles.

17.3.3.1 Incident detection

With automatic incident detection, emergency agencies and traffic controllers can get real-time alerts whenever traffic is hindered in any way. For example, applications can detect when vehicles

Figure 17.29 (a) Detection of traffic going the wrong way. (b) Detection of pedestrian in forbidden area.
Source: Citilog, Arcueil, France

drive too slowly, stop for too long, or stop in a forbidden area, when the road is congested, when a car is moving erratically or in the wrong direction, and when there are pedestrians or debris on the road. Traffic controllers get visual and audible alerts so they can quickly take adequate measures, such as alerting ambulances and warning other road users to slow down. This minimizes the risk of secondary accidents and helps prevent injuries and deaths, as well as infrastructure damage. Figure 17.29 shows examples of incident detection.

17.4 HYBRID ANALYTICS

Hybrid analytics applications have multiple uses. Security officers, investigators, and business managers can use them both for real-time event monitoring and for reporting purposes. For example, border control officers can use license plate recognition to detect wanted vehicles trying to cross a border, and police can use it to investigate which vehicles were involved in a crime. Facial recognition can be used to open doors for people with access to a building or to identify the perpetrator of a street robbery.

17.4.1 PTZ autotracking

PTZ autotracking is intelligence that automatically controls a PTZ camera to keep, for example, a person in sight. The advantage is that the camera can zoom in automatically to give a better view of the person. If only one PTZ camera covers the area, the disadvantage is that it might be pointing in the wrong direction at the time of a particular event. A hybrid approach would remedy that issue. Fixed cameras would find the events, and a PTZ camera would do the tracking. Another issue with PTZ autotracking is that only one object can be tracked at a time, and some PTZ tracking systems get confused if more than one object is moving in the view.

There are two different types of automatic tracking:

- *Automatic selection–automatic tracking:* The camera locks onto the first moving object until it loses that object. The camera will then find another moving object. This solution is useful in low-activity environments, such as parking lots and hallways. It provides a view of the object without the need for on-site security staff.
- *Manual selection–automatic tracking:* The surveillance operator selects an object to track, and the camera follows it. This setup helps the operator focus on the object instead of being distracted by operating the camera.

17.4.1.1 Gatekeeper

A PTZ camera with gatekeeper functionality can react to movement and automatically start panning, tilting, and zooming. Typically, the gatekeeper is used to monitor a gate or a specific area. When objects or people move in the gated area, the gatekeeper can follow them as they move away, zoom in on them, or move the camera lens to a preset position (see Figure 17.30). The gatekeeper can also trigger other actions such as reading a license plate (see section 17.4.2) or recording video, from the same camera, in other cameras, or in the VMS.

17.4.2 License plate recognition

License plate recognition (LPR), sometimes referred to as automatic license plate recognition (ALPR), has a variety of uses, ranging from access control to watching out for particular vehicles. For example, in an access control application, only vehicles with registered number plates are

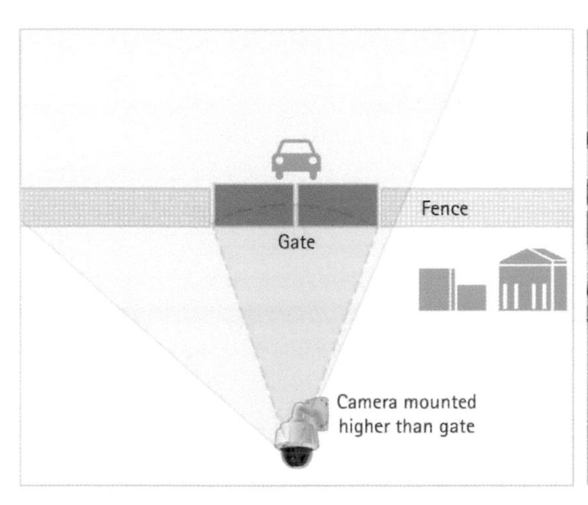

Figure 17.30 An illustration and a screenshot of the gatekeeper functionality. When an object enters the gated area, the gatekeeper zooms in and follows it.

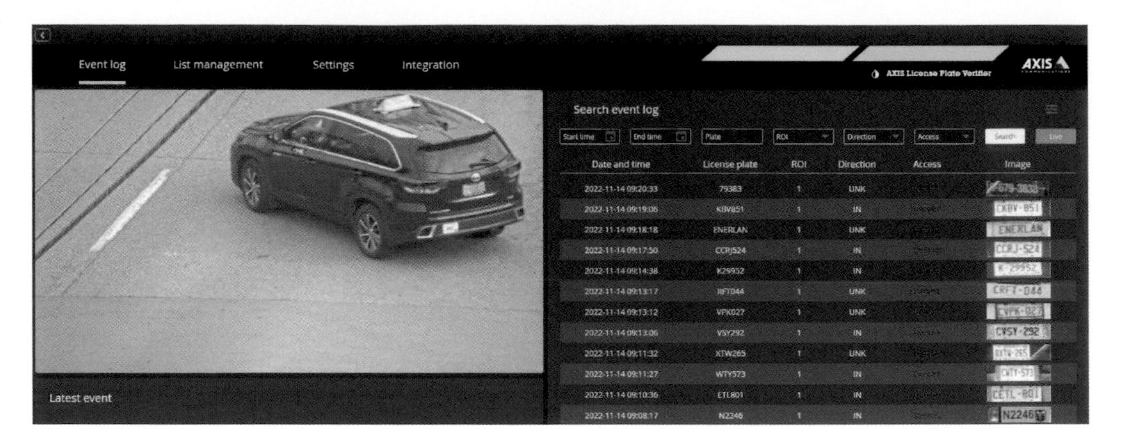

Figure 17.31 An example of an LPR application.

allowed access to a facility. In a criminal investigation, LPR can automatically look for certain sets of license plate characters to find suspect vehicles.

In a parking lot or parking building, LPR can automatically track the vehicles that enter, eliminating the need for a more expensive parking ticket infrastructure. LPR can also automatically monitor how long a particular vehicle stays parked. For example, a store might want to reserve its parking lot for its customers or to get an alert when a vehicle seems to have been abandoned.

In retail environments, LPR can identify cars whose drivers often shop in certain stores. This information can be used to, for example, analyze where these shoppers come from or to design direct marketing programs that reach only consumers in the correct geographical area.

An analytics application that performs LPR does so in several steps. In the car identification example, the steps include the following:

1. *Find the car:* The first step consists of finding the car in the image. For moving cars, video motion detection is a key element in making this work. For parked cars, it is a matter of recognizing the outline of a car in an image.
2. *Isolate the license plate:* Once the car is found, the next step is to isolate the license plate from the rest of the image. This is based on algorithms that define what a license plate looks like and where it might be mounted on a vehicle.
3. *Extract the characters:* The next step consists of extracting the letters and the numbers from the license plate using image analysis.
4. *Identify the characters:* Optical character recognition (OCR) transforms the characters from a collection of pixels into a stream of letters and numbers.
5. *Process the characters:* The final step consists of processing the resulting string of letters and numbers by storing it in a database or comparing it with existing entries.

Some challenges are especially difficult for LPR to deal with. Bad weather, blinding headlights, and dirty or damaged license plates can affect results. Sometimes, materials used in the manufacture of the plate can cause problems. For example, a glossy protective surface may cause reflections, making the capture difficult. And as license plate designs vary from region to region, the analytics application should be adapted to local conditions and fine-tuned to the specific implementation.

License plates on rapidly moving vehicles can be difficult to capture due to motion blur. In such cases, the camera needs to be configured correctly so that useable images can be captured in all possible lighting and weather conditions.

Figure 17.32 Specialty cameras are often used to ensure that a clear capture of the number plate can be provided at all times of the day.

To provide the best images possible for the LPR application, specialty cameras are often used (Figure 17.32). If your application requires license plates to be captured from vehicles moving at speed, the camera should have a global shutter. This technique exposes all the pixels on the image sensor at exactly the same instant, instead of row by row as in a rolling shutter camera. The advantage of the global shutter is that it reduces artifacts in the image and gives distortion-free straight lines.

From an implementation design point of view, LPR is a type of video analytics where the benefits of a distributed approach are very clear. When deploying an LPR application in a network camera, you can limit the transmission of data to the letters and numbers of a license plate – with perhaps just a snapshot of the vehicle – which drastically reduces the network load compared with a centralized implementation of an LPR system.

17.4.3 Speed monitoring

Also useful for monitoring traffic are speed-monitoring applications, which combine video and radar data in order to visualize measured vehicle speeds directly in the camera feed. Color-coded overlays can quickly identify speeding vehicles. A speed-monitoring analytic can also send the speed to public view monitors to display real-time messages asking drivers to slow down. The speed data can also be used to trigger other events, such as activating strobe lights, triggering alarms, and starting camera recordings.

Data can be exported for creating comprehensive graphical overviews. For example, about the number of vehicles per day and their speeds to assist traffic planners when deploying speed bumps or similar measures.

17.4.4 Facial recognition

Facial recognition has gotten a lot of attention, and its uses are many and varied. Police want to get alerts when certain individuals are seen in public or sensitive areas. Allowing only certain people to enter specific areas can enhance access control. For forensic purposes, the need can arise when searching for individuals in recorded videos. In casinos, managers may want to discover unwanted individuals. Customs can use automatic searches to find certain individuals and improve the precision of border and passport control.

The facial recognition process is similar to the LPR system described earlier. However, one difference is that it is possible to define beforehand what the system should look for in a number plate – that is, a string of letters and digits grouped in a certain order. To know what to look for, facial recognition depends on the initial step of building a database of known faces. Examples of such databases are passport photo databases and police registers. Sometimes, getting that data becomes the biggest challenge in a facial recognition system.

Figure 17.33 A speed-monitoring application can show live vehicle speeds directly in the camera feed.

Beyond that, the facial recognition steps include the following:

1. *Find the face:* Also called face detection, this means isolating the face from the rest of the body. In some applications, face finding is sufficient. For example, in an airport, it can be used to measure the queue time from entering and exiting a check-in point. In this case, the actual identity of the individual is of no interest. The system should merely be able to separate one individual from another.
2. *Face key point detection:* This involves identifying key points of a face such as the eyes, nose, mouth, etc. Some systems use these key points to first rectify the face image and then use a deep learning–based algorithm to process the image and generate a feature vector that encodes the specific aspects of a face, such as skin tone, facial features, and other unique aspects. In a monitoring situation, it may be enough for a guard to be presented with individual faces for matching against registered faces from access control badges.
3. *Match faces:* The final step involves matching the extracted face with signatures from a database to recognize individuals and make positive identifications.

As for LPR, the challenges for facial recognition systems are substantial. Even if lighting conditions are perfect, people generally move around and sometimes block each other, and faces change and age over time. Appearances can also easily be changed by a pair of glasses, a change of hair color or length, the growing of a beard, or even through surgery. Another particular challenge is the fact that people rarely look straight into a camera unless they are prompted to do so. To deal with this problem, developers have built three-dimensional (3D) facial recognition software, which extracts three-dimensional information from video streams and matches it with a database.

17.5 SOUND IDENTIFICATION

Many cameras have built-in audio detectors that can detect if a sound reaches a particular level. Some advanced algorithms can imitate the human ear and make a distinction between different sounds, although these can be difficult to train to an acceptable level. Sound analytics are better

Figure 17.34 An example of face-matching software.

Source: FaceFirst Inc.

Figure 17.35 Privacy can be maintained by blurring out personal features (left) or by representing movement as a colored area (right).

than video analytics at providing early warnings of aggressive situations, gunfire, breaking glass, or car alarms. When the application detects a sound, based on the defined parameters, it can trigger a camera to start recording and send an alarm to a security officer. For more on analytics in audio, see Chapter 7.

17.6 VIDEO ANALYTICS AND PRIVACY

Analytics applications, especially those that focus on identifying human features, are sometimes seen as being an invasion of privacy. However, those concerns can be overcome by not storing pictures or videos of faces. Once data is collected from, for example, a people-counting application, there is no reason to retain the pictures or video of individuals.

In some ways, video analytics can actually enhance privacy. For example, some applications can find and mask out all the people in a surveillance recording (see Figure 17.35). These applications

Figure 17.36 Blurring out vehicle details helps protect privacy.

support live, full frame rate motion-based masking, where people and objects moving against a set background are masked.

Analytics for privacy can be motion- or AI-based. Motion-based analytics are fairly trigger-happy, but moving objects that become stationary, for example, a person sitting down, will eventually merge into the background. When using AI-based masking, this object will never be merged to the background (always assuming the object was correctly detected and classified), which means that, in some situations, AI-based masking is more secure.

Some applications include functions for anonymizing vehicles by blurring out cars and their license plates. See Figure 17.36.

17.7 REALISTIC EXPECTATIONS ON VIDEO ANALYTICS

Although the applications described here can indeed perform well in a controlled environment, many will struggle to be robust and efficient enough to function optimally in real-life situations. Some systems will only work after tremendous tailoring efforts to meet every possible challenge in every situation.

Generally, the analytic applications that enhance traditional camera functionality – such as video motion detection and camera tampering – have matured well and can now be used in most situations without major tuning or modifications. Specialized applications such as people counting, LPR, and crossline detection are also used more widely, although typically, these require a fair amount of calibration for each installation.

Caught up in the enthusiasm of the potentials of analytics, the surveillance market initially had extremely high expectations on vendors and software suppliers to fulfill that potential. In retrospect, these expectations were too high compared to what is actually possible or cost-effective. With the advent of deep learning processing at the edge, many more applications can now be executed reliably and in a scalable way.

17.8 BEST PRACTICES

On a practical level, the key to success is to have a clearly defined use case. That is, the conditions that should (or should not) trigger an alarm should be known and be possible to describe using simple rules.

For example, tripwires generally perform better than, for example, motion detection. This is because they are placed in controlled environments, typically along a fence, where there are very few alternative explanations as to why an object would cross the fence without posing a threat.

Object-left-behind, on the other hand, is a much more challenging algorithm. This is because an object might not be left behind at all, and most objects left behind pose no real threat. For example, garbage thrown onto train tracks is not a bomb but would still be seen as a threat. To pinpoint the right analytics system, the right camera setup, and the right parameters, you must also ask yourself the right questions, for example, what it is you are trying to detect. You need to arrive at a good balance between false alarms and missed alarms.

Ideally, analytics should be able to differentiate between true-threat events and posing-as-threat events. For example, in a retail scenario, it would be ideal if the same application could not only detect that a customer put something in their bag but also that they went through the self-checkout without scanning and paying. It is not until then that they can be accused of shoplifting.

On a technical level, four important factors must be present for a video analytics application to work accurately:

- *Sufficiently high video image quality and light sensitivity*
- *Efficient video analytics algorithms*
- *Sufficient computer processing power, preferably using cameras with DLPUs*
- *Configuration and fine-tuning*

17.8.1 Video image quality

Getting the right video from the cameras involves the following considerations:

- *Lighting and the ability to see in darkness:* Any type of video analytics application is limited to what the camera can see. Is there sufficient light, and does it come from the right direction? What happens if the camera is blinded by headlights? Can the camera counteract bad lighting with WDR or low-light capabilities? If detection is needed in complete darkness, a thermal camera is likely the best option.
- *Frame rates and resolutions:* Contrary to popular belief, most analytics do not need high frame rates and high resolutions. In fact, five to ten frames per second at 1080p resolution is sufficient for many applications. Some specialized applications, such as LPR and facial recognition, may work with even lower frame rates, although not if the objects are moving rapidly. Both these types work better at higher resolutions.
- *Camera position:* Positioning cameras is critical to getting accurate results from analytics. For example, most people-counting applications work best with cameras positioned overhead at 90° because this allows the algorithms to efficiently separate individuals. Applications such as LPR and facial recognition work best with frontal views, that is, with the camera looking straight at the license plate or face.
- *Type of video:* Because compression always results in data loss, the best results are achieved when processing uncompressed video, as is often the case when running video analytics on the edge, inside the network camera. Running analytics based on data from thermal cameras is perfect when you need to tell living objects from dead objects, detect activity in complete darkness, or when you need to cover distances greater than those covered by conventional cameras. Note that there may be challenges for thermal imaging in areas where the ambient temperature is close to body temperature.

17.8.2 Efficient video algorithms

Video analytics applications are built on complex mathematical algorithms that process video and still images, each consisting of a myriad of details. The quality of an application depends on how accurately the algorithm performs these calculations and how robustly it deals with variations in the input (that is, the video stream). The only sure way of assessing the quality of an analytics algorithm is to field-test the application under realistic conditions to see how fast it is and how many correct responses and false alarms it generates.

In general, 90% accuracy is achievable for a modern analytics system. Reaching 95%, however, is very complicated, and 99% or beyond is extremely difficult in a real-world situation. From a user's point of view, the demand for accuracy depends on the following: How critical is the analytics application for the safety and security of people and property? How many errors are acceptable? How many false alarms can the system be allowed to generate before it is unusable? How many true positives (situations that should generate alarms) can the system be allowed to miss? Can the video analytics be combined with radar, PIR sensors, or audio analytics to increase accuracy? The cost of a video analytics system must also be weighed against other alternatives, such as employing more security staff.

17.8.3 Computer processing power

Some analytics applications are optimized to run on small embedded systems and perform well in a distributed system, whereas others require a powerful centralized server to be reliable. Because analytics applications are mathematically complex and, therefore, computer-power intensive, which is particularly true for modern deep learning algorithms, performance depends on the processor and available memory. Whatever the case, the more processing power available to an analytics application, the better and faster it will be, and deep learning processors will perform better than regular processors.

17.8.4 Configuring and fine-tuning the system

Video analytics and algorithms are designed to handle a large variety of situations. Each installation requires configuration and fine-tuning to match its own scenario. No system is perfect, and configuration is a balance between not missing critical situations and reducing false triggers. Optimizing a system can take anywhere from a day up to several weeks.

As a rule of thumb:

- The system will never be 100% accurate.
- The more parameters that can be adjusted in an application, the longer it takes to optimize.
- The configuration should be monitored and adjusted over at least one 24-hour period, as changes in lighting will impact results.
- Using thermal cameras can increase accuracy dramatically in some applications, such as crossline detection.
- Combining analytics with other systems such as access control, audio analytics, or radar can increase accuracy.

Chapter 18

System design considerations

One of the main benefits of a network video system is flexibility and scalability: the freedom to mix and match the most appropriate components from different vendors and the power to optimize or expand the system to any size. As with anything else, having freedom and power also requires knowledge. To build a truly flexible and scalable system, you need to know what the different components are, how they work, and how they interact. It is essential that you can select the right camera, install and protect it properly, configure it to match the scene complexities, and get it to stream live video or to record at the right time, in the right format, and with the right quality. At the same time, the appropriate network and storage solutions depend greatly on the selected cameras and their settings (such as resolution, compression, and frame rate) and the number of cameras. The budget might mean compromises have to be made. For example, complex scenes demand more bandwidth, so you may have to choose between spending more on storage or else reducing the frame rate, lowering the resolution, or increasing the compression.

This chapter discusses the most important aspects of designing and installing a network video system: how to select, install, and protect a camera and how to calculate the storage and network bandwidth. Many design tools are available, some of which are based on formats such as Autodesk® Revit® and Bluebeam Revu® which will also be discussed. There are also legal aspects to consider, some of which are mentioned later in this chapter.

18.1 SURVEILLANCE OBJECTIVE

To best position a camera, you need to know what kind of image you need. For example, to track people or objects moving to and from many positions in several directions, you probably want an overview image, as it gives the best chance of spotting such events. After finding and purchasing a suitable overview camera, you must install it in a position that achieves the purpose.

To identify a person or object, the camera must be positioned or focused so that it captures the level of detail needed for identification. As seen in Figure 18.1, the greater the angle to the object, the more difficult it becomes to recognize facial features. An angle of 10–15° gives the best view for facial identification. Placing a camera higher up puts it out of reach for vandals, but again, the angle makes identification of faces or details, such as license plates, more difficult.

There are tools that can help find the best position for a camera. Local law enforcement authorities may provide video surveillance guidelines.

Some vendors also provide design tools and plug-ins for various diagram and 3D software. These tools can help with placement of cameras, calculating view angles and coverage, and finding blind spots and items that block the view (see section 18.7).

DOI: 10.4324/9781003412205-18

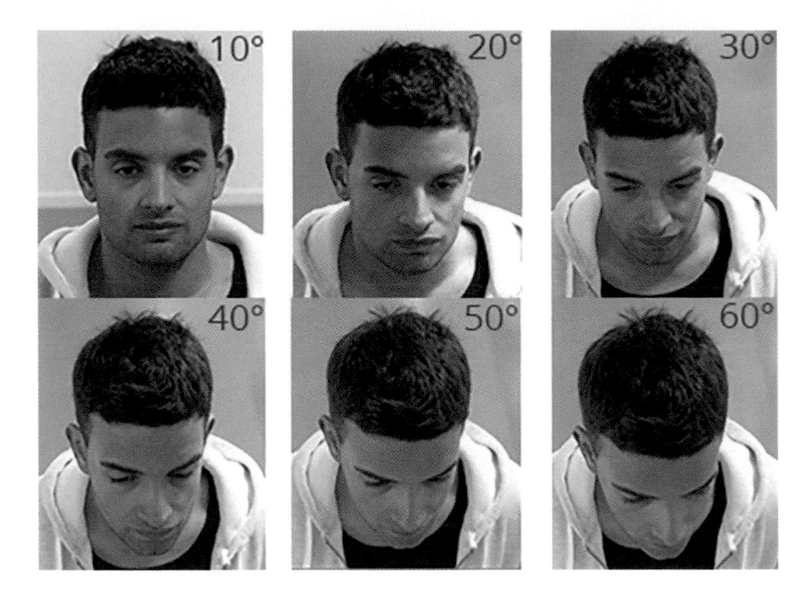

Figure 18.1 The greater the angle to the object, the more difficult it becomes to recognize facial features.

18.2 SELECTING A NETWORK CAMERA

In markets such as the video surveillance and physical security markets, new vendors are always appearing, bringing new products that potentially also have new capabilities. There are many different network camera vendors in the marketplace today. Because network cameras include much more functionality than their analog counterparts, choosing the right camera becomes not only more important but also more difficult. When choosing the technology to use, you need to consider that many physical security systems remain in operation for five to ten years, during which time they need to be maintained and serviced. This section outlines what to keep in mind when selecting a network camera. This includes the type of camera, image quality, resolution, compression, networking, and other functionalities, as well as the vendor.

18.2.1 Type of camera

To determine which types of network cameras are suitable and how many cameras are needed to cover an area, the scene, environment, and purpose must first be determined. Considerations include the following:

- *Purpose of the installation:* Determine the field of view or the kind of image that will be captured: an overview (viewing a scene in general or looking at the general movements of people) or high detail for the identification of persons or objects (for example, face or license plate recognition, point-of-sales monitoring).
- *Overt or covert surveillance:* This will help in selecting cameras that offer an openly visible or discreet installation.
- *Area of coverage:* For a given location, you should determine the areas of interest, how much of these areas should be covered, and whether the areas are located relatively close to each other or far apart. For example, if there are only a few small areas of interest close to each other, a single high-resolution camera with a wide-angle lens or a panoramic camera could be used, instead of multiple cameras with lower resolution.
- *Complete darkness and perimeter protection:* One way to manage this situation is to use outdoor cameras with built-in IR illumination. This will give you quality images even in total darkness. For detecting movement in darkness and other difficult conditions, you could also choose a thermal camera, which generally can detect movement at greater distances than optical cameras.

(a)

(b)

Figure 18.2 Two similar cameras recording the same scene can give images with very different quality. (a) This image has a lot of noise but provides enough detail to make positive identification. (b) This is more pleasant to look at, but the details have been lost due to motion blur.

- *Analytics:* Camera type and placement affect the successful use of most analytics. Also, consider whether the system will work best with edge-based or server-based analytics. For more information on analytics, see Chapters 16 and 17.
- *Tamper- or vandal-proof and other special housing requirements:* Proper protection against water, dust, temperature, and vandalism is essential. For more information, see section 18.4.
- *Indoor or outdoor camera:* If placing the camera outdoors, install it in an appropriate protective housing or use an outdoor-ready camera.
- *PTZ or fixed camera:* PTZ cameras with a large optical zoom factor can give high-detail images and survey a large area. Keep in mind that to make full use of the capabilities of a PTZ camera, an operator needs to control the movements, or alternatively, an automatic guard tour must be set up. For surveillance recordings without live monitoring, fixed network cameras are normally more cost-effective.
- *Light sensitivity and lighting requirements:* Consider adding external white lights or specialized lighting such as IR (infrared) lamps. Day-and-night functionality means you can get images in conditions that would otherwise be too dark. The light sensitivity levels of a camera are important and should be evaluated. Do not go by the measurements on a datasheet, as vendors measure this in different ways. For more information on light sensitivity measurements, see section 4.1.2.

18.2.2 Image quality

Your main concern when assessing image quality should be selecting a camera that meets your surveillance objective. Although image quality is one of the most important aspects of any camera, it is difficult to choose the right camera based on a datasheet. The reality is that many aspects of image quality cannot be quantified or measured. To illustrate the challenge, consider the two images in Figure 18.2. These images were taken from two different cameras with the same resolution and similar specifications under the same conditions and illumination. The cameras cost about the same and come from brand-name vendors. The conclusion is that the best way to determine image quality is to install different cameras and look at the results. Keep in mind that although a camera may provide good still images, the images might not be equally good when a lot of motion is introduced into the scene.

Many factors affect image quality. For example, white balance and a camera's ability to adapt to different lighting conditions – from fluorescent, high-pressure sodium to LED lights – is important to ensure color fidelity. Low light, backlight, dynamic range, and other extreme lighting conditions present challenges the camera needs to handle. Typically, a high-resolution camera is less light-sensitive than a lower-resolution camera. In other words, you may need to consider sacrificing resolution for better low-light performance or use a camera with a sensor and processing algorithms that are specially designed to meet these challenges. For more information about how light sensitivity, image processing, scanning techniques, sensor size, and other factors affect image quality, see Chapter 4.

18.2.3 Resolution

Best practices have emerged regarding the number of pixels required for certain video surveillance operations. For regular network cameras using visible light, the DORI (Detection, Observation, Recognition, Identification) system (based on the IEC EN62676-4: 2015 International Standard) defines the pixels per meter (PPM) required for the following:

- *Detection:* This level simply allows the detection of objects (e.g., persons or cars) in the camera view. Details cannot be seen. 25 PPM (10/ft) are required for detection.
- *Observation:* The level provides some general details of an object, such as distinctive clothing. 62 PPM (20/ft) are required.
- *Recognition:* This level can determine whether an individual is someone that has been seen before. 125 PPM (38/ft) are required.
- *Identification:* This level of 250 PPM (75/ft) enables an individual to be identified beyond reasonable doubt.

By using these PPM values, it is possible to rate a camera by the sensor/lens combination and verify that it will provide the performance needed in each application. Manufacturers often display the minimum DORI results on their cameras datasheets.

Note that in many video analytics applications, the pixel density requirements may be different to those specified by the DORI system. The requirements may be application-specific or even use-case-specific but should always be specified by the vendor of the application.

For an overview image, 70–100 pixels are generally enough to represent 1 m (20–30 pixels/ft) of a scene. For operations that require detailed images, such as face recognition, this can rise to as many as 500 pixels/m (150 pixels/ft). This means that to get positive identification of people passing through an area 2 m wide × 2 m high (7 × 7 ft), the camera needs to provide a resolution of at least 1 megapixel (1,000 by 1,000 pixels).

Figure 18.3 illustrates the field of view of a fixed dome camera with a 2.8 mm lens. The camera is mounted 3 m (9 ft 10 in) above the ground. The green area gives a resolution of 125 pixels/m (38 pixels/ft), as required for recognition, and the yellow zones gives up to 250 pixels/m (76 pixels/ft), which is required for identification.

The maximum distance for identification and recognition using cameras with different lenses is shown in Figure 18.4. Again, there are tools that help calculate resolution and field of view based on camera model and position (see section 18.7).

18.2.3.1 Determining the required resolution

The required resolution for video surveillance images depends on the size of and the distance to the objects under surveillance. To illustrate this, consider the surveillance of an airport entrance where people need to be identified any time there is an incident.

Say the entrance area is 20 m (65 ft) wide. The estimated vertical field of view needed to identify a person is 2 m (6.5 ft). But because network video offers a multitude of resolutions and formats, it is better to define resolution requirements based on pixel density in the horizontal dimension. The most typical object that needs to be recognized is a human face. The variations in face width are less than those of body height or width, so there is a smaller margin of error. Depending on conditions, you need 40–80 pixels across the face for positive identification of a person. Some manufacturers and organizations, such as the Swedish National Laboratory of Forensic Science (SKL), recommend

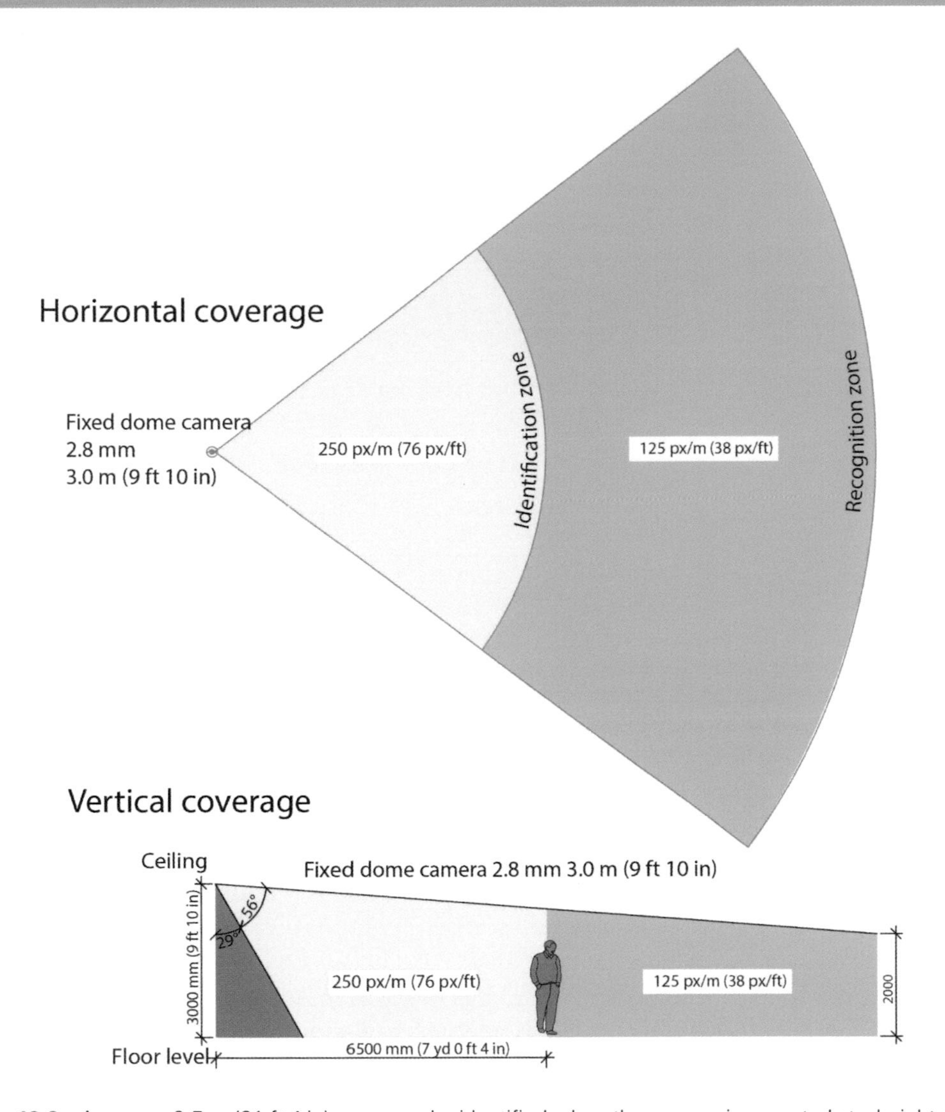

Figure 18.3 A person 6.5 m (21 ft 4 in) away can be identified when the camera is mounted at a height of 3 m (9 ft 10 in). The horizontal coverage of the camera is shown above, and the vertical coverage is shown below.

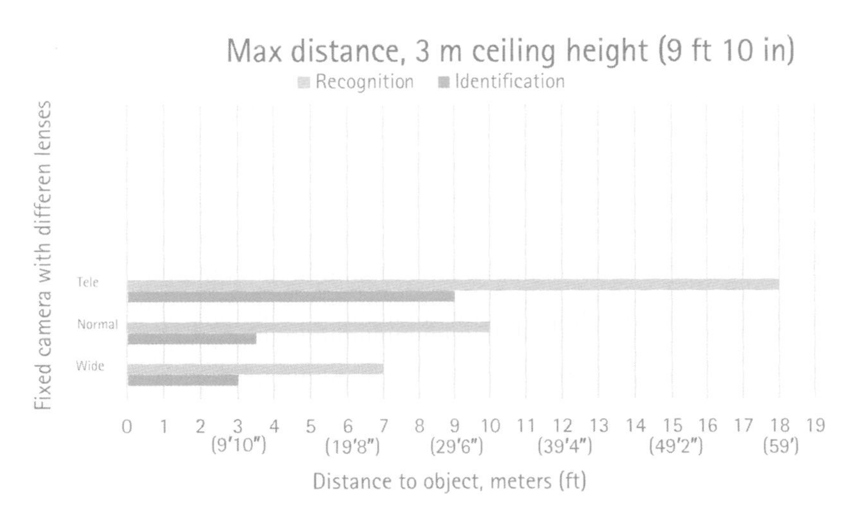

Figure 18.4 Example maximum distances for identification and recognition with different lenses. For example, a camera with a tele-lens enables identification at up to 9 m (29 ft 6 in) and recognition at up to 18 m (59 ft).

Figure 18.5 These images show what 20 (left), 40 (middle), and 80 (right) pixels across the face can look like.

Table 18.1 Required number of pixels per scene, given that the width of a typical face is 0.16 meters (½ ft)

Alternative	Face resolution, width	Scene, W × H	Scene resolution, W × H
A	20 pixels	20 m × 2 m	2500 × 250 pixels
B	40 pixels	20 m × 2 m	5000 × 500 pixels
C	80 pixels	65 ft × 6 ft	10400 × 960 pixels

Table 18.2 Number of cameras needed to cover the scene

Alternative	Scene resolution, W × H pixels	720p (1 MP) 1280 × 720	1080p (2 MP) 1920 × 1080	4K (8 MP) 3840 × 2160
A	2,500 × 250	2	2	1
B	5,000 × 500	4	3	2
C	10,000 × 1,000	8	6	3

80 pixels across a face for identification, while others, such as the international standard published by CELENEC, suggest that 40–100 pixels/ft is sufficient. Counting pixels across the face is a convenient measuring method because all adult faces are about the same width, about 0.16 m, or a little more than ½ foot. Therefore, in the following examples, we focus on the horizontal resolution. The vertical resolutions that correspond to the 2 m vertical field of view are included for reference.

Figure 18.5 shows images of a face using three different resolutions, measured in the number of pixels across the face. You can clearly see the impact of higher resolutions (see also Table 18.1 and Table 18.2).

If you know the number of pixels needed across the face (the face resolution) and the width of the surveillance area (the scene width), you can calculate the total number of pixels needed to cover the width of the scene (the horizontal resolution). Divide the scene width by the face width (0.16 m or ½ ft) and then multiply by the face resolution:

$$Horizontal\ scene\ resolution = \frac{Scene\ width}{Face\ width} \times Face\ resolution$$

In Table 18.1, we use the measurements from the airport scenario earlier.

Calculation in meters, alternative B:

$$5{,}000\,pixels = \frac{20\,m}{0.16\,m} \times 40\ pixels$$

Figure 18.6 Example of a built-in pixel counter. To change the size of the selection frame, either drag the corner or set the required width and height in pixels. Are there enough pixels to identify a person when they stand in this part of the scene?

Calculation in feet, alternative C:

$$10{,}400\,pixels = \frac{65\,ft}{\frac{1}{2}\,ft} \times 80\,pixels$$

With the resolutions determined, you can calculate the number of required cameras:

$$Number\ of\ cameras = \frac{Horizontal\ scene\ resolution}{Camera\ resolution}$$

Table 18.2 shows some possible combinations and the benefit of megapixel cameras.

Although somewhat hypothetical, the examples in Table 18.2 show that there is a lot to gain by calculating how many cameras you need. If the aim is to get high-resolution coverage of this scene with alternative C, you need eight 720p cameras. This drops to three cameras if you instead use 4K cameras. Likewise, if the demands on resolution are moderate, there is little point in going for a 4K camera if a 720p camera will suffice.

Some cameras have a built-in pixel counter, which makes it easier to verify that the camera's resolution is high enough for the scene and to make a positive identification (see Figure 18.6). You could ask a person to stand as a model when you calculate the pixels, but you could also use any object with a known width. If the object is too small to fit in the pixel counter frame and you cannot accept a smaller field of view, you need a camera with a higher horizontal resolution.

18.2.4 Compression

The dominant video compression methods today are H.264 and H.265. Some manufacturers may also include support for legacy formats, such as Motion JPEG.

Designing the network and the storage system largely depends on the compression standard. The following are some basic recommendations:

- *Motion JPEG:* A common compression method in the early days of network video but little used today. Suitable for systems with limited retention and systems that only need very limited frame rates, as in time-lapse applications. Motion JPEG consumes more bandwidth and storage, so is often used in low–frame rate scenarios to capture high-quality images, not video, such as in LPR applications where the vehicle is moving at a high speed.
- *H.264:* One of the dominant video compression standards in video surveillance today. Compared to MJPEG, H.264 reduces network bandwidth and storage. It captures video at a higher rate with more efficiency. The Main and High profiles make H.264 more efficient and give improved image quality.
- *H.265:* Provides high resolution at lower bitrates, lowers the minimum bandwidth requirements, and reduces storage needs as compared to H.264, especially at higher resolutions. However, it requires high-performance network cameras with sufficient computing power and efficient decoders on the video monitoring side of the system. Network video products with H.265 compression are readily available on the market. The benefits are typically greater for higher resolutions, but there may be extra requirements put on the viewing station, as the video decoding is more resource-intensive.
- *Smart codecs:* The H.264 and H.265 standards refer only to the decoder, whereas there are opportunities to improve the encoding. Some proprietary codec technologies can filter out areas of low interest and compress these more aggressively while still recording the details of interest and motion at higher quality. This switching between a maximum group of pictures (GOP) value and a lower value drastically reduces the bitrate and the requirements on storage and bandwidth. Some technologies apply the same principles to the frame rate, making it change dynamically, which also saves on bandwidth and storage. For more information on compression, see Chapter 6.

18.2.5 Networking functionality

In the same way that high-quality images are essential, networking functionality is also important. Besides the obvious Ethernet connectivity, a professional network camera should also support the following capabilities:

- *Power over Ethernet (PoE):* This means that the camera can receive power through the same cable as for the data, thus eliminating the need for separate power cable runs. This means you can save hundreds of dollars per camera in installation costs alone. Make sure the camera complies with the Power over Ethernet IEEE 802.3af/802.3at Type 1 (up to 15 W) or 802.3at Type 2 (up to 30 W) standard. This will give you the freedom to select from a wide variety of network switches. Outdoor PTZ cameras may require more power (up to 60 W or even more in extreme environments). These will require a switch or midspan that conforms to 802.3bt, which provides higher levels of power. For more information about PoE, see section 9.6.
- *DHCP*: Used by many organizations to manage IP addresses. A DCHP-enabled switch or router automatically gives each connected device an IP address. Advantages include making it quicker and easier to deploy or switch cameras, stability through periodic renewal, reservations, and failover. With less advanced or poorly managed DCHP servers, the disadvantages include the risk of temporarily losing IP addresses if the server goes down or resets, IP addresses changing over time, or that someone accidentally connects the server to a router that is open to the internet. Some users may prefer the predictability of static IP address and the ability to define and use IP addresses per their own requirements.
- *HTTPS encryption:* Used for secure communication. Many video management systems use HTTPS as the default for communication. See Chapter 12 for more on cybersecurity.
- *SNMP:* Helps IT administrators monitor network conditions and see if connected devices need attention.

- *IP address filtering:* Restricts access to the camera to predefined IP addresses only.
- *IPv6:* This is the most recent version of the Internet Protocol (IP). IPv6 uses 128-bit addresses that are divided into eight groups of four hexadecimal digits. Its development was driven by the limitations of IPv4, which uses 32-bit addresses.
- *Wireless technology:* Wireless is a good option if running a cable to the camera is impractical, difficult, or expensive. A non-wireless camera can use wireless if first connected to a wireless bridge. Wireless technology can be useful, for example, in historic buildings where the installation of cables would damage the interior or in facilities where cameras are often moved to new locations, such as in a supermarket or in outdoor installations. Wireless technology can also be used to bridge sites without having to install expensive cabling. For more about wireless networks, see Chapter 11.
- *802.1X:* Enhances the security of local area networks through a port-based authentication framework.

Always weigh in the opinion of the IT department. They should be able to determine if the camera provides adequate networking functionality and security. For more information on networking technologies, see Chapter 9.

18.2.6 Other functionalities

Network cameras have many other functionalities apart from simply providing video. When selecting a camera, it is important to evaluate these functionalities, which may include the following:

- *Audio for communication and detection:* Users can listen in on an area and communicate instructions, orders, or requests to visitors or intruders. When microphones detect sounds above a certain level, they can trigger alarms or cameras to start recording. Consider whether one-way or two-way audio is required. Microphones and speakers can be built-in or external. Consider the legal aspects of monitoring and recording audio in your region. Some regions forbid the use of audio for surveillance purposes, except in police interview rooms and other government-controlled facilities. For more information about audio, including how to use audio (and I/Os) in cameras that lack these features, see Chapter 7.
- *Network radar:* Radar uses electromagnetic waves to detect movement and is not sensitive to things like moving shadows, lights, rain, or insects. A motion detector based on radar is a good complement to cameras, as the device calculates distance, velocity, and size of objects in relation to the detector. Like a camera, a radar detector can trigger an alarm when it detects an object, and it can also trigger a recording for visual verification.
- *Analytics:* Many network cameras offer edge-based intelligence, such as object detection, video motion detection, tampering detection, license plate recognition, and people counting. Edge analytics make the system more scalable and help reduce bandwidth and storage requirements because the camera itself decides when to send and process video. Analytics require a lot of processing power, and if the processes are performed on the server rather than in the camera, the server can quickly become overloaded. For more information about analytics, see Chapters 16 and 17.
- *Input/output (I/O) connectors:* Connecting input devices to a camera (such as a door contact, infrared motion detector, glass-break sensor, or shock sensor) enables the camera to react to an external event, for example, by sending and recording video. In many cases, as in scenarios where the goal is to capture the identity of a person at an entrance, there is no need for the camera to continually send video. Through the input port, the camera knows when the door opens and only captures and sends video from that event. Outputs enable the camera or a remote operator to control external devices, for example, alarm devices, door locks, or lights.
- *Alarm management functions:* Network cameras can perform alarm management tasks, such as processing and linking input, output, and other events. For example, if the level of audio in a room passes a defined threshold, the camera can send an output signal that turns on the lights

and send video to the video management software. Pre- and post-alarm buffers in a camera can record video before and after an alarm occurs. A camera can also send notifications, for example, via email, messaging services, or push notifications. Communication with other devices and VMS systems can also occur via the messaging protocol Message Queuing Telemetry Transport (MQTT). See section 9.8 for more information on MQTT.

- *Other physical security devices:* In an enterprise environment, it is favorable to take a holistic approach to system requirements. When you evaluate a video surveillance system, also look at other systems such as access control, intercom, audio, and intrusion detection to determine if there is a way to construct an integrated system that covers all your physical security needs.

18.2.7 Vendors

Choosing a video surveillance vendor and partner can be confusing when there are so many to choose from, each with their own range of products and solutions. Narrow down the selection by isolating just a few vendors. Here are a few tips that may help when selecting a camera vendor:

- *Wide product portfolio:* Go with those who maintain a full product camera line, including fixed cameras, domes, PTZ cameras, thermal cameras, panoramic cameras, etc. This way, one or two companies can satisfy current and future needs for system expansion and functionality upgrades, such as audio, radar, and body-worn cameras. If analog video products must be integrated into a network video system, make sure the selected company's product portfolio also includes video encoders and decoders.
- *Multi-application support and ease of integration:* Even if you decide to go with a single vendor, make sure you select cameras that have open application programming interfaces (API) and that can integrate with several video management software applications so you don't get tied to a single supplier. Open multi-vendor video management systems give users the most flexibility.
- *Standardized interface:* A standardized interface for IP-based security devices allows integrators and users to combine products from different vendors in a single solution. Standardization helps device manufacturers ensure interoperability with products from other manufacturers, and software developers can ensure that their products support multiple device brands. One organization that works to facilitate interface standardization and interoperability between products is ONVIF (Open Network Video Interface Forum), a non-profit organization with a membership largely made up of companies manufacturing IP-based physical security products. Standardized interfaces have the downside of not always covering all functionality in some advanced network cameras.
- *Tools for managing large deployments:* Like all intelligent network devices, network cameras have an IP address and built-in firmware. Upgrading the firmware is usually easy, and many vendors provide upgrades free of charge. When making a purchase decision, consider the cost of setting IP addresses and future upgrades of all the cameras in the system. The camera vendor should have tools to manage these processes, and their estimates for cost and downtime should be clear and measurable upfront. The vendor should also provide management software that can locate all network video devices automatically and monitor their status (health monitoring). Any such tool should also provide support for implementing and maintaining cybersecurity policies.
- *Tools and support for system design:* Enterprise-level systems have tremendous capabilities and flexibilities. This also means that to get the most out of the system it needs to be properly designed. Does the vendor have the support and tools to help with designing and documenting the system?
- *Tools for system maintenance:* One of the key strategies for maintaining cybersecurity for networked systems is to keep all devices such as IP cameras updated with the latest firmware. It is, therefore, key that the vendor provides tools for maintaining the devices, whether the system has 5 or 5,000 devices.

- *Warranty:* Video surveillance systems are a substantial investment for most organizations and have a life expectancy of several years. Make sure that the vendor has a reasonable warranty on the selected cameras and check if they offer extended warranty services.
- *Networking knowledge:* In the analog video surveillance world, there was little need to evaluate the IT knowledge of the vendors. In the world of IP-based security and surveillance systems, the case is very different. Not only is networking functionality important – even more so are the technologies available to provide adequate cybersecurity.
- *Long-term partner:* Select a vendor that has the potential to be a long-term partner. Remember that your system needs to be operational, maintainable, and perhaps expandable for five to ten years. Does the company have a large base of installed cameras? Is their focus on network camera technology? Can they offer local representation and support? Is the company a global player? Because needs change and grow, it is important to choose a vendor where there is innovation and long-term plans for support, upgrades, and product paths. Look at the prospects of future growth and the need for added features and functionality.

After deciding on the camera, it is a good idea to purchase a single unit and test its quality and real-world performance before ordering large quantities.

18.3 INSTALLING A NETWORK CAMERA

How a network camera is installed is just as important as the process of purchasing it. Next are some recommendations on how best to achieve high-quality video surveillance based on camera positioning and environmental considerations.

18.3.1 Add light if needed

The most common reason for poor quality images is lack of light. An easy, cost-effective way to improve the lighting conditions and get better images is to add more light in both indoor and outdoor situations. The positioning of extra lighting is as important as the positioning of the camera. You want to avoid reflections, shadows, or blinding the camera or people moving in the area.

In basic terms, there are three situations in which adding light can help:

- When the scene is too dark for the camera to produce useable images.
- When the existing lighting is not good enough for the camera to produce high-quality or bright images without motion blur or noise, the latter which also consumes bandwidth and storage.
- When the lighting conditions of the scene are challenging. For example, if the scene includes both bright and shadowed areas, or the scene includes backlighting. An evenly-lit scene is always easier to deal with, no matter if it is dim or very bright.

The positioning of a white-light illuminator relative to the camera's position is also important, especially when different areas in the scene are reflecting differently, depending on factors such as the weather. For example, worn light-gray asphalt becomes very dark when wet.

Smooth and textured surfaces reflect light differently. Uneven surfaces bounce light in all directions because of their inherent irregularities. When the light hits the object squarely, the reflection tends to be stronger. So typically, the camera's position should be beside the illuminator and directed straight at the target. This also helps the camera deal with shadows in the scene.

With smooth surfaces, the angle of reflected light is equal to the incoming light. In these cases, the camera and the illuminator should be placed so they have the same angle.

18.3.1.1 Use IR light when white light is impracticable

For discreet or covert security, or in areas where artificial light is unwanted, choose an IR-sensitive day-and-night camera. An IR illuminator, which provides near-IR light, can be used in conjunction to enhance the camera's ability to produce high-quality video in low-light or nighttime conditions.

Figure 18.7 In cameras with advanced built-in IR LEDs, the angle of illumination can automatically follow the camera's field of view.

To maintain good image quality and prevent focus shift at night when using IR illuminators, make sure that the camera has an IR-corrected lens. For more about IR illuminators, see Chapter 4.

There is also the option of using cameras with built-in IR LEDs. Previously, built-in LEDs had a very short lifespan, and they were known to generate heat that would increase the noise in the image. Technology advancements have resulted in better mechanical designs, in which heat can escape more easily, a longer lifespan that mirrors that of the camera, with gaskets that prevent IR light from polluting the images, and autofocus that follows the camera's focus and field of view (see Figure 18.7). With built-in LEDs, it is possible to configure the camera to turn the lights on automatically at a specific time, when the amount of ambient light reaches a specific level, or when triggered by an event. Using built-in IR LEDs is straightforward because of the seamless design and ease of installation.

18.3.2 Avoid direct sunlight and glare

Always avoid direct sunlight into the camera, as it blinds the camera and can reduce the performance of the image sensor. If possible, position the camera with the sun behind the camera. Some cameras have sunshields that help reduce the impact of direct sunlight and make placement easier.

Sunlight is not the only issue. A camera can usually deal with viewing a self-illuminated object during the day, but at night, the ambient light level is lower and the contrast between high and low light levels is greater. Therefore, objects such as car headlights or illuminated signs appear brighter than they do in daylight (see Figure 18.8). Such glare is problematic for cameras and human eyes alike. To help the camera get better video, increase the ambient light level in the scene by using additional lighting.

18.3.3 Avoid backlight

Backlight creates various problems for all conventional cameras. Next are some of the most common problems and suggestions on what to do about them.

- *Avoid the brightest areas:* Pointing the camera at the brightest areas in the scene causes problems with over- and underexposure. The bright areas become overexposed (bright white), while other objects appear too dark. This typically occurs when attempting to capture an object in front of a window. To solve this problem, reposition the camera or block the light with curtains or blinds. If neither is possible, add frontal lighting.

Figure 18.8 Two images of the same scene in different lighting conditions. (a) During the day, the colors are vibrant. (b) At night, the contrast between light and dark makes it difficult to see the area surrounding the light-box. Adding some lighting would improve the image quality.

- *Reduce the dynamic range:* In outdoor environments, viewing too much of the sky results in too wide a dynamic range. The camera will self-adjust to achieve a proper light level for the sky. Consequently, the object or landscape of interest will appear too dark (see Figure 18.9). One way to solve this problem is to mount the camera higher up, using a pole if needed.
- *Adjust the camera settings:* It may be necessary at times to adjust settings for brightness, sharpness, white balance, and WDR for different environments (indoor, outdoor, and fluorescent) to get an optimal image. Some areas have multiple light sources, for example, daylight plus fluorescent lighting. Make sure the camera can deal with this. Cameras with support for high (or wide) dynamic range are better at handling backlit scenarios (Figure 18.10). As discussed in Chapter 4, there are several different types of WDR technologies.
- *Change the exposure time:* Use short exposure times for rapid movement or when a high frame rate is required. A long exposure time makes images look nicer but probably lowers the total frame rate and results in increased motion blur. In cameras with automatic exposure, the frame rate increases or decreases with the amount of available light. It is only as the light level decreases that artificial light or prioritized frame rate or image quality becomes an issue. Thanks to improvements in sensor technology and image processing, modern cameras now have better low-light abilities, and the trade-offs between motion and noise are less than they used to be.

18.3.4 Lens selection

The fastest way to calculate the required lens and field of view is to use an online lens calculator, preferably one that also considers pixel density, as available from most camera and lens manufacturers. For more information on lenses, see section 4.2.

Outdoors it is best to use an auto-iris lens because it automatically adjusts the amount of light that reaches the image sensor. This optimizes the image quality and protects the sensor from being damaged by strong sunlight. Even better than an auto-iris lens is a P-Iris lens, especially when using high-resolution cameras. P-Iris lenses are less prone than standard auto-iris lenses to producing diffraction in their images. Diffraction is an issue, especially when the light is strong. And the higher the resolution of the sensor, the more prominent the diffraction. For more information about the differences between auto-iris and P-Iris lenses, see section 4.2.7.

Long-distance coverage usually requires longer lenses than those included with the camera, so check if lens replacement is possible. If installing the camera in an outdoor housing, will the replacement lens fit in the housing?

Figure 18.9 (a) The image shows the sky and has a lot of backlight. (b) The camera is moved to improve the image.

Figure 18.10 Advanced cameras include a feature that compensates for backlight, which in this example is most apparent in the part of the image showing the sky.

18.4 PROTECTING A NETWORK CAMERA

Surveillance cameras are often placed in very demanding environments. In outdoor installations, protection against varying weather conditions is necessary. In industrial settings, cameras may require protection from hazards such as dust, acids, or corrosive substances. In vehicles such as buses and trains, cameras must withstand humidity, dust, and vibrations. Cameras may also require protection from vandalism and tampering.

Manufacturers of cameras and camera accessories use various methods to meet environmental challenges. Solutions include placing cameras in protective housings, designing special-purpose cameras for each type of environmental challenge, and using intelligent algorithms that can detect and alert users to a change in a camera's operating conditions.

The level of protection provided by enclosures, whether it is a one-fits-many or an integrated solution, is often indicated by IP, NEMA, and IK ratings. All electronic devices, cameras, and encoders must also fulfill the emission, immunity, and electronic safety requirements of the region and the environment in which they are used.

The following subsections discuss such topics as coverings, positioning of fixed cameras in enclosures, environmental protection, vandal and tampering protection, types of mountings, and protection ratings.

18.4.1 Camera enclosures in general

A camera's operating conditions depend on its materials and components. When the conditions of the environment are beyond a camera's operating conditions, the camera needs to be protected by a housing (also called an enclosure). Some cameras are designed and supplied ready for use outdoors and in other demanding conditions. However, to use an indoor camera outdoors, you need a separate enclosure in which to place the camera.

Camera enclosures come in different sizes and qualities; some are made of metal, others of plastic. Most cameras today come with the protective enclosure as an integrated part of the camera. When selecting the appropriate enclosure for a specific camera, several things must be considered:

- *Mount:* What kind of mounting bracket do you need (wall, pole, corner, parapet)?
- *Cable runs:* How can the cables be run? How are you going to manage them? How much cable do you need? What kind of shielding is required? What quality cable do you need for your data? What kind of conduits do you need?
- *Operating temperature and environment:* Do you need heaters, fans, sunshields, or wipers? Do you need or dustproof or waterproof materials and seals?
- *Power supply:* How much power does the camera need? Does the housing need power? What is the available power? Is Power over Ethernet sufficient? Do you need higher voltage (12, 24, 110 volts)?
- *Vandal resistance:* How much physical force does the housing have to withstand? Which impact rating (IK class) does that correspond to?
- *Vibrations:* Are there vibrations in the area? For example, will the camera be placed near railroad tracks or a busy freeway? Will it be placed in a bus?

And for fixed camera (box camera) housings:

- *Opening:* A side or a slide opening can make a difference when opening the camera for maintenance.

And for dome and PTZ camera housings:

- *Dome:* Do you need a clear, smoked, or mirrored dome?

Some housings also have peripherals such as antennas for amplifying the signal to wireless cameras. An external antenna is only required if the housing is made of metal. Provided the signal is strong enough, a wireless camera inside a plastic housing usually works without the use of an external antenna. Some vendors offer extra powerful access points that provide the level of connectivity required by wireless network cameras, such as wide coverage and signals strong enough to reach the cameras and to ensure the stability of the network.

Figure 18.11 (a) Housings for fixed cameras. (b) A camera mounted on the bottom cover of a housing, which is connected to an external power supply that supplies power to the heater and the camera.

Figure 18.12 Some enclosures can be repainted to blend in with the facade of a building.

Sometimes it is preferable to have the protective enclosure in the same color as the surroundings to make the camera less noticeable or more aesthetically pleasing. Some cameras are designed so that the enclosure can easily be repainted.

18.4.2 Transparent coverings

To see out from a housing, the camera lens needs a window. This is usually made of high-quality glass or durable polycarbonate plastic. Because windows act like optical lenses, they need to be of high quality to minimize their effect on image quality. Imperfections that compromise the clarity of the window material result in lower-quality images.

Windows in housings for PTZ and dome cameras need to meet even higher demands. Not only must the windows be specially shaped in the form of a dome but they must also have high clarity, or else, imperfections or specks of dirt may be magnified. When the zoom factor is high, even the tiniest particles become problematic. Although smoked domes give a more discreet installation than clear domes, keep in mind that they also reduce the amount of light available to the camera and, therefore, affect its light sensitivity. See Figure 18.13 for examples of clear and smoked domes. Some

Figure 18.13 Examples of domes for PTZ and fixed dome cameras. (a) Clear dome. (b) Smoked dome. (c) Partially smoked dome. Although the smoked dome makes it difficult to see in which direction the camera is pointing, it also reduces the amount of incoming light and, therefore, the image quality

manufacturers also offer mirrored domes. A smoked or mirrored dome acts much like sunglasses do in reducing the amount of light that can pass through the covering. Therefore, the camera might have problems with the depth of field and with adjusting to f-stop changes. To remedy some of these issues, it is best to use auto-iris lenses. With a fixed dome camera, you can use a dome that is partially smoked and partially clear. This makes it more difficult to see where the camera is pointing, but because the lens looks through a clear window, image quality is not compromised.

18.4.2.1 Overcoming the limitations of conventional domes

Camera vendors have long tried to solve the shortcomings of the dome manufacturing process. Until recently, molding technology made it impossible to make a dome in one piece without introducing flaws. To make a dome with the required level of clarity, manufacturers must join two parts, a half-sphere and a cylinder. This joining of parts always results in a transition, which is a problem, as it affects part of the camera's view, resulting in blurry images. Therefore, the discussion has been limited to where to place the transition and whether it should be smooth and wide (large, slightly blurred area) or sharp and thin (small, very blurred area). When monitoring areas with varying altitudes, such as escalators, hilly roads, or steep arena stands, the transition is particularly troublesome because the camera cannot see clearly above its horizon. The more the camera tilts, the blurrier the image. This phenomenon is called mirroring (see Figure 18.17).

As explained earlier, a window or dome acts as an extra lens. With conventional domes, the varying distance between the camera and the dome usually causes issues with reflections, distortions, and other optical effects. This is another motivation for trying to solve the manufacturing challenges and spawn a new generation of domes.

Imagine that a camera's dome is a sphere rather than an elongated half-sphere. To get the best possible image quality, the camera block should be placed at the center point of the sphere, that is, at the zero point on all the axes. However, in most cases, the camera is placed lower on the vertical axis to avoid the error of refraction caused by the transition of the dome. The degree of vertical misalignment is called the L-value (see Figure 18.14). This gives the camera a greater tilt range but also a drop in image quality.

Some PTZ cameras on the market today solve the L-value challenge by using two half-spheres that are joined together and tilted at an angle (see Figure 18.15 and Figure 18.16).

The tilt makes it possible to place the camera block at the center point of the sphere, where the L-value is 0. Because of its position, the camera block remains consistently at the same distance from the dome wall. This means that refractions and other optical effects are kept to a minimum.

Thanks to its advanced mechanics, this dome rotates with the camera block, which makes it possible to maintain optimal image quality in all pan and tilt positions and to identify with certainty

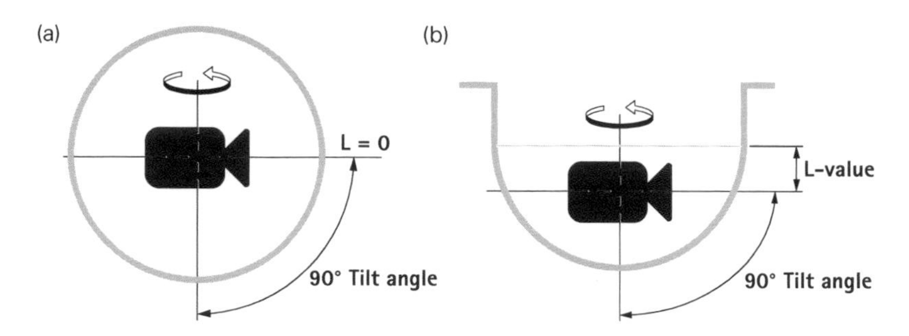

Figure 18.14 (a) The optimal L-value is zero. (b) In most domes, the L-value ends up being well over that.

Figure 18.15 A PTZ camera with a type of spherical dome that solves the L-value challenge.

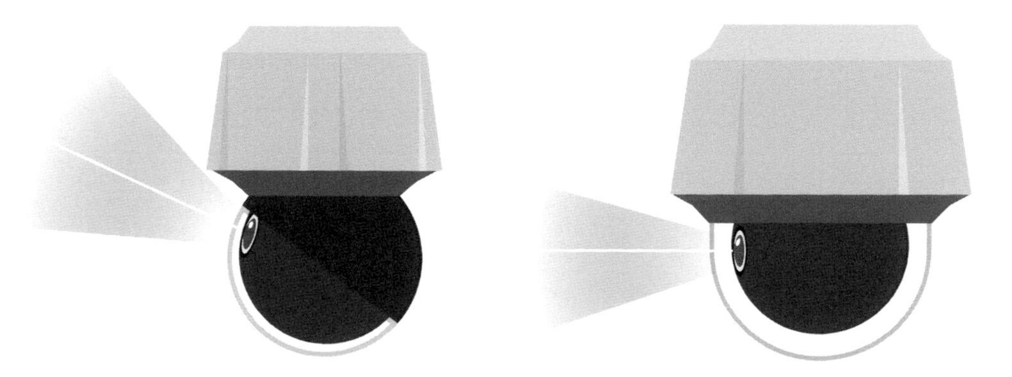

Figure 18.16 (a) A spherical dome. (b) A conventional dome.

objects as much as 20° above the camera horizon (see Figure 18.17). In other words, cameras with this type of dome are much more suited for monitoring uneven terrain than cameras with conventional domes.

Another feature of this dome is its ability to shake itself off when wet, helping the camera to produce sharp images even in rainy weather.

When it rains or when cleaned with a hose, the dome starts rotating at high speed in alternating directions (see Figure 18.19). This breaks the water's surface tension, and the drops fall from the dome.

Figure 18.17 Two images taken at 20° tilt angle and 20× zoom. (a) The transition in a conventional dome causes a mirroring effect in the image. (b) With the spherical dome, the image is significantly sharper.

Figure 18.18 Two merged snapshots of the same rainy scene, before the dome shakes (left) and after the shaking (right).

18.4.3 Positioning of fixed cameras

When installing a fixed camera in an enclosure, it is important to position the lens of the camera right up against the window to prevent glare in the image, as caused by reflections from the camera and the background. To reduce reflections, special coatings can be applied on the glass used in front of the lens (see Figure 18.20).

18.4.4 Environmental protection

The main environmental threats to a camera, particularly outdoors, are cold, heat, water, and dust.

Housings with built-in heaters and fans (blowers) can handle environments with low and high temperatures. Enclosures that have active cooling with a separate heat exchanger help cameras cope with hot environments.

To withstand water and dust, housings are carefully sealed. In situations where cameras might be exposed to acids, such as in the food industry, housings made of stainless steel are required (see Figure 18.21). Some specialized housings can be pressurized, submersible, bulletproof, or explosion-protected. Special enclosures may also be required for aesthetic considerations.

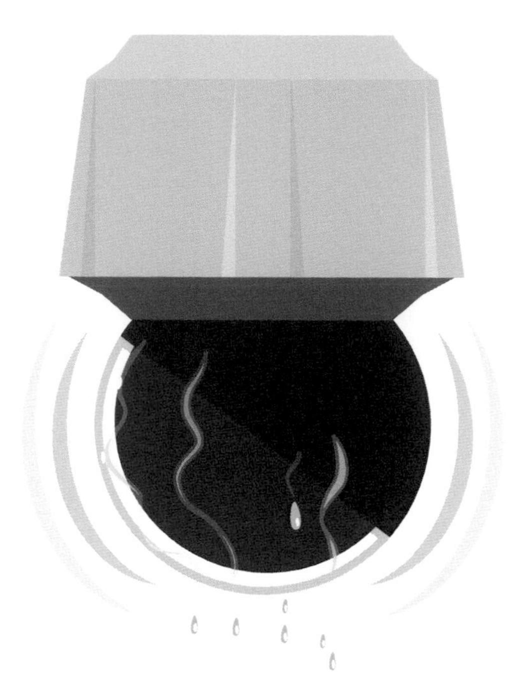

Figure 18.19 Water drops fall from the dome when it shakes.

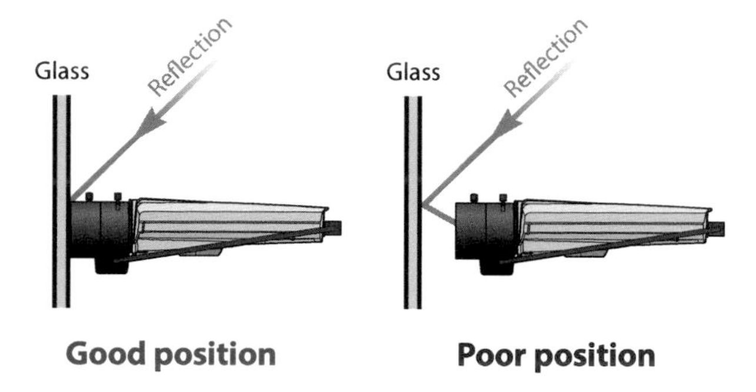

Figure 18.20 When installing a camera behind glass, correct positioning of the camera becomes important to avoid reflections.

Figure 18.21 A stainless steel PTZ camera.

Other environmental elements include wind and traffic. To minimize vibrations, particularly on pole-mounted camera installations, the housing should be small and securely mounted. Electronic and mechanical image stabilization is an important feature for these installations (see Chapter 4).

For housings, the terms *indoor housing* and *outdoor housing* often refer to the level of environmental protection, although these terms do not always correspond to conditions at the installation site. An indoor housing is used primarily to prevent the entry of dust, and it does not normally include a heater or fan. But for installation in, for example, a freezer room, a room with sprinklers, or a parking garage, which might be cleaned with high-pressure washers, a camera needs to cope with condensing humidity, or it might need a heater.

18.4.5 Vandal and tampering protection

In some surveillance applications, cameras are at risk of violent attack. Transportation vehicles, schools, prisons, and retail environments are examples of areas where vandals or criminals may try to tamper with, redirect, destroy, spray paint, or remove cameras. For examples of cameras for vehicles and prisons, see Figure 18.22 and Figure 18.23.

Although a camera or housing can never guarantee protection from destructive behavior in every situation, there are several measures available that can help security managers deal with vandalism. Considerations to keep in mind include camera and housing design, mounting, placement, and intelligent alarms.

Figure 18.22 An example of a compact, fixed dome camera specially designed for installations in mass-transit vehicles, such as buses and trains. It can withstand vibrations, dust, high humidity, and fluctuating temperatures.

Figure 18.23 A corner-mounted anti-ligature camera.

Figure 18.24 Examples of fixed camera housings. The housing to the left is classified as vandal-resistant.

Anti-ligature cameras are designed to resist almost any attempt at violence, whether the intent is self-harm, the harming of others, or destruction of the camera or the facilities. Many anti-ligature cameras fit snuggly into corners or have rounded edges so there are no surfaces to grip (see Figure 18.23). Typically, you find these cameras in prisons and psychiatric wards or in small spaces, such as interview rooms and elevators.

18.4.5.1 The goals of vandal protection

Many features and best practices can be implemented to increase protection against vandalism. The important goals of vandal protection, regardless of actual technical implementation, include the following:

- *Making it difficult:* Tampering with a video surveillance camera should be difficult. Perhaps even more important is that their design and placement should make them look like they are difficult to tamper with. A vandal should think twice before trying to interfere with a camera.
- *Creating uncertainty:* If vandals do attack a camera, they should remain uncertain as to whether they actually succeeded in destroying the camera or stopping the recording.
- *Delaying:* Even if it is not possible to fully protect a camera from a determined attack, it is still worthwhile making it time-consuming for a vandal to redirect or destroy the camera. Every second gained increases the chance of discovery or that the vandal gives up.
- *Detecting and sending alarms:* A camera with built-in analytics can detect that someone is tampering with it and can send a notification. This allows operators to quickly alert staff to clean, adjust, or replace the camera, to stop the vandal from finishing the attack, or in some other way, to fix the problem.

18.4.5.2 Mechanical design

Casings and related components made of metal provide better protection than ones made of plastic. The shape of the housing or camera also matters. A housing or a fixed camera that protrudes from a wall or ceiling is more vulnerable to physical violence than housings or casings for a dome camera, which are often more discreet. A dome's smooth, rounded surface makes blocking the camera's view more difficult. For example, hanging a piece of clothing over a dome camera is nearly impossible. The more a housing or camera blends into an environment, the better its protection against vandalism. For examples of vandal resistant cameras, see Figure 18.24, Figure 18.25, and Figure 18.26.

For improved vandal-resistance, domes can be made of a durable, transparent material, such as polycarbonate plastic, which is the material used to create bulletproof glass. Increasing the thickness of a dome improves its ability to withstand heavy blows. However, the thicker the dome, the greater the risk of flaws in the material. At high zoom levels, these flaws can be magnified and make

Figure 18.25 Examples of fixed dome cameras. All are vandal-resistant.

Figure 18.26 A vandal-resistant PTZ camera.

the image blurry. Increasing the thickness can also create unwanted reflections and refraction, which have a negative impact on image quality.

Special coatings can also be applied to cameras and housings to minimize the impact of graffiti. Dirt and drops of water on the dome or housing window cause image distortion. Therefore, some housings have built-in wipers that keep them clean.

18.4.5.3 Mounting

The way cameras and housings are mounted also affects the level of protection. A camera mounted so that it is accessible from the exterior is more vulnerable to attack. It is more exposed than a camera that uses a recess mount, where only the transparent part of the camera or housing is visible (Figure 18.27)

One thing to consider is if it is worth sacrificing the protection that tamper-resistant fasteners offer for the flexibility of standard fasteners. Nonstandard screws can make it more challenging

Figure 18.27 An example of a PTZ camera with a recessed mount. When the camera is mounted, you can see the dome and the trim ring (left) but not the mounting bracket (right).

for unauthorized persons to remove cameras and housings. The more unusual the screws, the better protection they provide. However, all authorized tasks that involve mounting, dismounting, or moving the cameras become more difficult and expensive because staff need special tools.

When making plans for mounting cameras and protecting the system from vandals, always include the cable runs. Running the cable directly through the wall or ceiling behind the camera provides the best level of protection. A metal conduit is also a good alternative when trying to protect cables from attack.

18.4.5.4 Camera placement

You can also deter vandals by placing cameras out of reach. A camera mounted high up on a wall or in the in the ceiling is less likely to attract spontaneous attacks. The downside may be the field of view, which to some extent can be compensated for by selecting a different lens.

18.4.5.5 Protecting cameras with analytics

Analytics in network cameras and video management systems can help protect cameras against vandalism. One basic function is where cameras visually detect if they are redirected or obscured, in which case, they can send alarms to control rooms or to staff in the field. Other algorithms can detect abnormal sounds picked up by a connected microphone. Cameras can also be fitted with a tampering switch that reacts when the camera is opened, and an accelerometer can detect if the camera is shaken or moved.

Without this type of shock sensors and analytics, keeping track of the proper functioning of hundreds of cameras in demanding environments is too difficult. In systems where no one is actively watching live video, analytics simplify automatic surveillance by notifying staff when someone interferes with a camera. For more information about analytics and their applications, see Chapters 16 and 17.

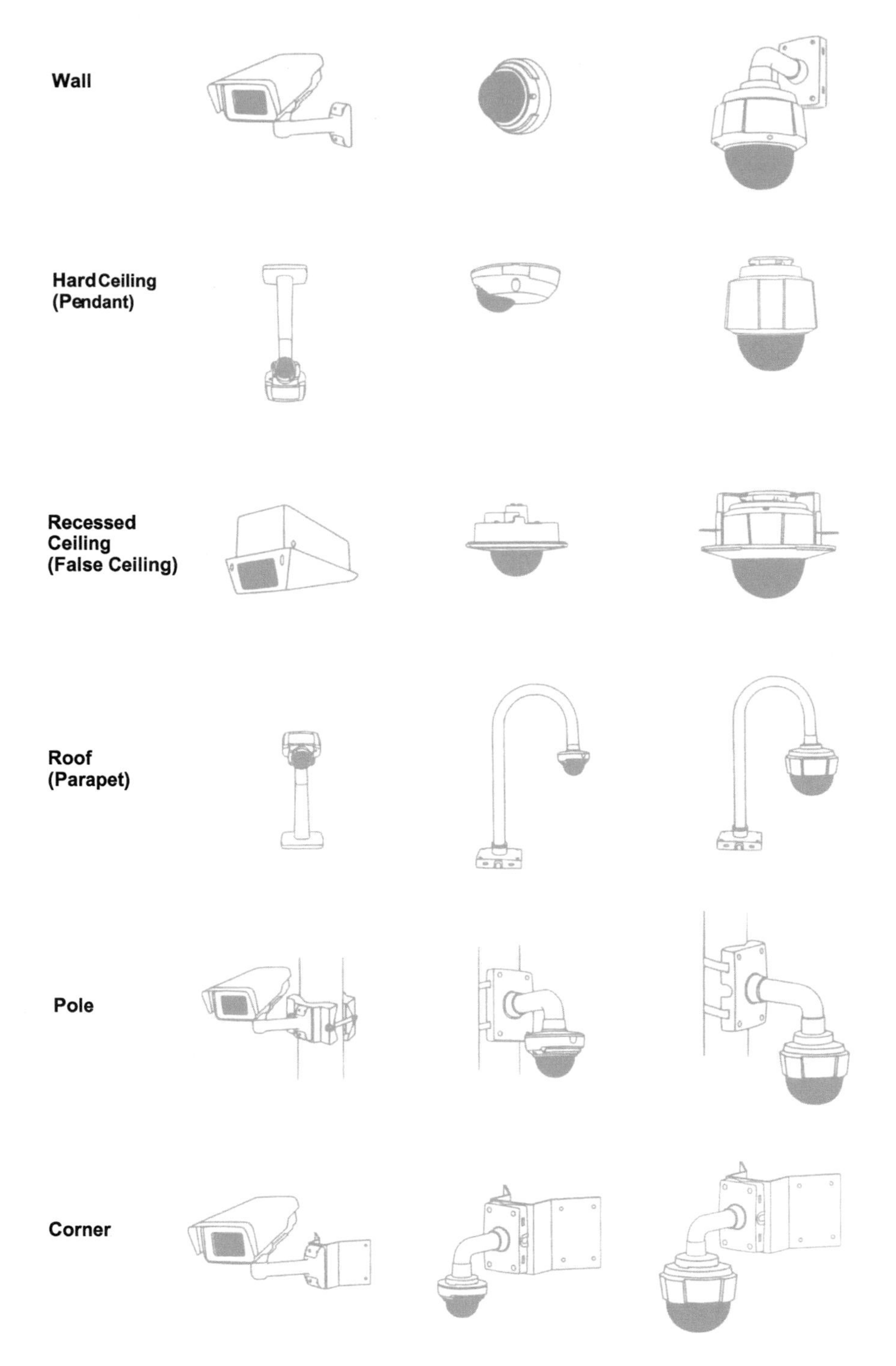

Figure 18.28 Examples of common mounting types.

18.4.6 Mounting types

Because surveillance is not limited to a specific type of space, there needs to be a wide range of mounting options. To minimize vibrations, always make sure the camera mount is stable. Because PTZ cameras move, this very action can cause image interference if the camera mount is not properly secured. In outdoor situations, sturdy mounting equipment is necessary to avoid vibrations

Figure 18.29 From left to right: A flush mount, a surface mount, and a pendant mount.

Figure 18.30 Examples of a regular wall mount (left) and a corner wall mount (right).

caused by strong winds. If the mount is not strong or stable enough, the worst-case scenario is that the camera falls and damages people or property.

18.4.6.1 Ceiling mounts

Ceiling mounts are used primarily indoors. The enclosure itself can be any of the following:

- *Flush mount:* Mounted inside the ceiling with only parts of the camera and housing (usually the dome) visible. This mount is also known as a recessed mount or drop-ceiling mount.
- *Pendant mount:* Hung from a ceiling like a pendant.
- *Surface mount:* Mounted directly on the surface of a ceiling or wall and, therefore, fully visible. This is also known as a hard-ceiling mount.

Figure 18.29 provides examples of these mounting types.

18.4.6.2 Wall mounts

Wall mounts are used to mount cameras inside or outside a building. The cable can often be routed through the arm, and many mounts have an internal cable gland or gasket to protect the cable.

18.4.6.3 Pole mounts

Pole mounts (Figure 18.31) often hold PTZ cameras in large outdoor areas, such as parking lots, roads, and city squares. This type of mount is usually designed to minimize the effects of wind and

Figure 18.31 An example of a pole-mounted PTZ camera.

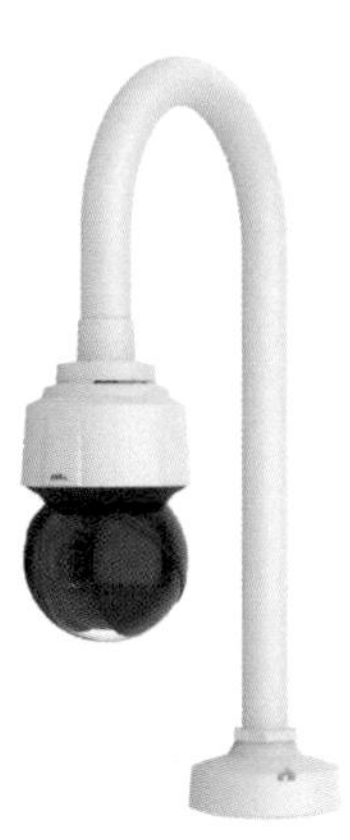

Figure 18.32 An example of a parapet-mounted camera.

ground vibrations. The pole and the mount must be able to absorb these vibrations and limit their impact on the camera. When calculating the sway, consider the height and diameter of the pole, as well as the material. Concrete poles sway less than metal and wooden poles. Factor in the weight and dimensions of the equipment the pole needs to carry. This is especially important for PTZ cameras and cameras with high optical zoom. If the pole is under-dimensioned, you risk an unreasonable amount of motion blur in the images. More advanced PTZ cameras have built-in electronic image stabilization to limit the effect of wind and vibrations. However, heavy cameras can cause serious injuries if they should fall. As with wall mounts, the cable can usually run inside the pole, and cable inlets and outlets must be sealed properly.

18.4.6.4 Parapet mounts

Parapet mounts (Figure 18.32) are used to mount cameras on rooftops or to raise the camera for a better angle of view. A benefit of parapet mounts is that the camera is cheaper and easier to service than if it hung from a wall mount. Because the arm can swing inwards, maintenance staff can access the camera from the rooftop rather than having to use lifts or other types of aerial work platform.

18.4.6.5 Special mounts

Some installations require special mounting solutions. When a camera needs to hang from the corner of a building, a corner adapter on a standard wall mount provides the solution.

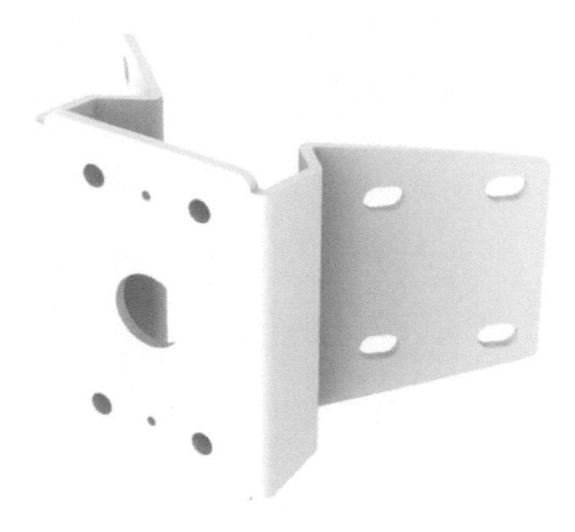

Figure 18.33 A corner adapter for fixing a standard mount kit to a corner. See also Figure 18.30 for a corner wall-mounted installation.

Figure 18.34 A telescopic pendant mount that swings if accidentally pushed.

A telescopic pendant mount allows a camera to hang from a high ceiling. If it has a ball joint, the pendant mount can hang from a sloped ceiling regardless of the angle. It can also swing to the side if it should get hit by a forklift or a tall slow-moving vehicle (Figure 18.34). Other special mounts include pipe adapters for mounting cameras on standard threaded pipes and recess mounts for mounting cameras on soffits. Some mounts work with both PTZ cameras and fixed dome cameras, although they sometimes require adapter kits.

Many outdoor camera installations require a cabinet or connection box for peripherals, such as power supplies or media converters (for connection to fiber-optic cables). Rather than a standard

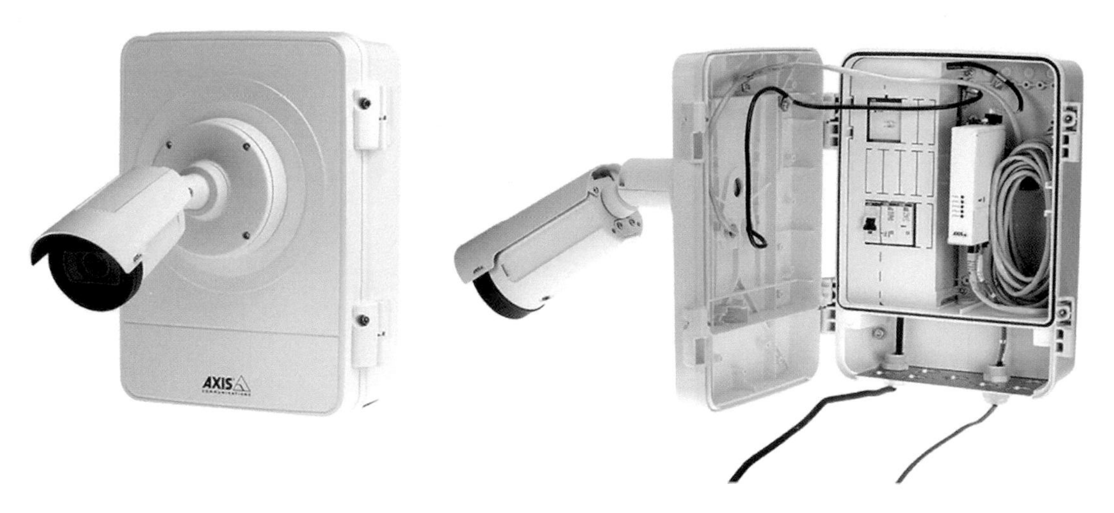

Figure 18.35 A cabinet with a bullet-style camera mounted. There is room inside for media converters, surge protectors, power supplies, switches, or other electrical peripherals.

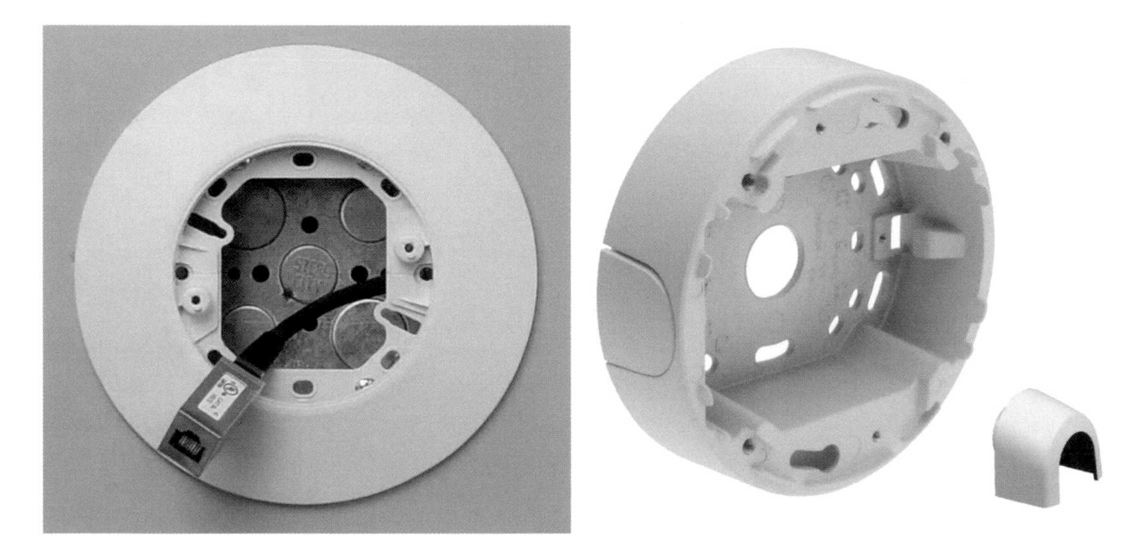

Figure 18.36 (a) Junction-box plate that fits a single gang box, double gang box or 4 inch square junction box, or a 4 inch octagon junction box. This plate is used to mount a small fixed dome camera. (b) Conduit back box that can be mounted on a hard, flat surface or a junction box. This is used to mount a fixed dome camera and fits a ¾ inch conduit or a conduit adapter, such as the one shown beside the back box.

connection box, which tends to spoil the visual appearance, you can use a cabinet with a built-in camera mount, which is more aesthetically pleasing and easier to install, as shown in Figure 18.35. Such cabinets can include modular attachment options, such as DIN rails and clips.

Sometimes, connections need to be housed in a junction box. These boxes make the connections neat and practical because they are accessible behind the front panel, whereas their concealment protects them from tampering. More importantly, they can contain any sparks or heat arising from loose connections and short circuits. Some camera vendors can supply a complete solution, including a wide range of modular mounting systems, cabinets, and conduit adapters, as well as back boxes and junction-box plates that fit both the gangs of standard junction boxes and the hole patterns of the camera mounts. For examples, see Figure 18.36.

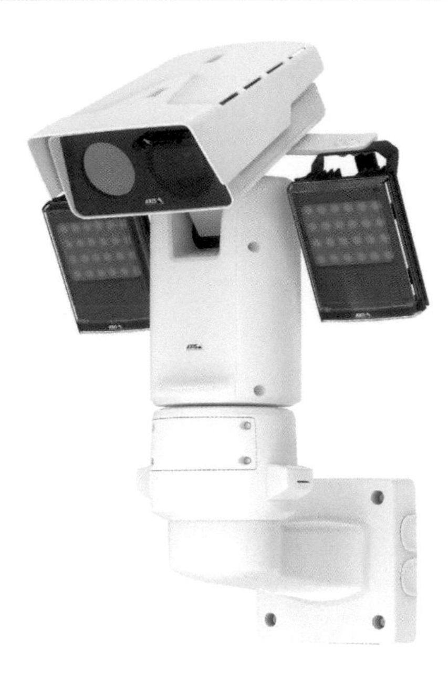

Figure 18.37 A pan-tilt motor with a bi-spectral camera and IR illuminators.

Pan-tilt motors bring remote pan and tilt functionality to fixed cameras. The motor itself is usually mounted on a pole or a wall. Sometimes the pan-tilt motor is the base for a bi-spectral camera (a thermal and a conventional camera in one) and IR illuminators, see Figure 18.37.

18.4.7 Electromagnetic compatibility (EMC)

All electric and electronic devices release electromagnetic energy, also known as radio frequency emission or RF emission. Emission is a byproduct of electrical or magnetic activity. Unfortunately, the emissions from one device can interfere with others, which can lead to data loss, image quality degradation on monitors and cameras, and can cause other equipment to malfunction. This is known as electromagnetic interference (EMI) or electromagnetic compatibility (EMC) problems.

The other main concept of EMC is immunity, which is a measure of how good a system is at rejecting interference from other devices. The opposite of immunity is susceptibility. It refers to the tendency of a product to malfunction or break down when exposed to emissions.

Before determining how to install a product in a particular environment, always take both emission and immunity into consideration.

18.4.7.1 EMC standards

All network video manufacturers must declare the electromagnetic compatibility of their products. In most of the world, this includes emission as well as immunity.

In Europe, electromagnetic compatibility is included in the CE mark (see Figure 18.42), which in turn is included in the EU's harmonization legislation. Conformance with the EMC directive is mandatory for network cameras, but manufacturers can self-certify their products. Usually, they show compliance through the following standards:

- *Emission*: EN 55022 and sometimes EN 61000-6-3 or EN 61000-6-4
- *Immunity*: EN 55024, EN 61000-6-1, and EN 61000-6-2

The standards that begin with EN 55022 and EN 55024 are product standards that are valid specifically for information technology equipment (ITE). They are harmonized with international standards

(CISPR 22) and, for example, the corresponding standards in Canada (ICES-003) as well as Australia and New Zealand (AS/NZS CISPR 22). The EN 61000–6 standards are generic standards.

In the USA, the Federal Communications Commission (FCC) stipulates the rules and regulations for telecommunication devices. Network cameras are included in the set of rules called CFR 47, Title 47, Part 15 Radio Frequency Devices, Subpart B–Unintentional Radiators. However, the FCC rules only refer to emission and not immunity.

Harmonic current emissions and flicker are covered by separate EU standards (EN 61000-3-2 and EN 61000-3-3). Moreover, there are additional standards for specific types of digital products or applications. For example, railroad applications should follow EN 50121-4 and IEC 62236-4, audio and audio-visual equipment EN 55130-4, and alarm systems EN 50130-4. There are also standards for wireless equipment.

As of October 2013, manufacturers can use the standard EN 55032 (CISPR 32) to show EMC compliance. This standard covers IT, audio, and video equipment, which is all grouped as multimedia equipment (MME). The corresponding standard for immunity for MME is EN 55035 (CISPR 35).

18.4.7.2 Emission

Emission refers to the ability of equipment to function satisfactorily without emitting too much electromagnetic energy that can disturb other equipment in that environment.

When manufacturers declare a product's compliance with emission requirements, they state which environmental category (Class A or Class B) the product can be used in. Although the limits are not identical in all regions, the relevant standards use the same categories:

- *Class A digital devices*: Intended for commercial, industrial, or business environments. Due to the higher level of radio frequency interference already present in commercial markets, the emissions requirements are less stringent than Class B devices.
- *Class B digital devices*: Intended for residential areas. Because these products are used in home environments, they must comply with stricter emission requirements than Class A devices.

18.4.7.3 Immunity

Immunity is a measure of the ability of electronic products to tolerate the influence of electromagnetic phenomena and electrical energy (radiated or conducted) from other electronic products. For example, any system connected to the AC mains power line must be immune to transient surges. The required immunity level depends on the type of appliance and the environment it is intended for.

The requirements for a product to withstand electrical disturbances or interference are the opposite of emission requirements. That is, the immunity requirements are higher in an outdoor, industrial, or similar environment than in a residential environment.

18.4.7.4 Shielded or unshielded network cables

In the USA, Class A devices can be connected by unshielded twisted-pair (UTP) network cables. This is because the FCC rules disregard immunity, and as mentioned earlier, the emission limits are less stringent for Class A devices than Class B devices.

When products have been tested against both the FCC Class A and the FCC Class B requirements, you can choose between the following:

- *Unshielded (UTP) network cables* – fulfilling the FCC Class A limits for emission.
- *Shielded (STP) network cables* – fulfilling the FCC Class B limits for emission.

In demanding electrical environments, it is best to use an STP network cable. Outdoor environments always fall into that category, so always use STP cables for outdoor cameras. Even indoor environments can be electrically demanding, for example, where the network cable runs parallel with the electrical mains supply cables or where large inductive loads such as motors or contactors are close to the camera or its cable. It is also mandatory to use an STP cable with an indoor camera or encoder if the cable runs partially or completely outdoors (see section 18.4.13).

In summary, most indoor surveillance deployments in the USA can use UTP cables and still fulfill the regulatory requirements. For the customer, UTP network cables are attractive because they are cheaper, less bulky, and allow simpler termination of the connector.

In Europe and most other countries, the limits for immunity require both Class A and Class B devices to be connected by STP cables.

For more information about shielded and unshielded cables, see Chapter 10.

18.4.8 Media converters

In some video surveillance installations, there may be a need to switch from one type of transmission media to another or to simply use the existing infrastructure and not replace it.

18.4.8.1 PoE over coax

One such case is when older installations include existing coaxial cable runs that were originally installed for analog cameras, instead of ripping out the old cabling connectivity and power to your network cameras over the coax cable. These devices are available in several different PoE power classes.

18.4.8.2 Ethernet-to-fiber converters

Ethernet can also be transmitted over fiber optic connections, which are typically used in city surveillance, airports, and other long-distance installations. A network camera can connect via RJ45 to a media converter switch that provides one or more SFP slots for connecting the camera to the network over long distances.

18.4.9 Safety of electrical equipment

The low voltage directive (LVD) provides broad objectives for the safety of electrical equipment. Its purpose is to align the certification of electrical products so that they are safe to use in all EU member states. Just like the EMC directive, LVD is included in the CE mark, and manufacturers can self-certify their products.

For information technology equipment (ITE), compliance with LVD is usually shown through the EN 62368–1 standard and its international equivalent, IEC 62368–1. IEC stands for the International Electrotechnical Commission, an organization that publishes standards for all electrical, electronic, and related technologies. In the USA, the Occupational Safety and Health Administration (OSHA) has given Underwriters Laboratories (UL) the right to perform certification of information technology equipment. This standard is known as UL 62368–1.

The standards' identical numbers indicate that they are harmonized, which means that in essential areas, they are equal. However, there are regional differences. Therefore, products that are marketed and sold on multiple markets need to show all the required certifications. In the USA, Underwriters

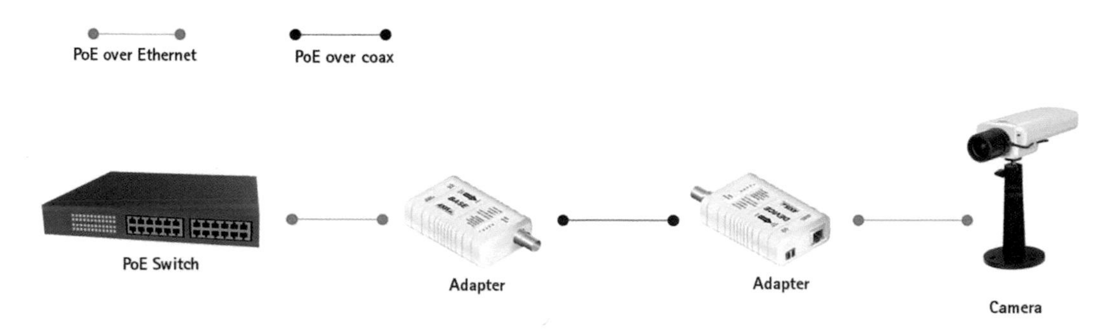

Figure 18.38 A PoE-over-coax converter kit provides power and connectivity over existing coaxial cable.

Laboratories (UL) performs certification of information technology equipment to ensure compliance with the requirements of both the construction and general industry OSHA electrical standards. In Canada, the equivalent certification body is the Canadian Standards Association (CSA), but usually, the European or American certifications are sufficient for this market.

IEC/EN/UL 62368-1 applies to network cameras, encoders, and their power supplies. Outdoor products must also comply with IEC/EN/UL 60950-22. Products with built-in LEDs also need to comply with EN 62471, which among other things includes exposure limits to prevent hazards to eyes and skin.

As mentioned earlier, the purpose of the safety standards is to ensure that products are safe to use without risk of personal injury or property damage caused by hazards, such as electric shock, fire, dangerous temperatures, and mechanical instability.

In outdoor or demanding electrical environments or if cables are routed outdoors, devices must be protected against power surges. There are several ways to ensure that a power surge has a path to ground. One way is to use power sourcing equipment (a midspan, a network switch, a power supply, or other end device) that is properly grounded and to use STP network cables to connect all devices. If the camera has a grounding screw, both ends of the grounding wire must be in contact with their respective grounding surfaces. For power supplies, the safety standards include a specific set of conditions, such as limited power source (LPS) and safety extra low voltage (SELV). Simply put, if a power supply fulfills the requirements for SELV, this means it is built in such a way that its voltage stays within safe values.

Figure 18.39 An Ethernet-to-fiber converter.

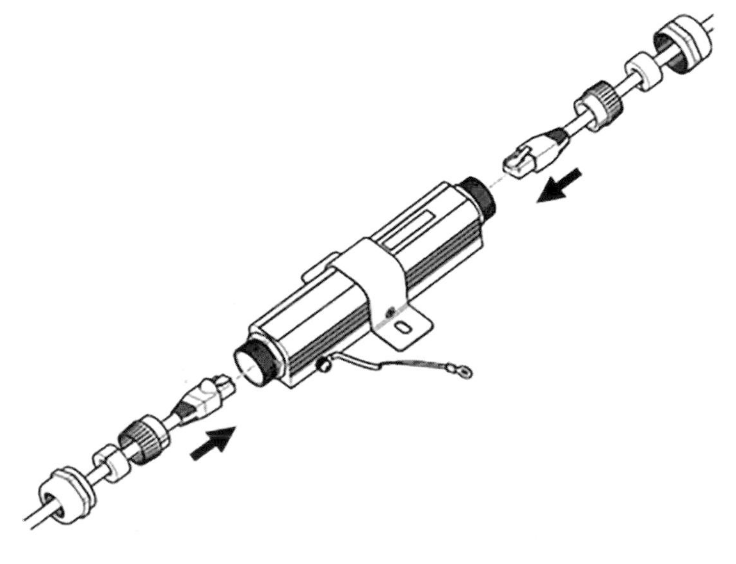

Figure 18.40 An example of an in-line surge protector.

An installation should follow the safety standards and the national electrical code both to protect the investment (limiting the risk of damage through power surges) and to make the electrician's job easier, faster, and cheaper.

18.4.10 Environmental ratings

Various standards groups have defined different classes of protection, so there are several different environmental classifications that manufacturers can label their products with. The most common ratings are IP, NEMA, and IK, which are explained next.

18.4.10.1 IP ratings

Ingress protection (IP) ratings define the level of protection an electrical appliance has against the intrusion (ingress) of solid objects and liquids. IP is sometimes interpreted as standing for international protection. The ratings are based on the harmonized international and European standard IEC/EN 60529. Like the safety standards for safety of electrical equipment, the IP standard is connected to the low voltage directive, and manufacturers can self-certify their products. Although companies must ensure that they fulfill the basic requirements of the standard, some subject their products to tougher tests than others. For example, an IP test is more difficult to pass if the product has first been impact-tested.

Figure 18.41 shows an example of an IP66-rated enclosure designed for housing system devices. It allows a non-IP-rated device such as a PoE switch or a surge protector to be installed outdoors and still be protected against the elements. Many network cameras are supplied as outdoor-ready, meaning they, too, have an IP rating, which is often IP66.

An IP rating consists of the letters IP followed by two digits (e.g., IP66) and sometimes two extra letters. The first digit indicates the level of protection an enclosure provides against access to hazardous parts and the ingress of solid foreign objects. Examples of hazardous parts are electrical conductors or moving parts. Examples of solid objects are fingers, tools, and dust. The higher the number, the better the protection. The second digit indicates the level of protection against intrusion by liquids. Again, the higher the number, the better the protection. For example, an IP66 rating

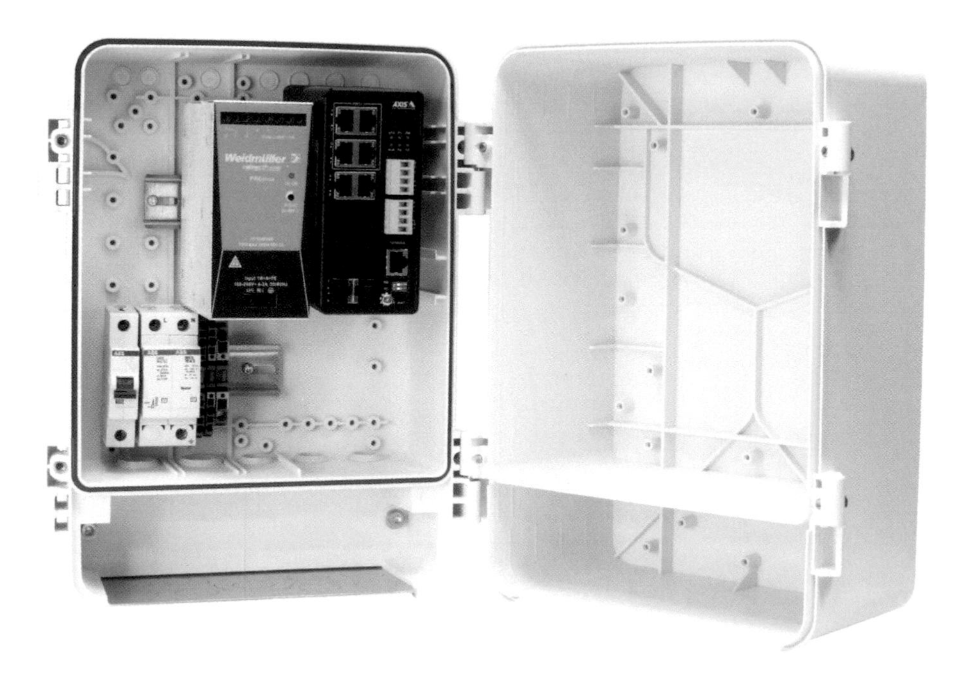

Figure 18.41 An example of an IP66-rated protective enclosure for system devices.

means it is dust-tight and protects against ingress from powerful water jets. Products intended for outdoor use should have an IP rating of at least IP44, although most outdoor installations require IP66. When there is no protection rating given for either solid objects or liquids, the letter X is used (for example, *IP2X*). See Table 18.3 and Table 18.4 for explanations of the first two digits. The extra two letters are optional and rarely used, so are not included in the tables.

If the second digit is 6 or lower, this implies compliance also with the requirements for all levels below it. So an IP65 enclosure is automatically approved for IP55 or IP64 levels of protection. However, equipment that is has a second digit of 7 or 8 should not be used where it might be exposed to water jets. Therefore, never assume that an IP67-rated product can also withstand IP66 environments unless it is dual-coded, that is, IP66/IP67.

18.4.10.2 NEMA ratings

The National Electrical Manufacturers Association (NEMA) is a US-based association that provides standards for electrical equipment enclosures. NEMA has adopted and published a harmonizing IP

Table 18.3 IP ratings-first digit: Foreign solid objects

Level	Protected against	Effective against
0	Not protected	No protection.
1	Objects larger than 50 mm	A large surface of the body such as back of the hand but no protection against deliberate contact with a body part.
2	Objects larger than 12.5 mm	Fingers or other objects can penetrate as far as 80 mm if safe from hazardous parts. Objects with a diameter of 12.5 mm cannot penetrate fully.
3	Objects larger than 2.5 mm	Objects, such as tools and thick wires, cannot penetrate at all.
4	Objects larger than 1 mm	Objects, such as wires and screws, cannot penetrate at all.
5	Dust-protected	Ingress of dust is not completely prevented, but dust does not enter in sufficient quantity to interfere with satisfactory operation of the equipment.
6	Dust-tight	No ingress of dust.

Table 18.4 IP ratings-second digit: Liquids

Level	Protected against	Effective against
0	Not protected	No special protection.
1	Dripping water	Dripping water (vertically falling drops) has no harmful effect.
2	Dripping water when tilted up to 15°	Vertically dripping water has no harmful effect when the enclosure is tilted at any angle up to 15° from its normal position.
3	Spraying water	Water falling as spray at an angle up to 60° from the vertical has no harmful effect.
4	Splashing water	Water splashed against the enclosure from any direction has no harmful effect.
5	Water jets	Water projected from a nozzle against the enclosure from any direction has no harmful effect.
6	Powerful water jets	Water from heavy seas or water projected in powerful jets cannot enter the enclosure in harmful quantities.
7	Brief immersion in water	Ingress of water in a harmful quantity cannot be possible when the enclosure is immersed in water under defined conditions of pressure and time.
8	Continuous submersion in water	The equipment is suitable for continuous submersion in water under conditions which shall be specified by the manufacturer. The conditions must be harsher than for IP 7.
9	Water from high pressure and steam jet cleaning	Water directed at the housing from any angle under very high pressure has no harmful effect.

standard, ANSI/IEC 60529, through the American National Standards Institute (ANSI). However, they also have their own standard, NEMA 250, which they have launched successfully on the global market.

Like the IP standard, NEMA 250 addresses ingress protection, but it also considers other items such as corrosion resistance, performance, and construction details (see Table 18.5). Therefore, it is safe to say that a NEMA type is comparable to an IP rating, but it would be wrong to state the opposite. This table **cannot** be used to convert IP ratings to NEMA types, and equivalence should always be verified through testing.

The UL standards for enclosures, UL 50 and UL 50E, are based on the NEMA 250 standards. The major difference between them is that while NEMA allows self-certification, UL enforces compliance by demanding that products pass third-party testing and inspection.

18.4.10.3 IK ratings

Many security cameras are placed in environments where they are subjected to various kinds of impacts. The most obvious causes of impact are vandalism and other physical attacks, but falling branches, debris caught in the wind, climbing animals, and perching birds can also cause damage. Also, even an experienced installer can drop a camera.

For enclosures, there is a standard that specifies degrees of protection against external mechanical impact. It was originally approved in 1994 by the European Committee for Electrotechnical Standardization (CENELEC) as the European standard EN 50102. When it was adopted as an international standard in 2002, it changed numbers to IEC/EN 62262. The standards are identical, and both are still valid.

Similar to the IP ratings, the degrees of protection against impact are indicated by a code that consists of the letters IK followed by two digits (see Table 18.6).

Tests are used to demonstrate an acceptable level of robustness when assessing the safety of a product. The main concern is to make sure that the inner workings of a product are properly protected by the enclosure. Although the product inside the enclosure needs to be safe from accidental or intentional probing after impact, it does not necessarily have to be operational. During the test, each exposed surface is hit five times, evenly distributed over the surface. The same point, or area around it, cannot be hit more than three times. The points to which impact should be applied is specified in the relevant product standard. The product standard may also include exceptions to the rule of maximum five hits. After the test, the product needs to be evaluated: Is the damage admissible? Is the product still safe and reliable? So really, an IK10 rating does not mean that the product is resistant to impact but is instead more a measure of robustness.

Manufacturers who are serious about providing high-quality products go further than the requirements of the standard. To ensure that the product retains its level of robustness throughout its lifespan, they may test the weakest part of the camera rather than the strongest. As mentioned in the section on IP ratings, they may also extend their efforts by doing the IP tests after they have performed the IK tests. Others may check that the product is still operational after the IK test. This does not imply that it retains its IP rating, although a well-engineered product could be assumed to tolerate such an environment.

18.4.10.4 Explosive atmospheres

An explosive atmosphere is defined as an area where flammable substances, such as liquids, vapors, gases, or combustible dusts, are likely to occur and mix with air in such proportions that excessive heat or sparks might make them explode. Examples include gas stations, oil platforms and refineries, chemical processing plants, printing industries, gas pipelines and distribution centers, grain handling and storage, aircraft refueling and hangars, and hospital operating theaters.

It is the explosive environment that must be protected against potential igniters from the camera and other equipment. In other words, the camera must be explosion-protected, explosion-proof, or

Table 18.5 NEMA ratings for enclosures in nonhazardous locations

NEMA	IP	Indoor	Outdoor	Protected against
Type 1	IP10	•		Access to hazardous parts and ingress of solid foreign objects (falling dirt). No protection against liquids.
Type 2	IP11	•		Access to hazardous parts and ingress of solid foreign objects (falling dirt). Ingress of water (dripping and light splashing).
Type 3	IP54	•	•	Access to hazardous parts and ingress of solid foreign objects (falling dirt and windblown dust). Ingress of water (rain, sleet, snow). Will be undamaged by the external formation of ice on the enclosure.
Type 3R	IP14	•	•	Access to hazardous parts and ingress of solid foreign objects (falling dirt). Ingress of water (rain, sleet, snow). Will be undamaged by the external formation of ice on the enclosure.
Type 3S	IP54	•	•	Access to hazardous parts and ingress of solid foreign objects (falling dirt and windblown dust). Ingress of water (rain, sleet, snow). The external mechanisms remain operable when ice-laden.
Type 3X		•	•	Access to hazardous parts and ingress of solid foreign objects (falling dirt and windblown dust). Ingress of water (rain, sleet, snow). Provides an additional level of protection against corrosion. Will be undamaged by the external formation of ice on the enclosure.
Type 3RX		•	•	Fulfills the requirements of NEMA 3R and NEMA 3X.
Type 3SX		•	•	Fulfills the requirements of NEMA 3S and NEMA 3X.
Type 4	IP56	•	•	Access to hazardous parts and ingress of solid foreign objects (falling dirt and windblown dust). Ingress of water (rain, sleet, snow, splashing water, and hose-directed water). Will be undamaged by the external formation of ice on the enclosure.
NEMA 4X	IP56	•	•	Access to hazardous parts and ingress of solid foreign objects (windblown dust). Ingress of water (rain, sleet, snow, splashing water, and hose-directed water). Provides an additional level of protection against corrosion. Will be undamaged by the external formation of ice on the enclosure.
Type 5	IP52	•		Access to hazardous parts and ingress of solid foreign objects (falling dirt and settling airborne dust, lint, fibers, and flyings). Ingress of water (dripping and light splashing).
Type 6	IP67	•	•	Access to hazardous parts and ingress of solid foreign objects (falling dirt). Ingress of water (hose-directed water and the entry of water during occasional temporary submersion at a limited depth). Will be undamaged by the external formation of ice on the enclosure.
Type 6P	IP67	•	•	Access to hazardous parts and against ingress of solid foreign objects (falling dirt). Ingress of water (hose-directed water and the entry of water during prolonged submersion at a limited depth). Provides an additional level of protection against corrosion. Will be undamaged by the external formation of ice on the enclosure.
Type 12	IP52	•		Without knockouts. Access to hazardous parts and ingress of solid foreign objects (falling dirt and circulating dust, lint, fibers, and flyings). Ingress of water (dripping and light splashing).
Type 12K	IP52	•		With knockouts. Access to hazardous parts and ingress of solid foreign objects (falling dirt and circulating dust, lint, fibers, and flyings). Ingress of water (dripping and light splashing).
Type 13	IP54	•		Access to hazardous parts and ingress of solid foreign objects (falling dirt and circulating dust, lint, fibers, and flyings). Ingress of water (dripping and light splashing). Spraying, splashing, and seepage of oil and noncorrosive coolants.

flame-proof. These terms are used synonymously and mean that the camera will contain any explosion originating within its housing and will not generate conditions that could ignite vapors, gases, dust, or fibers in the surrounding air. This term does not mean that the camera itself will withstand an exterior explosion.

Table 18.6 IK ratings

Level	IK01	IK02	IK03	IK04	IK05	IK06	IK07	IK08	IK09	IK10	IK10+
Impact energy (Joule)	0.14	0.2	0.35	0.5	0.7	1	2	5	10	20	50+ (a)
Mass (kg)	0.2					0.5		1.7	5		
Drop height (mm)	56	80	140	200	280	400	400	300	200	400	

(a) IEC/EN 62262 provides for a maximum resistance of IK10 at 20 J with the possibility of extending the impact energy up to 50 J. Some types of equipment need more protection. Therefore, the market has extended the test beyond what the standard provides. These IK ratings are known as IK10+. When using such a rating, the manufacturer should indicate the impact energy, mass, and drop height of the striking element.

Figure 18.42 ATEX and CE marks.

18.4.10.5 IECEx and ATEX certifications

When a camera is installed in a potentially explosive environment, its housing must meet very specific safety standards. The international standard is known as IECEx or *International Electrotechnical Commission System for Certification to Standards Relating to Equipment for Use in Explosive Atmospheres.*

In the US, explosion-protected electrical equipment is classified according to the class/division system described in the *National Electrical Code (NEC), article 500.* The Canadian equivalent is the *Canadian Electrical Code (CEC), section 18.*

In Europe, products intended for these environments must comply with the ATEX directive, which stands for *appareils destinés à être utilisés en atmosphères explosibles* (in English, *equipment for potentially explosive atmospheres).* ATEX-certified products bear the hexagon-shaped Ex mark and CE mark (see Figure 18.42).

The rest of the world uses a zone system described in the IEC 60079 set of standards for the IECEx certification.

Becoming ATEX- or IECEX-certified is a complicated process. Some camera manufacturers self-certify their own enclosures, and it may be advantageous to obtain both camera and enclosure from the same supplier. However, most manufacturers usually submit their explosion-protected products for testing and certification by a notified body (NB), which is appointed by the member state, or an *Ex Certification Body* (ExCB), which is appointed by IECEx. ATEX has an NB exception for gas and dust atmospheres that are classified as low-risk atmospheres – that is, areas in which an explosive mixture is not likely to occur in normal operation, and if it does occur, it will exist only for a short time.

Explosive atmospheres are divided into categories, gas and dust, where each category is divided into zones. See Table 18.7. There are also two categories of equipment, one for mining (I) and one for surface industries (II).

18.5 STORAGE AND SERVER CONSIDERATIONS

Depending on system size and requirements, setting up the video management part of a network video system can be anything from a simple five-minute task to a very complex and time-consuming activity. Designing a server and storage system begins with a few basic decisions: How much storage is needed? Is redundancy a requirement? Do you prefer an on-premises setup, or should some functions be cloud-based? The required performance of the servers, possible system

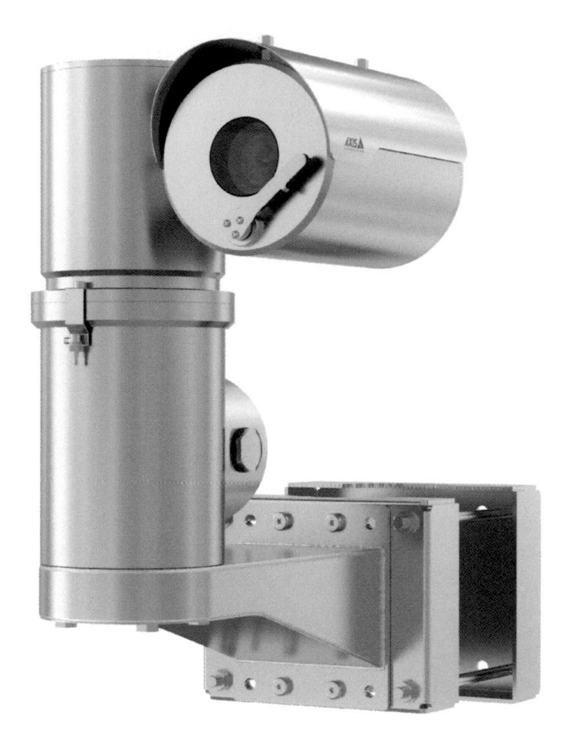

Figure 18.43 Example of a PTZ camera designed and certified for use in potentially explosive atmospheres and harsh environmental conditions. The stainless-steel housing makes it suitable for offshore, onshore, marine, and heavy industrial environments.

Table 18.7 Zone divisions of explosive atmospheres

Zone (IECEx and ATEX)	Atmosphere	Definition	Class and division (NFPA 70)
Zone 0	Gas	An area in which an explosive mixture is continuously present or present for long periods.	Class I Division 1
Zone 1	Gas	An area in which an explosive mixture is likely to occur in normal operation.	Class I Division 1
Zone 2	Gas	An area in which an explosive mixture is not likely to occur in normal operation, and if it occurs, it will exist only for a short time.	Class I Division 2
Zone 20	Dust	An area in which an explosive mixture is continuously present or present for long periods.	Class II Division 1
Zone 21	Dust	An area in which an explosive mixture is likely to occur in normal operation.	Class II Division 1
Zone 22	Dust	An area in which an explosive mixture is not likely to occur in normal operation, and if it occurs, it will exist only for a short time.	Class II Division 2

architecture, and suitable storage setups also affect which video management software you use. For more information on storage and servers, see Chapter 13.

18.5.1 System component distribution

Video surveillance systems can be set up in several different ways. Exactly how this is done depends a lot on what you require from the system. For example, for a small system, it may suffice to install everything locally on your own hardware. On the other hand, even if your system is small, you might still require the benefits of other system types where some components are located elsewhere, in the cloud, for example, for greater convenience or for recording redundancy. The larger your system, the more likely it is that you would require these benefits, especially when dealing with multiple sites.

- *On-prem(ises):* Whether the system is for a small retail store or a large and complex installation at an airport, there are pros and cons of having all the equipment on the premises. One of the main benefits is that no video needs to be transported over the internet for recording. There might also be a requirement to keep the system air-gapped (disconnected) from other systems, from a cybersecurity point of view. In small systems, storage can be edge-based (in the camera), on a server, or on a network-attached storage (NAS) device. Larger systems connect cameras to NVRs with or without built-in PoE ports. System management (the VMS) and viewing can be provided by a simple mobile or desktop app.
- *Hybrid solutions:* Hybrid means a combination of on-prem and cloud-based solutions. The video is normally viewed and managed from a client rather than from the recording server. The system is often distributed so that some components, the VMS or the storage (some or all of it), are located elsewhere, usually on a cloud server. This type of solution may be offered by the camera manufacturer, or there may be a third-party hosting company involved. Another type of hybrid system is one in which recording of video is on-prem (edge-based or on local server/NVR) but where all recordings are also copied to cloud servers for redundancy.
- *Cloud solutions:* Most or all the components possible to deploy remotely are located on cloud servers. The only components at the site might be the cameras and their associated hardware. All system access is via mobile and desktop apps. Managing large amounts of data and bandwidth requires multiple high-performing and reliable servers dedicated to video management tasks. This setup requires IT staff for a range of tasks: load balancing, scaling up the system as required, or performing maintenance without disrupting the service. For many users, maintaining this type of infrastructure themselves is not an option, so leaving it to a supplier that provides cloud-based services gives them a dependable and professional surveillance system that fulfills their requirements while leaving them to manage their core business.

18.5.2 Multi-site systems

In an enterprise installation, such as a global company, a large hospital, or a school district with many buildings, there are often several video surveillance systems located at different facilities. Some of those systems might be small, some midsized, and some large. In a federated enterprise system, these subsystems are all tied together into one system, meaning that all cameras and systems can be managed, monitored, and maintained from one location (Figure 18.44). Each site records and stores the video from local cameras.

18.5.3 Provisioning the server

A server used for video management should be properly provisioned to handle the current camera system and should be possible to scale up as and when needed. Each server has a certain baseline of how many cameras it can handle, based on its CPU (central processing unit), network interface, internal RAM (random-access memory), and hard drives. You also need to factor in the total number of cameras, their resolution and frame rate, and the retention goal of the system.

Depending on the task and the load it creates, some physical servers use only a fraction of their processing power and memory. To save on hardware, reduce rack space, and limit power consumption, many systems use virtual servers, where the same server supports several systems and applications. But it can also mean that several physical servers are merged into a single more powerful machine to handle an increased load. A dedicated server running a single application is always faster than a server running multiple applications.

The server racks should be well-organized (see Figure 18.45), and the surveillance system should not share cables with other systems such as point-of-sale (POS) systems.

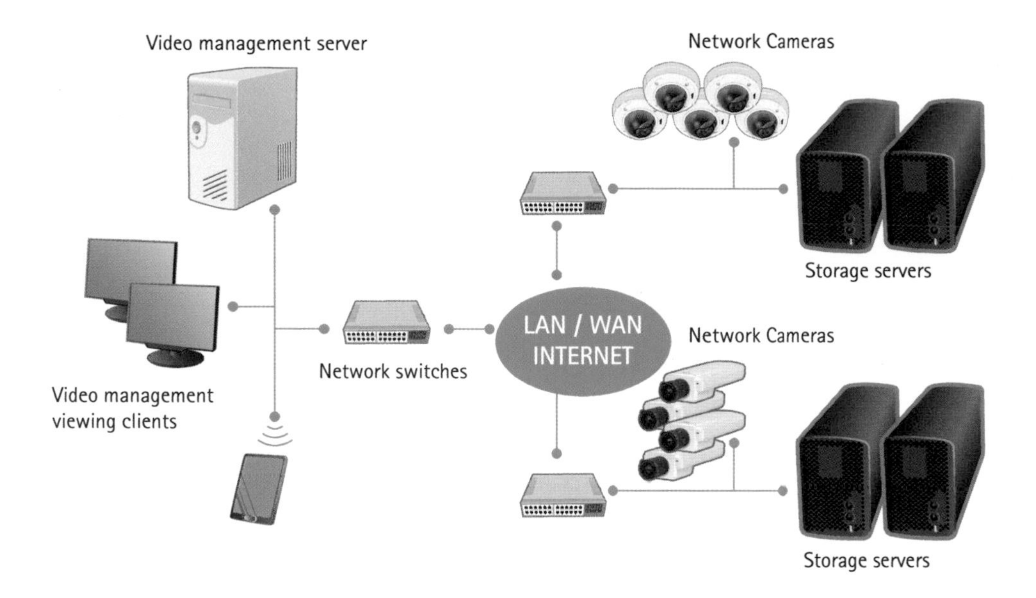

Figure 18.44 A multi-site system that is also federated into one system.

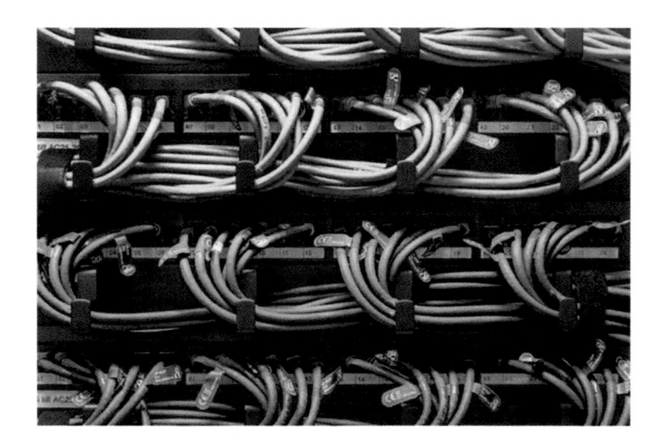

Figure 18.45 Network cables in a server rack.

18.5.4 Calculating storage

Calculating the appropriate amount of storage is a very important task when designing a video surveillance system. This is not an exact science because the size of video files depends on the complexity and amount of motion in a scene. Some guidelines, along with an example of how the amount of storage can be limited by changing various system parameters, are provided here.

Factors to consider when calculating storage needs:

- *Number of cameras*
- *Number of hours per day the camera will be recording*
- *How long the data must be stored – also known as the retention time*
- *Event-triggered recordings only or continuous recording*
- *Other parameters such as frame rate, compression, image quality, and complexity*

Note that the sample values provided here do not take into consideration any overhead or other technical issues that could result in greater file sizes. Nor do they consider any storage space

Table 18.8 Sample bitrates and storage requirements for H.264 (24 h recording)

		No smart codec		With smart codec	
Resolution	Frames per second	Bitrate (Mbit/s)	Storage GB/day	Bitrate (Mbit/s)	Storage GB/day
HDTV 1080p	5	0.33	3.60	0.31	3.34
	15	0.93	10.0	0.44	4.79
	30	1.75	18.9	0.52	5.58
4 Mpixel	5	0.63	6.77	0.58	6.22
	15	1.75	18.9	0.83	8.92
	30	3.30	35.7	0.96	10.4
8 Mpixel	5	1.20	13.0	1.09	11.8
	15	3.36	36.3	1.56	16.9
	30	6.37	68.8	1.81	19.6

Table 18.9 Sample bitrates and storage requirements for H.265 (24 h recording)

		No smart codec		With smart codec	
Resolution	Frames per second	Bitrate (Mbit/s)	Storage GB/day	Bitrate (Mbit/s)	Storage GB/day
HDTV 1080p	5	0.22	2.35	0.20	2.14
	15	0.61	6.54	0.29	3.09
	30	1.14	12.4	0.36	3.62
4 Mpixel	5	0.41	4.37	0.37	3.94
	15	1.13	12.2	0.53	5.68
	30	2.13	23.0	0.61	6.64
8 Mpixel	5	0.76	8.24	0.68	7.32
	15	2.13	23.0	0.97	10.5
	30	4.04	43.6	1.14	12.3

required for the operating system or video management software. Also consider that formatting the drive reduces the amount of available storage. For example, a 2 TB drive might yield 1.82 TB of available memory after formatting.

18.5.4.1 Bitrate and storage for H.264/H.265

The values in Table 18.8 and Table 18.9 show sample storage calculations for the video compression formats H.264 and H.265, with values supplied for both with and without smart codecs applied. Because of the number of variables that affect average bitrate levels in H.264 and H.265, these calculations are provided as guidelines only and should not be taken as absolute values.

The formula for calculations is:

$$\frac{Approximate\ bitrate}{8} \times 3600\ s = \frac{MB\ per\ hour}{1000} = GB\ per\ hour$$

$$GB\ per\ hour \times hours\ of\ operation\ per\ day = GB\ per\ day$$

$$GB\ per\ day \times retention\ time\ in\ days = Storage\ need$$

The amount of motion in a scene can have a significant impact on the amount of storage required. The values in Table 18.8 and Table 18.9 are based on the use of a high-end network camera performing continuous recording of an indoor scene at a retail store for 24 hours per day, with an image compression level of 30. The samples with a smart codec use the *medium* setting.

18.5.4.2 Bitrate and storage for Motion JPEG

Calculating the same requirements for Motion JPEG is more clear-cut, as this format consists of multiple individual image files. Motion JPEG recordings vary depending on the frame rate, resolution, and level of compression. For most surveillance applications, the amount of bandwidth and storage using Motion JPEG would be many times higher than with H.264/H.265 and is thus rarely used, so no samples are provided here, only the formula.

Calculating bitrate and storage for Motion JPEG:

$$Image\ size\ x\ frames\ per\ second\ x\ 3600s = \frac{KB\ per\ hour}{1{,}000{,}000} = GB\ per\ hour$$

$$GB\ per\ hour \times hours\ of\ operation\ per\ day = GB\ per\ day$$

$$GB\ per\ day \times retention\ time\ in\ days = Storage\ requirement$$

18.5.5 Bits or bytes?

There is often confusion as to what is meant by various terms such as Mb, MB, Mbit/s, MB/s, etc. Which is for capacity? Which is about speed?

18.5.5.1 Data sizes

Starting off at the lower end; the smallest unit of digital information is one binary digit, more commonly known as a *bit* (b), which can have one of two possible values, often represented as 1 or 0 but sometimes as on/off, true/false, yes/no, +/-.

As bits are such small increments of data, we use multiples to talk about larger values:

- 1 thousand bits = 1 kilobit (kbit)
- 1 million bits = 1 megabit (Mbit)

Bits can also be represented in groups, by using the *byte*. Although there are exceptions and variants, the modern standard as defined by ISO/IEC 2382–1:1993 is that 1 byte (B) = 8 bits (b), where historically, these eight bits are the minimum required to represent a single character of text.

The same system of prefixes applied to bits is also used for bytes:

- 1 million bytes = 1 megabyte (MB)
- 1 billion bytes = 1 gigabyte (GB)

When working with multiples of bits and bytes, it is useful to know that computer architectures are binary systems and not decimal – that is, they work with base two instead of base ten. This means that the exact totals in the two systems will differ, even if the various prefixes (kilo, mega, giga, etc.,) are often used for both decimal and binary values.

For example, to multiply 1 byte by 1,000:

- In base ten (decimal), this would be 10^3 = 1,000
- In base two (binary), the nearest equivalent is 2^{10} = 1,024

Although slightly different, both these values are generally given as "1 kilobyte". This is because, for most people, it is much easier to count in decimal than in binary. Hard drive storage space is routinely specified using decimal values, where, for example, 1 GB is defined as exactly one billion bytes. However, as the computer using the drive runs on a binary system, a disk specified by

Table 18.10 Comparisons of data sizes

Unit	Size (binary)	Size (decimal/metric)
Kilobyte (KB)	1,024 Bytes	1,000 bytes
Megabyte (MB)	1,024 Kilobytes	1,000,000 bytes
Gigabyte (GB)	1,024 Megabytes	1,000,000,000 bytes
Terabyte (TB)	1,024 Gigabytes	1,000,000,000,000 bytes
Petabyte (PB)	1,024 Terabytes	1,000,000,000,000,000 bytes
Exabyte (EB)	1,024 Petabytes	1,000,000,000,000,000,000,000 bytes

a vendor at 10 GB will be seen by the computer as having only 9.31 GB, as in a binary system: 10 GB = 10,737,418,240 bytes.

As there are 8 bits in every 1 byte, it also follows that:

- 1 Mb = 0.125 MB
- 8 Mb = 1 MB

See Table 18.10 for further comparisons of data sizes.

18.5.5.2 Data transfer rates

Specifying the rate at which data is transferred is common practice. For example, an internet connection may be specified at 50 megabits per second (Mbit/s). Although technically, this is a measure of capacity (number of bits moved per second), values in Mbit/s are generally viewed as speeds (e.g., a high-capacity connection is perceived as being "fast").

For network video, remember that data is sent over a network "bit by bit" and not in bytes. This means that Mbit/s can be useful to describe a connection over which you might view a live video stream, that is, when not downloading and saving the video. A "fast" connection will give a smoother viewing experience (by reducing latency) and will better allow the viewing of high-resolution video. If, instead, you simply wish to download and save the video, then the connection speed is of less importance. A "slower" connection might take several minutes longer than a "fast" connection to download the video, but does that really matter?

Bytes, megabytes, and gigabytes are mostly used to specify data sizes, for example, the size of an individual file or the storage capacity of a hard drive. However, MB/s can also be used to specify data transfer rates, especially when downloading data, for example, when saving a video to disk. As one byte is the smallest useful amount of data in computer systems, it makes more sense to specify these rates in MB/s.

Also relevant to this discussion is that bits transferred over a network do not always arrive in the correct order nor do they necessarily originate from the same source. This is not a problem, as the application receiving the data bits assembles them in the correct order, but it serves to further highlight the fact that specifying bits for transmission speeds across networks is often more useful than bytes.

Finally, remember that manufacturers and vendors of IT equipment and services may promote the numbers that would appear to give the best value for money (speed in this case). This can mean that a value is given in bits rather than bytes. For example, an internet service provider (ISP) specifies a particular broadband service at 100 Mbit/s. If you use this connection to download a file of 100 MB, this will take eight seconds to complete, as the write speed converted to bytes/s is 12.5 MB/s (100 megabytes = 800 megabits).

18.6 PROVISIONING NETWORK BANDWIDTH

When designing a network video system, it is important to properly design the network and associated bandwidth. In the early days of network video, bandwidth was limited and network design was a major challenge. Today's networks, where we count in tens or hundreds of gigabits, make this less challenging because, if designed correctly, they can easily cope with today's large amounts of network video. Network video products use network bandwidth based on their configuration. Bandwidth usage, as with storage, depends on image resolution, compression, and frame rate, as well as the complexity of the scene.

18.6.1 Limiting the bandwidth

There are several technologies available that enable the management of bandwidth consumption, including the following:

- *Switched networks and VLANs:* By using virtual local networks (VLANs) on a switched network – a common technique today – the same physical computer and video surveillance network can be separated into two autonomous networks. Although these networks remain physically connected, the network switch logically divides them into two virtual and independent networks. Fiber interconnections between switches are ideal for systems with multiple sites where intermediate distribution frames (IDFs) and main distribution frames (MDFs) can interconnect through a fiber backbone.
- *Quality of service (QoS):* Using QoS will guarantee a certain bandwidth to a specific application or to a certain camera in a video surveillance system.
- *Multicast:* With multicast, you can reduce bandwidth on the network with multiple clients requesting a video stream. Multicast requires that all the networking equipment also supports multicast.
- *Event-driven frame rate:* A frame rate up to 25 or 30 fps on all cameras all the time is not always required by all applications. With the configuration capabilities and built-in analytics of network cameras and video encoders, frame rates under normal conditions can be set lower – for example, 5 fps – which will dramatically decrease bandwidth consumption. In the event of an alarm (e.g., if motion detection is triggered), the recording frame rate can be increased automatically. In many cases, it is sufficient to have the camera send video over a network only if the video is worth recording; the rest of the time, nothing needs to be sent. Some smart codecs include dynamic frame rate functionality, which automatically provides an event-driven frame rate.

18.6.2 Network and system latency

In a video surveillance system, it may be important to consider latency. This is especially true when the video is being monitored live and when a PTZ camera is used. A well-designed network should have very little latency, typically in the hundred-millisecond range. In a network with many hops, latency can be one second or more, which can present problems.

More important in a surveillance system is to measure the total end-to-end system latency, from live image to viewing on the monitor. Factors that may affect latency include shutter speed, encoding, cable infrastructure, network, video management software, server hardware, client hardware, and decoding.

18.6.3 Network cabling

Never underestimate the importance of the cables in a wired network. Poorly or incorrectly installed network cables can cause numerous problems. Even the smallest cabling issue can have serious effects on the operation of the network. A kink in a cable can cause a camera to respond intermittently, and a poorly crimped connector may prevent Power over Ethernet (PoE) from functioning properly. Consider the following when installing cables:

- Use the correct wiring standards.
- Do not combine the wiring standards T568a and T568b on the same cable.

- Use high-quality cables with the relevant CAT rating. This is most often CAT6 or CAT7 for devices and CAT8 or fiber in server rooms.
- Where possible, ensure the cable system is certified for its category rating via electronic testing to ensure distance and interference fall within acceptable parameters.

Cables are categorized according to the data rates that they can transmit effectively. The specifications also describe the material, the connectors, and the number of times each pair is twisted per meter. Nowadays, the most used cable type is CAT6 or better. Make sure that the cabling you use matches the requirements of the installation.

- CAT3 with 16 MHz bandwidth (no longer used)
- CAT5e with 100 MHz bandwidth (little used)
- CAT6 up to 250 MHz
- CAT6a up to 500 MHz
- CAT7 up to 600 MHZ
- CAT7a with a frequency range through 1,000 MHz

For gigabit connectivity and a future-proof installation, even if your existing network switches and routers only support 100 Mbps, it is still a good idea to use CAT6 cabling or better. This way, you avoid having to purchase and reinstall the cabling infrastructure when the bandwidth requirements increase.

Cabling for the video surveillance system should be done by certified construction contractors. The work should include the following:

- Pulling the cables in the ceiling.
- Installing racks and patch panels.
- Terminating (patching the cables) and labeling them.
- Making sure that at least 30 cm (12 in) of cable hangs in the ceiling at each point where cameras will be installed
- Certifying the cables to warrant the work

18.6.3.1 Tips for better network cabling

- The maximum network cable run between devices is 100 m (325 ft).
- If using sockets, take the distance between the socket and the computer into account. A good rule of thumb is 90 m (300 ft) for horizontal runs and 10 m for the patch cabling at both ends of the channel.
- Use the same connector and cable type, such as STP, for the whole length of cable.
- To avoid interference, keep the network cable runs separate from electrical mains cabling. Different voltages in mains cables require different standoff distances.
- To avoid fire hazards and violations of building codes, do not suspend network cabling from ceiling tiles.
- Use shielded network cables outdoors to protect products from surges and interference or if required by local regulations (see section 18.4.11.4).
- Because network cabling typically uses solid wires, cabling should not be twisted or bent into a tighter radius than 4 times the diameter of the cable. Where cable strain is expected, for example, in elevator traveling cables, consider using stranded cabling instead.
- When fastening cable runs, do not use metal staples or adjust cable ties too tightly.
- Avoid using a daisy chain network topology.
- Always pull at least 30 cm (1 ft) extra cable to use as a service loop in case of termination failures that require re-termination.

Figure 18.46 Always leave an extra length of cable – a service loop – for possible future re-termination.

18.6.3.2 Preparing the network cable

A network cable consists of four pairs of twisted wires. The wires are color-coded (orange, green, blue, and brown). The cable specification has been designed for high-speed data transfer and very little crosstalk.

- Do not untwist more of the cable ends than recommend by the manufacturer. If you do, this could cause problems such as near-end crosstalk, which will have a detrimental impact on your network.
- Keep the pairs together.
- Wire the plug correctly, making sure that each wire is properly connected to its respective pin at both ends.
- Use the correct connectors. Network connections use RJ45 connectors specifically designed for either stranded or solid cable. Match the connector with the cable type, such as STP or UTP.
- Use the correct crimping tool for the specific type of connector.

18.6.3.3 Certifying the cable installation

Installers who need to prove to the network owner that the installation has been done correctly and meets TIA or ISO standards need to certify their work. Network owners who want to guarantee that the infrastructure can handle the video bandwidth can use a tester to certify the network infrastructure. In some cases, testers are employed to pinpoint specific problems. Certification tests are vital if there is any disagreement between the installer and network owner after the installation is completed.

In twisted-pair copper wire networks, copper cable certification is achieved through a thorough series of tests in accordance with standards set by the Telecommunications Industry Association (TIA) or the International Organization for Standardization (ISO). These tests are done using a certification testing tool, which gives a pass-or-fail indication.

A cable tester is used to verify that all the intended connections exist and that there are no unintended connections in the cable being tested. When an intended connection is missing, it is "open". When an unintended connection exists, it is a "short" (a short circuit). If a connection leads to the wrong place, it is "mis-wired", meaning that it has two faults: It is open to the correct contact and is shorted to an incorrect contact.

Generally, the testing is done in two phases. The first phase – the *opens test* – makes sure each of the intended connections is good. The second phase – the *shorts test* – makes sure there are no unintended connections.

18.7 TOOLS FOR SYSTEM DESIGN

Designing a network video system requires making a lot of choices and fine-tuning. A modern video surveillance system is complex, includes many components, and is based on best-of-breed technology. Therefore, it is crucial to have tools that make choosing the right components easier. System designers need to determine which are the correct:

- camera types
- enclosures and environmental protection
- lenses to provide the desired field of view
- camera locations and coverages
- camera mounts and accessories
- cabling infrastructures
- network infrastructure, bandwidth, and security
- servers for running the VMS and connecting the storage
- sizes and performances of the storage units
- rack designs for the server room
- video management software solutions

There are many different tools available for different stages of the design. Some are basic and help you choose a few of the components, while others are comprehensive and advanced tools for total system design. The following sections cover component selection tools as well as system design tools.

18.7.1 Calculators and component selection tools

To validate the field of view, there are lens calculators (see also Chapter 4). To assist in choosing the correct product for the installation, there are product selection guides available, as well as guides for selecting the right accessories for the camera. Some are available both online (see Figure 18.47) and as smartphone apps.

18.7.2 Comprehensive system design tools

Demands for accessible and visually appealing yet inclusive design tools are growing, especially now that we have become accustomed to a world with smart devices and easy-to-use apps. System design tools make picking the right camera and recording solution for any surveillance scenario quick and easy. You can import your own floor plans or maps to further simply the positioning of each camera. When the project design is finished, you get a bill of materials that includes everything you need for the installation.

You can keep track of multiple projects, and templates provide a quick way to get new projects started. Set up your scenario and get a shortlist of cameras, mounts, accessories, and recording solutions that meet your needs.

Many types of system architecture benefit from a comprehensive toolset, and some vendors provide very advanced system design packages. These include all the features in the tools described in

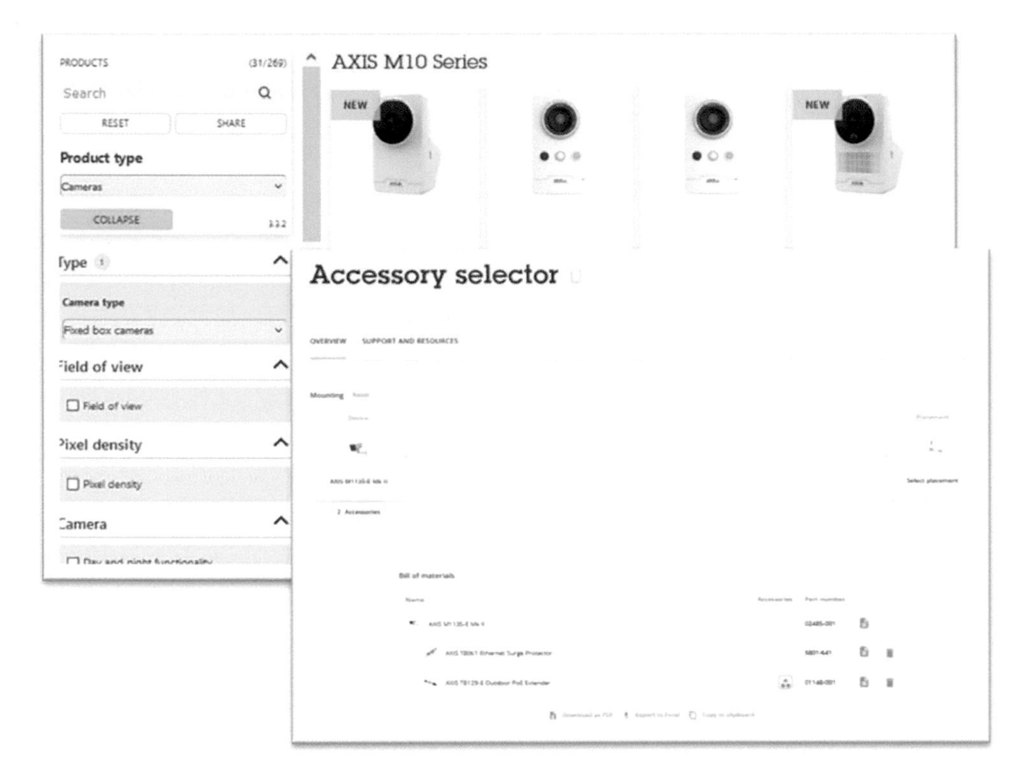

Figure 18.47 Examples of online product and accessory selectors.

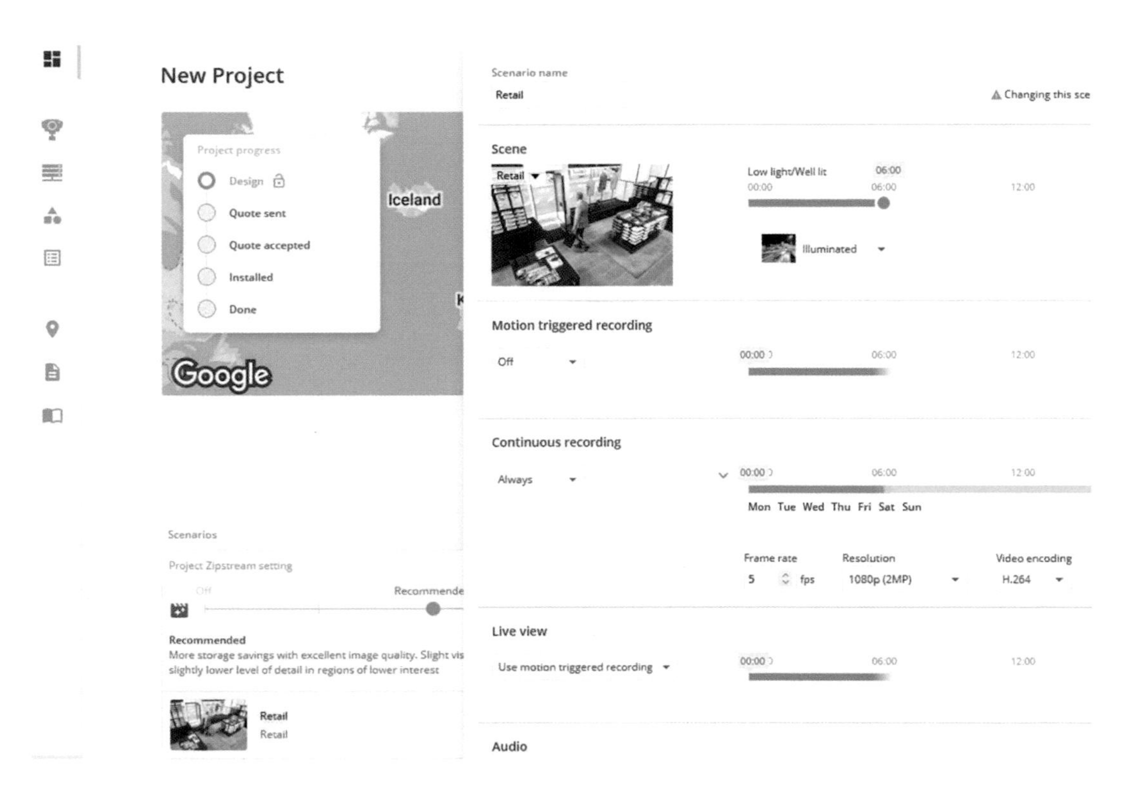

Figure 18.48 System design tools gather several product-selection tools and system calculators into a single intuitive and user-friendly tool.

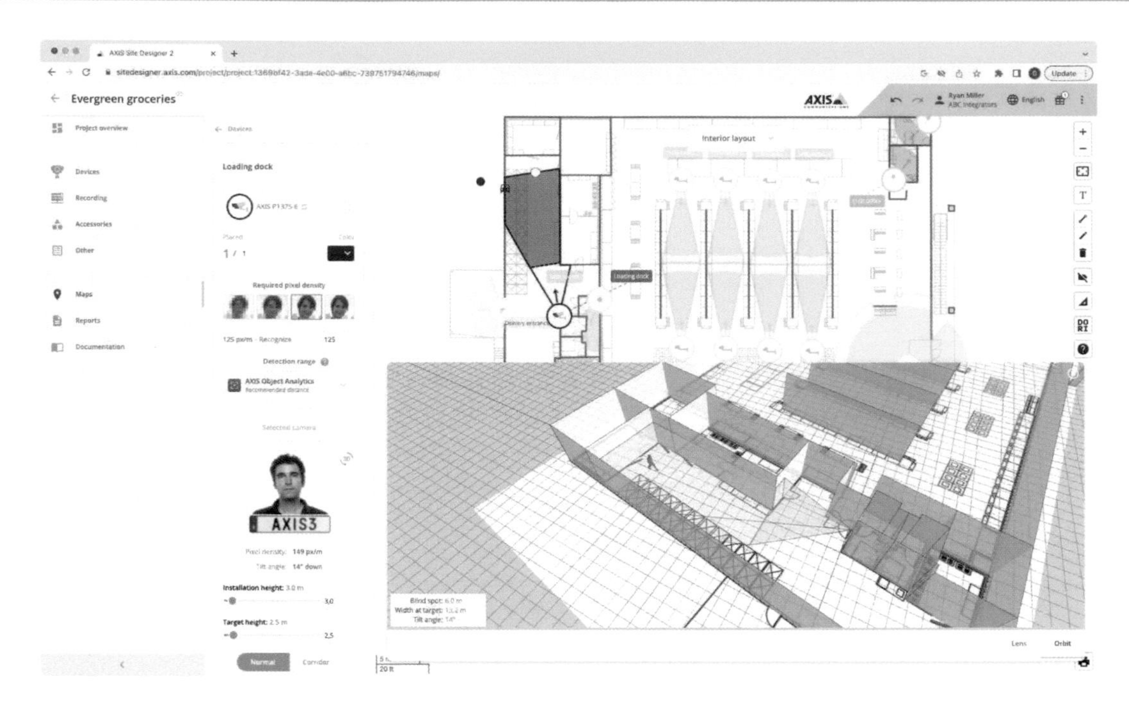

Figure 18.49 An advanced system design tool offers a wide range of options and great flexibility.

this chapter, and they connect the data from the design process. In this way, human error can be reduced and efficiency greatly increased.

Beyond recommending cameras based on the user-defined data, an advanced design tool allows the user to define requirements such as indoor, outdoor, WDR technology, and form factor. It can provide virtualization of the cameras and their intended environment, as well as field-of-view representation (see Figure 18.49).

Based on camera selections, a design tool can calculate and recommend server and storage solutions. Some of the design tools on the market can also provide a bill of materials (BOM) and the possibility to gather and compile associated product documents, such as datasheets, technical specifications, and other conformity documents.

Some platforms can build commonly used outputs for requests for proposal (RFP), including floor plans with field-of-view representations and camera legends.

A truly comprehensive design tool even includes tools for designing server rooms, allows for drag-and-drop placement of appliances in vertical racks, and generates a 2D or 3D design, complete with calculations for power.

18.7.3 CAD software plugins

Based on market positioning and maintenance costs, each design company has their own preference for 2D or 3D CAD software. Some manufacturers supply plugins for these CAD platforms, such as Autodesk® Revit® and Bluebeam Revu®.

To complete the system design, the software using the plugin can provide a detailed camera listing and a bill of materials (BOM) that includes all the products in the design.

18.7.3.1 Autodesk® Revit®

System designers using Revit® can work with different plugins with interactive 3D security camera models. You can place security cameras directly in the building layout and visualize each camera's coverage, which together with the 3D camera view helps you design a project that covers critical

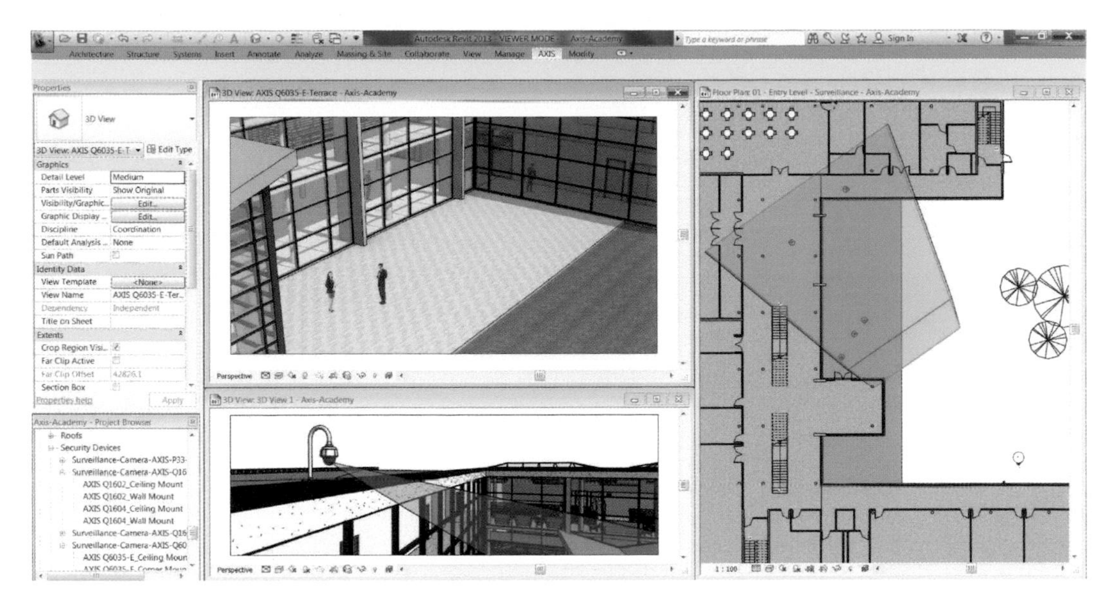

Figure 18.50 Installing 3D CAD models of cameras makes it easy to integrate video surveillance into a building project.

areas and avoids blind spots (see Figure 18.50). This type of plugin provides the metadata needed for building information modeling (BIM), which makes it easy to integrate surveillance system planning into different Revit® based projects, as well as other BIM compatible design tools.

18.7.3.2 Bluebeam Revu®

Tools such as Bluebeam Revu provide a great collaborative platform for streamlining work on projects of many sizes. The program lets users put together a centralized collection of all the materials required in a project, including surveillance details such as camera symbols and their corresponding field-of-view areas. The product-specific properties that design companies provide as part of their Bluebeam tools offering allow you to show the coverage ratio of each camera in the design and help you accurately visualize the distance to the objects in your drawing. You can work with bill of materials details and create export files for use in other design tools.

18.8 LIFECYCLE MANAGEMENT

Lifecycle management is, by definition, an all-encompassing plan and is executed throughout the life of a system. For the user, lifecycle management commonly starts with design and specification and extends to the inevitable phase of technology refresh or decommissioning. From a manufacturer's perspective, the cycle starts much earlier – in design and engineering – and includes components and materials selection. Manufacturers are increasingly responsible for planning for the future recycling and decommissioning of their products, in line with evolving legislation on sustainability. Some manufacturers have been early adopters of various green initiatives such the UN Global Compact.

It quickly becomes apparent that the term "lifecycle" can imply various scales of time, but if we focus on the user's perspective, where there are requirements for protecting assets and people, then a typical lifecycle might look like the example in Figure 18.51.

Offering reliable and comprehensive tools for system design is an important role of the manufacturer. These tools are just one part, although an important part, of a comprehensive lifecycle management offering. As mentioned earlier, this is often where the journey to a new system begins for the customer and their consultant or integration partner.

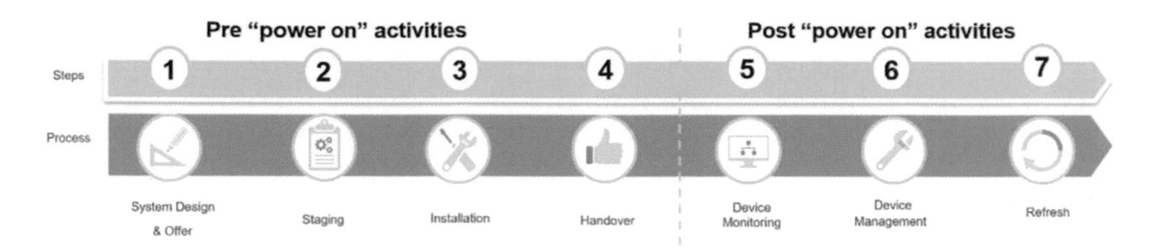

Figure 18.51 An example of a system lifecycle.

18.8.1 System management

Planning and putting in place tools, processes, and resources for managing a surveillance system is critical for its performance, stability, and security. The parts of a system are often referred to as being either IT (information technology) systems or OT (operational technology) systems. The IT system is responsible for availability, stability, performance, and cybersecurity. Tools designed for the IT tasks are for monitoring, maintaining, and managing the system components. They often include device status dashboards, inventories, updating, and configuration capabilities. Their main purpose is to assure optimal performance and high availability and to keep the system as cybersecurity as possible. The OT system refers to the part of the system responsible for productivity – where operators do the work. Examples of OT systems are the Video Management Systems (VMS), reporting/incident management tools, document creation tools, etc.

In some cases, the same resources may be responsible for both using the system and for maintaining it, and many VMSs include tools for managing and maintaining the overall system. This is common where the solution is not overly complex and where resources or budgets are limiting factors. But even in these systems, there is value in using tools that are separate from the VMS, as this allows maintenance work to be performed without disrupting operations. Some maintenance and management tools offer the possibility for remote connectivity, enabling those responsible to maintain, diagnose, or correct issues without having to travel to the site.

On the other hand, there are large-scale systems that may be geographically distributed and have higher complexity. In these scenarios, it is common for different parties to have different roles, that is, essentially splitting the OT and IT responsibilities. The management and maintenance of the system(s) will likely be performed with separate tools and often by individuals without access to the VMS.

Whether the system is simple or complex, small or large, or if different resources have different responsibilities, it is good practice to separate maintenance tools and the VMS. Due to the importance of the VMS being the evidence and incident management system, access to it should be limited to those with proper credentials.

Established manufacturers usually offer a comprehensive suite of tools, both as needed for device configuration during installation and as needed for ongoing management and maintenance. Separate installation tools, independent of the VMS, are critical during installation, as infrastructure and cameras are often installed at different times. The tools include capabilities such as updating to the latest operating system (OS) or firmware during staging, cutting down time spent in the field, and allowing technicians to focus on installation and connectivity. You can also make initial security settings, such as passwords and cyber-hardening, and devices can be prepared to operate on the network infrastructure. The installation tool can then be used to verify operation after installation, independent of the VMS.

Having installation activities done separately allows technicians to focus on configuring and setting up the user preferences in the VMS, including operator training and system testing, without being distracted by device configuration. Lastly, the IT system tools that will help maintain and manage the devices can be installed. The tasks for these tools include regular updates, maintaining

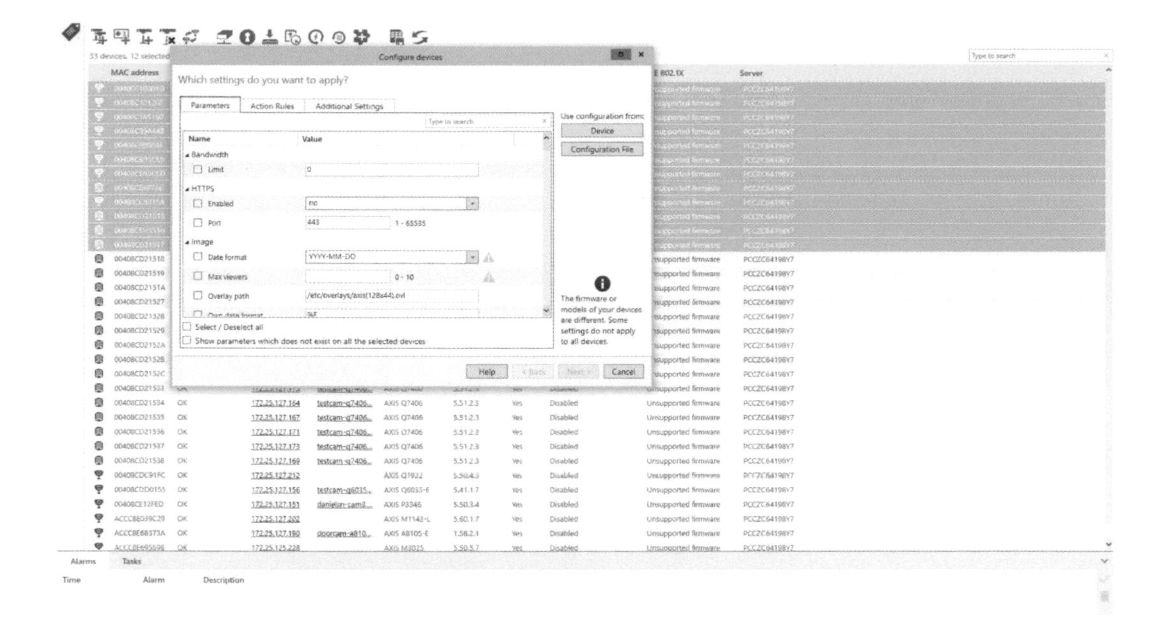

Figure 18.52 Example screenshot from an advanced management tool.

polices for cybersecurity, and having an overview of all the assets. The dashboards within these tools offer assurance and control of the environment. They also provide insights into warranty, end-of-life and recommended replacement products, required updates, and the capability to perform quick system audits as requested by other stakeholders within the organization.

18.9 MAINTENANCE

Any video surveillance system, however expensive or advanced, is going to require maintenance. Things get broken, parts wear out, software gets outdated, etc. Preparing for coming maintenance and, even better, factoring it in during the system design phase will be advantageous when it comes to dealing with the inevitable tasks to come.

Maintenance can generally be divided into two categories: planned activities, for example, regular firmware updates, and reactive activities brought about by unplanned events, for example, the repair or replacement of a camera accidentally moved or damaged.

18.9.1 Planned maintenance

The successful operation of a modern video surveillance system depends on continued operation of its individual components, or at least most of them, and especially those not covered by backup or redundancy measures. Seeing over all these components and planning activities to keep them in optimum shape is vital if you expect to rely on your system to deliver results around the clock, all year round.

By their very nature, planned maintenance activities will occur according to schedules, some of which can be built around information provided by the vendors of the system components. For example, a camera manufacturer should provide a schedule of upcoming firmware releases and when to expect them. Another source of input for planning could be the mean time between failure (MTBF) value specified for a hard drive or other system device.

Listed next are examples of activities that can included in system maintenance plans:

- *On-site cleaning and checkup:* Changes to the camera view due to the buildup of dirt and the activities of birds and insects can be managed by a regular cleaning program, which also provides an opportunity to inspect and possibly adjust the camera. The cleaning program can also include the clearance of vegetation or other unwanted objects in the field of view.

- *Checking the camera view remotely:* In many cases, no one will be looking at the view from a camera unless there is an incident that warrants it, and clearly, this is not the time to discover that the view has been compromised or changed. Simply checking the view on a regular basis can remove the need for more drastic measures later, although a check of this type should not replace the cleaning program. Note that there are analytics applications that can assist in detecting changed camera views. See Chapter 17 for more information.
- *Firmware upgrades:* Running up-to-date firmware versions in your cameras and other devices mitigates risks, as the latest versions will include patches for known vulnerabilities that attackers may try to exploit. Regularly checking for applicable firmware updates should be a feature of any maintenance program and to keep installations secure. For more on cybersecurity, see Chapter 12.
- *VMS updates:* As well as improvements and bug fixes, updates to video management software can sometimes include useful new features that add value to the security system overall.
- *Backups:* Creating backups of your cameras and other devices should occur on a regular basis, according to a set schedule. It is also recommended to make regular backups of the VMS installation, as well as the recordings created in the system. Device backups will assist you in replacing a device or restoring it after a cybersecurity event.
- *Other system documentation:* As well as thoroughly documenting the system during the installation phase and creating regular device backups, all new devices and changes to existing devices or configurations not already registered elsewhere need to be recorded throughout the system's lifespan. Examples of such information can include the following:
  - The angle and field of view for a camera – this can be documented by taking a screenshot
  - Custom modifications to hardware or firmware
  - The physical location of the device, including details of how it is mounted, etc.
- *Replacement of recording media:* All data storage devices are subject to wear and tear, and manufacturers should be able to specify the expected lifespan of a hard drive or an SD card. Be sure to schedule the replacement of these devices to mitigate the risk of data loss.
- *Replacement of system hardware:* Servers, switches, and surveillance cameras all have a certain period of warranty from the manufacturer, typically one, three, or five years. Some organizations have a policy to proactively replace all IT-related equipment before the warranty expiries.
- *Batteries and power supplies:* Some system devices contain batteries, which will need replacing at some point. Device power supplies connected to mains power are subject to wear due to heat generation and should be checked on a regular basis.

18.9.2 Unplanned maintenance

Apart from the usual repairs, upgrades, and replacements to be expected during the normal operation of a video surveillance system, there is also the ever-present risk of damage or disruption due to cybersecurity events. Regular and comprehensive maintenance activities will aid you enormously should you ever need to restore the system after, for example, a cyberattack. For more information on restoring a system, see section 12.9.

18.10 LEGAL ASPECTS

Video and audio surveillance can be restricted or prohibited by laws. Each region or country has its own sets of rules that regulate video surveillance. For example, notifying the public of the existence of video surveillance is usually mandatory. These notifications do not need to be intrusive but should nevertheless be visible (see Figure 18.53). Always check the local laws and regulations before installing a video surveillance system.

Figure 18.53 A sticker or placard is a commonly used method of notifying the public that an area is under video surveillance.

The legislation or guidelines may cover the following:

- *License:* It may be necessary to register or obtain a license from an authority to conduct video surveillance, particularly in public areas.
- *Purpose of surveillance:* Is the equipment permitted by the laws in the area?
- *Position or location:* Is the equipment located in such a way that it only monitors the spaces the equipment is intended to cover? If unintended areas are covered, consultations with the owners of such spaces might be required. In some areas, video surveillance can be prohibited, for example, restrooms and changing rooms in retail stores.
- *Notification:* Signs that warn the public that they are entering a zone covered by surveillance equipment might be necessary. The sign may have to be of a particular type, or it may need to follow specific guidelines.
- *Quality of images:* There may be rules regarding the quality of images, which can affect what may be permitted or acceptable for use as evidence in court.
- *Video format:* Police authorities may require the video format to be one that they can handle.
- *Information provided in recorded video:* Video recordings may have to be stamped with the time and date.
- *Processing of images:* There may be rules regulating how long images should be retained, who can view the images, and where recorded images can be viewed.
- *Drawings:* There may be requirements for drawings of where cameras are placed.
- *Personnel training:* There may be regulations that require operator training in security and disclosure policies as well as privacy issues.
- *Access to and disclosure of images to third parties:* There may be restrictions on who can access the images and how they can be shown. For example, if video will be disclosed to the media, images of individuals may have to be anonymized, for example, by blurring.
- *Monitoring and recording of audio:* A permit may be required for recording audio in addition to video.
- *Regular system checks:* There may be guidelines on how often and thoroughly a company should perform system checks to make sure all equipment is operating as it should.
- *Audit trail:* Meaning the ability to track who used the system, at which time, and for what purpose. In addition, proof of a video's authenticity may be required by methods such as watermarking.

Index

Note: Page numbers in **bold** indicate a table on the corresponding page.